Spatial Analysis in Geology Using R

The integration of geology with data science disciplines, such as spatial statistics, remote sensing, and geographic information systems (GIS), has given rise to a shift in many natural sciences schools, pushing the boundaries of knowledge and enabling new discoveries in geological processes and earth systems. Spatial analysis of geological data can be used to identify patterns and trends in data, to map spatial relationships, and to model spatial processes. R is a consolidated and yet growing statistical programming language with increasing value in spatial analysis often replacing, with advantage, GIS tools. By providing a comprehensive guide for geologists to harness the power of spatial analysis in R, **Spatial Analysis in Geology Using R** serves as a tool in addressing real-world problems, such as natural resource management, environmental conservation, and hazard prediction and mitigation.

Features:

- Provides a practical and accessible overview of spatial analysis in geology using R
- Organised in three independent and complementary parts: Introduction to R, Spatial Analysis with R, and Spatial Statistics and Modelling
- Applied approach with many detailed examples and case studies using real geological data
- Presents a collection of R packages that are useful in many geological situations
- Does not assume any prior knowledge of R; all codes are explained in detail
- Supplemented by a website with all data, codes, and examples

Spatial Analysis in Geology Using R will be useful to any geological researcher who has acquired basic spatial analysis skills, often using GIS, and is interested in deepening those skills through the use of R. It could be used as a reference by applied researchers and analysts in public, private, or third-sector industries. It could also be used to teach a course on the topic to graduate students or for self-study.

Chapman & Hall/CRC The R Series

Recently Published Titles

Rasch Measurement Theory Analysis in R: Illustrations and Practical Guidance for Researchers and Practitioners
Stefanie Wind and Cheng Hua

Spatial Sampling with R
Dick R. Brus

A Criminologist's Guide to R: Crime by the Numbers
Jacob Kaplan

Analyzing US Census Data: Methods, Maps, and Models in R
Kyle Walker

ANOVA and Mixed Models: A Short Introduction Using R
Lukas Meier

Tidy Finance with R
Christoph Scheuch, Stefan Voigt, and Patrick Weiss

Deep Learning and Scientific Computing with R torch
Sigrid Keydana

Model-Based Clustering, Classification, and Density Estimation Using mclust in R
Lucca Scrucca, Chris Fraley, T. Brendan Murphy, and Adrian E. Raftery

Spatial Data Science: With Applications in R
Edzer Pebesma and Roger Bivand

Modern Data Visualization with R
Robert Kabacoff

Learn R: As a Language, Second Edition
Pedro J. Aphalo

Spatial Analysis in Geology Using R
Pedro M. Nogueira

For more information about this series, please visit: https://www.crcpress.com/Chapman–HallCRC-The-R-Series/book-series/CRCTHERSER

Spatial Analysis in Geology Using R

Pedro M. Nogueira

CRC Press is an imprint of the
Taylor & Francis Group, an **informa** business
A CHAPMAN & HALL BOOK

First edition published 2024
by CRC Press
2385 NW Executive Center Drive, Suite 320, Boca Raton FL 33431

and by CRC Press
4 Park Square, Milton Park, Abingdon, Oxon, OX14 4RN

CRC Press is an imprint of Taylor & Francis Group, LLC

ISBN: 978-1-032-65032-6 (hbk)
ISBN: 978-1-032-65187-3 (pbk)
ISBN: 978-1-032-65188-0 (ebk)

DOI: 10.1201/9781032651880

Typeset in Palatino
by KnowledgeWorks Global Ltd.

This book is dedicated to my parents,

Raul and Margarida.

Contents

Preface....xvii
Acknowledgements....xxi
About the Author....xxiii

Part I Introduction to R

1. **R as a Geologist's Tool**....3
   References....6

2. **Getting Started**....8
   2.1 Downloading and Installing R and RStudio....8
   2.2 Understanding the RStudio Environment and Workspace....9
   2.3 Basic Syntax and Data Structures in R....10
   2.4 Operators in R....14
   2.4.1 Basic Operators....14
   2.4.2 Other Operators....14
   2.4.3 A Special Operator: The Pipe....15
   2.5 Flow Control....17
   2.5.1 If()-Else Statements....17
   2.5.2 For() Loops....18
   2.5.3 While() Loops....19
   2.6 Packages and Libraries....19
   2.7 An Overview of CRAN (Comprehensive R Archive Network)....21
   2.8 Popular Packages for Data Analysis and Visualisation....22
   2.9 R Packages for Geologists....23
   2.10 Versions and Updates....24
   2.11 Other Resources for Learning R....24
   2.12 Concluding Remarks....25
   References....26

3. **Reading and Writing Data**....27
   3.1 Data in CSV Files....27
   3.1.1 Reading CSV Files....27
   3.1.2 Writing CSV Files....28
   3.2 Retrieving Data from a Website....28
   3.3 Retrieving Data from a Web Server....33

3.4 Retrieving Data from Excel Files 35
3.4.1 Reading Excel Files 35
3.4.2 Writing Excel Files 35
3.5 Other Types of Data 37
3.6 Concluding Remarks 37
References 37

4. Data Cleaning and Manipulation 38
4.1 Know Your Data 38
4.2 Subsetting Data 40
4.2.1 Subset Numeric Data 41
4.2.2 Subset Text Data 44
4.3 Missing Values 46
4.3.1 Handling Missing Values 47
4.3.2 Replacing NAs 48
4.4 Transforming Variables in R 50
4.4.1 Converting Character to Numeric 50
4.4.2 Converting Continuous to Categorical 51
4.4.3 Converting Categorical to Continuous 51
4.4.4 Appending Rows 52
4.4.5 Appending Columns 54
4.5 Reshaping Data 56
4.5.1 Transposing a Data Frame 56
4.5.2 Long to Wide and Vice Versa 57
4.6 Concluding Remarks 59
References 60

5. Functions for Everyday Data Analytics 61
5.1 Arrays and Vectors 61
5.1.1 The sort() Function 61
5.1.2 The unique() Function 62
5.2 Strings and Texts 62
5.2.1 The paste() and paste0() Functions 62
5.2.2 The substr() Function 63
5.2.3 The gsub() Function 63
5.3 Data Frame-Related Functions 64
5.3.1 The nrow() Function 64
5.3.2 The ncol() Function 64
5.3.3 The head() Function 64
5.3.4 The tail() Function 65
5.3.5 The colnames() and rownames() Functions 65
5.4 The apply() Functions 66
5.5 Creating Functions 69
5.6 Concluding Remarks 70
References 70

6. Basic Statistics ... 71
6.1 A Primer in Exploratory Data Analysis ... 71
6.1.1 The 'Meuse' Dataset ... 71
6.1.2 Basic Functions ... 72
6.2 Handling Outliers ... 76
6.3 Correlating Variables ... 79
6.4 Inference Statistics ... 80
6.4.1 The T-test ... 80
6.4.2 Welch Two-Sample T-test ... 82
6.4.3 ANOVA ... 83
6.4.4 Regression Analysis ... 85
6.5 Concluding Remarks ... 89
References ... 89

7. Basic Data Visualisation ... 91
7.1 Bar Plots ... 91
7.2 Histograms ... 92
7.3 Box Plots ... 93
7.4 Scatter Plots ... 94
7.5 Organising the Plots ... 95
7.5.1 The par() Function ... 96
7.5.2 The layout() Function ... 97
7.6 Save the Plots ... 97
7.7 Concluding Remarks ... 99
References ... 99

8. Putting It All to Work: Part I ... 100
8.1 Reading the Data ... 100
8.2 Insights from the Data ... 102
8.3 Subsetting Data ... 104
8.4 Transforming Data ... 106
8.5 Plotting the Data ... 107
8.6 The Bureau of Mineral Resources Subset ... 109
8.7 Descriptive Statistics ... 113
8.7.1 Descriptive Parameters ... 122
8.7.2 Correlations Made Beautiful ... 124
8.8 Concluding Remarks ... 127
References ... 127

Part II Spatial Analysis with R

9. Fundamentals of Spatial Analysis ... 131
9.1 Key Concepts and Techniques ... 132
9.2 Applications in Geology and Earth Sciences ... 132
References ... 134

10. Spatial Objects ... 135
10.1 Vectorial Objects ... 137
10.1.1 Shapefiles ... 139
10.1.2 GeoPackages ... 140
10.1.3 GeoJSON ... 141
10.1.4 GPX Files ... 141
10.1.5 KML Files ... 142
10.2 Gridded and Raster Objects ... 143
10.3 Coordinate Reference Systems (CRS) ... 145
10.4 Spatial Analysis Packages ... 147
10.4.1 The 'sp' Package ... 148
10.4.2 The 'sf' Package ... 151
10.4.3 The 'raster' Package ... 153
10.4.4 The 'terra' Package ... 156
10.4.5 See Also ... 157
10.5 Concluding Remarks ... 158
References ... 158

11. Going Vectorial ... 159
11.1 Sources of Vectorial Spatial Data ... 159
11.2 Read and Write Spatial Data ... 161
11.2.1 Reading Spatial Data Files ... 161
11.2.2 Writing Spatial Data Files ... 162
11.2.3 Converting between Spatial File Types ... 163
11.3 OpenStreetMap: A Source of Geographic Data ... 167
11.3.1 The 'osmdata' Package ... 167
11.4 Visualising Spatial Data ... 172
11.4.1 A First on 'ggplot2' ... 173
11.4.2 A First on 'leaflet' ... 175
11.4.3 Plotting Spatial Points ... 177
11.4.4 Plotting Spatial Lines ... 179
11.4.5 Plotting Spatial Polygons ... 183
11.5 Concluding Remarks ... 188
References ... 189

12. Handling Spatial Data ... 190
12.1 Geometries and Projection ... 193
12.1.1 Defining Coordinate Reference Systems ... 193
12.1.2 Defining the Geometry ... 194
12.1.3 Reprojecting ... 197
12.2 Buffering ... 198
12.3 Subsetting Spatial Data ... 199
12.3.1 Cropping ... 200
12.3.2 Clipping ... 203

12.4 Spatial Queries (Logical) 207
12.4.1 The st_contains() Function 207
12.4.2 The st_touches() Function 209
12.4.3 The st_crosses() Function 210
12.4.4 The st_overlaps() Function 212
12.4.5 The st_within() Function 213
12.4.6 The st_intersects() Function 214
12.5 Spatial Calculations (Numeric) 216
12.5.1 st_centroid() Function 217
12.5.2 The st_area() Function 217
12.5.3 The st_length() Function 218
12.5.4 The st_distance() Function 219
12.6 Spatial Operations (Geometric) 219
12.6.1 The st_sym_difference() Function 220
12.6.2 The st_union() Function 220
12.6.3 The st_difference() Function 221
12.6.4 The st_intersection() Function 222
12.7 Concluding Remarks 224
References 224

13. Putting It All to Work: Part II Vectors 225
13.1 Setting the Data 225
13.2 Preparing the Maps 228
13.2.1 The ggplot() Map 230
13.2.2 The 'leaflet' Map 231
13.3 Subsetting the Stream Sediments 235
13.3.1 Subsetting by Geological Unit 236
13.3.2 Naming the Geological Units 238
13.3.3 Answering the Questions 238
13.4 Group Statistics 240
13.4.1 Descriptive Statistics by Geological Unit 240
13.4.2 Correlation by Geological Unit 241
13.4.3 Testing Independency from Geological Unit 242
13.4.4 Plotting the Independence 246
13.5 This Is Not My Fault 249
13.5.1 Where Are You From? 251
13.6 Concluding Remarks 255

14. Into the Grid with Rasters 257
14.1 Types of Raster Data 258
14.2 Types of Raster Files 258
14.3 Reading and Writing Raster Data 260
14.4 Converting between Raster Types 262
14.5 Retrieving Elevation 263

14.6 Retrieving Bathymetry....266
14.6.1 WFS Service (More Than Bathymetry)....270
14.7 Know Your Data, Again.......274
14.8 Basic Descriptive Statistics of a Raster....277
14.9 Concluding Remarks....278
References....278

15. Basic Raster Operations....279
15.1 Reprojecting Rasters....279
15.2 Cropping Rasters....281
15.3 Resampling Raster Data....281
15.4 Aggregating and Disaggregating....283
15.5 Filtering Raster Data....284
15.6 Masking Raster Data....286
15.7 Concluding Remarks....290
References....290

16. Terrain Operations....291
16.1 Slope of Terrain....292
16.2 Aspect of Terrain....293
16.3 Hillshade of Terrain....294
16.4 Terrain Ruggedness Index (TRI)....296
16.5 Concluding Remarks....297
References....298

17. Working with Satellites....299
17.1 Satellite Packages....299
17.1.1 Sign In ... Into a Satellite Journey....300
17.2 Working with Sentinel Images....301
17.2.1 Register to Download....302
17.2.2 Verifying the Available Images....303
17.2.3 Downloading the Data....308
17.2.4 Visualising the Results....310
17.3 Working with Landsat-8 Images....314
17.3.1 Download Landsat-8 Data....314
17.3.2 Know Your Landsat-8 Data....315
17.3.3 Read Landsat-8 Data....318
17.3.4 Plotting a Composite RGB Landsat-8 Image....318
17.3.5 The NDVI....320
17.4 Concluding Remarks....322
References....322

18. Putting It All to Work: Part II Rasters....324
18.1 Setting Up the Environment....324
18.2 Combining Bands....327

18.3 Calculating Indices ... 329
18.4 A Thematic Map ... 333
18.5 Geomorphology with Whitebox ... 338
18.5.1 The 'whitebox' Package ... 338
18.5.2 Installation of the 'whitebox' Package ... 340
18.5.3 A First on Geomorphological Analysis ... 340
18.5.4 Streams and Waterflow ... 346
18.5.5 Basins and Rivers ... 349
18.5.6 Geomorphs, a Collection of Geographic Features ... 349
18.6 Concluding Remarks ... 353
References ... 354

Part III Spatial Statistics and Modelling

19. Introduction to Spatial Statistics and Modelling Applied to Geology ... 357
19.1 The 'spatstat' Package ... 358
References ... 359

20. Point Pattern Analysis ... 361
20.1 Distribution Patterns ... 361
20.1.1 The K-Function ... 362
20.1.2 K-Function Interpretation ... 364
20.1.3 Clustered Points ... 365
20.1.4 Regular Points ... 367
20.1.5 Randomly Distributed Points ... 367
20.2 PPA for the Northern Territory Dataset ... 369
20.2.1 Point Density Analysis ... 371
20.3 Earthquakes Data ... 373
20.3.1 PPA of Earthquakes ... 374
20.4 The Fry Method ... 376
20.5 Nearest Neighbour and PPA ... 379
20.6 Concluding Remarks ... 380
References ... 381

21. Interpolating Data ... 383
21.1 The 'gstat' Package ... 384
References ... 386

22. Inverse Distance Weighting (IDW) ... 388
22.1 Timor-Alor Earthquakes ... 390
22.2 Concluding Remarks ... 395
Reference ... 396

23. Kriging 397
23.1 Types of Kriging 397
23.2 Variograms and Semi-Variograms 398
23.2.1 Clustered Pattern 400
23.2.2 Random or Uniform Pattern 403
23.2.3 Trend Pattern 405
23.3 Kriging Earthquakes 407
23.3.1 Ordinary Kriging 409
23.3.2 Universal Kriging 409
23.4 Creating the Rasters 411
23.5 Concluding Remarks 417
References 418

24. The Way Forward 419

Index 421

'no data was harmed in preparing this book'

This book is full of examples of code in R that can be used or adapted for your specific case or just tested for the fun of seeing the results. Anyway, some parts of code are not executable and are merely used to demonstrate some syntax or concept.

For easily grasping what are the code snippets that work or not, each piece of code is preceded by the friendly symbols to indicate:

Working code, please try it and be happy.

This is just a coding concept or a syntax example. Try it and you will receive a friendly Error message from the R console.

Depending on the level that you have as a computer programming person or as a data analysis expert, and this is highly subjective, the examples provided can be easy, interesting, or hard to grasp. As a tentative way of providing some feedback on how I felt about the difficulty of the example, a star coding is provided using one star (easy) to five stars (maybe I will read it again and out loud).

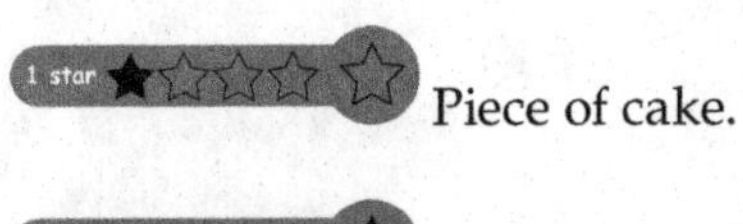

Piece of cake.

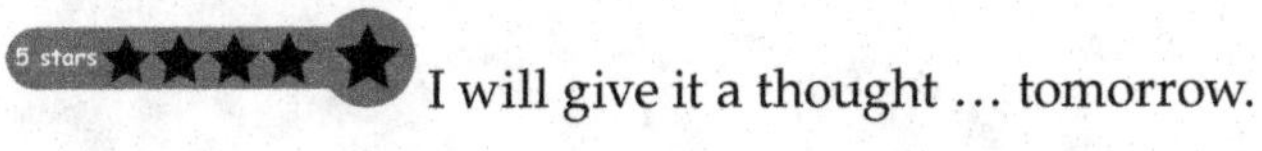

I will give it a thought ... tomorrow.

Preface

The law of the instrument[1] is illustrated by the popular phrase 'if all you have is a hammer, everything looks like a nail', referring to the idea that we conduct our work depending on the instruments that are available. As a geologist, if the only tool I have is a geologist's hammer, all I can do is to smash rocks and look at their guts. With this book I intend to provide some unconventional tools to the geologist's field equipment.

No, the tools suggested in this book do not replace field work, acute observation, laboratory experimentation, and reasoning; alas, most of the fun of a geologist's work comes from these tasks. These new tools are complements, new ways of gathering and visualising information, providing new insights to our knowledge of the earth systems, and, yes, they are also good fun.

Spatial analysis of geological data can be used to identify patterns and trends in data, to map spatial relationships, and to model spatial processes. R is a consolidated and yet growing statistical programming language with increasing value in spatial analysis often replacing, with advantage, geographic information systems (GIS) tools. With a free and open-source software environment that is well-suited for spatial data analysis, it has a wide range of statistical and graphical functions that are easy to learn and use.

This book is intended to provide a hands-on introduction to spatial analysis in geology using R. It is aimed at people who have acquired elementary spatial analysis skills often using GIS but who are interested in deepening their skills. The book will also be of interest to graduate students and researchers who are interested in using R for spatial analysis in geology.

Another important design feature of the text is to present a constellation of packages that are useful in many geological situations, might that be retrieving geological data from a server, accessing elevation data for a region, using bathymetric and geological information from the oceans, downloading and treating satellite images, or drawing contour curves and streams from a study region.

It is organised into three parts that are intentionally independent, complementary, and with escalating complexity. It is possible to read a single chapter to learn about a subject, an entire part to dominate a matter, or the entire book to develop a spatial analysis project, depending on its necessities. Selected chapters in the text – i.e., in Chapters 8, 13, and 18 – present a wrap-around to encapsulate the set of operations and functions learned. These chapters consist of guided problem-solving case studies.

Part I of the book introduces the R programming language, its libraries, methods, and functions (Chapter 2); on how to read and write data (Chapter 3); basic operation for data cleaning (Chapter 4); common R function

[1] For more details, please refer to https://en.wikipedia.org/wiki/Law_of_the_instrument

(Chapter 5); basic statistics with R (Chapter 6); and basic data visualisation with R (Chapter 7). The format of this part is mostly designed to be used as a reference text. This is neither a computer programming book nor a thorough description of all the possibilities of the R programming language; however, when convenient I recommend complementary readings. The operations and functions selected are the ones that are most commonly used in a typical R session from a geologist's point of view. It has a chapter dedicated to data visualisation, an important task in spatial data analysis.

Part II covers a range of spatial objects and methods (Chapters 10 and 11) used in spatial analysis; for those acquainted with GIS, some parts will sound familiar but often surprisingly twists on new ways to handle spatial data (Chapters 12 and 15) are presented. Special attention is paid on how to gain access to geological and geographical data, such as roads (Chapter 11), terrain elevation (Chapters 14 and 16), bathymetry (Chapter 14), or satellite images (Chapter 17), all of these with very high quality and open access. The creation and visualisation of static and interactive maps are detailed in Chapter 13, as maps are the utmost way of geologists communicating.

Part III is dedicated to what often is called geostatistics, including point pattern analysis (Chapter 20), spatial autocorrelation and spatial interpolation (Chapters 21 and 22), and spatial regression. It is founded on several case studies demonstrating the application of R and its advanced analysis tools.

The book does not assume any prior knowledge of R. All of the code in the book is explained in detail so that readers can follow along and learn how to use R for spatial analysis. The code is presented as code fragments or snippets named sequentially by chapter, exercise, and chunk (e.g. #04-20/01, for Chapter 4, exercise 20, chunk 01). We created a 'star' system to help the reader to recognise the expected difficulty of each code snippet.

My predilection for lists implies that sometimes the topics of a package or a function are presented as a list; this is not something to memorise but to come back to for reference when necessary.

Who Should Read This Book?

The book is intended in a user-friendly reading, for a wide audience, mostly geoscientists, including:

- Graduate students and researchers
- Academics and post-graduate students
- People who have learned spatial analysis skills using GIS and want to benefit from using a programming language
- Applied researchers and analysts in public, private, or third-sector

Why Use R for Spatial Analysis?

There are a number of reasons why you might want to use R for spatial analysis. R is a free and open-source software environment, which means that it is available to everyone, with a plethora of statistical and graphical functions. Additionally, R is very easy to learn and use, even if you do not have any prior programming experience.

What Will You Learn in This Book?

This book will teach you the basics of spatial analysis in R. You will learn how to:

- Understand the basics of the R programming language.
- Perform basic and advanced statistical analysis of data using R.
- Load and manipulate spatial data.
- Create graphs, maps, and interactive visualisations of geological data.
- Perform geostatistical analysis including point pattern and spatial interpolation analysis.

The book also includes a number of case studies that illustrate the use of R for spatial analysis in geology.

What Do You Need to Know Before You Read This Book?

This book does not assume any prior knowledge of R. However, it is helpful if you have some rudimentary knowledge of computer programming and spatial analysis. If you do not have any prior knowledge of spatial analysis, you may want to read an introductory textbook on spatial analysis before you read this book.

The book can be read from start to finish; however, it is also possible to read individual chapters or parts depending on interest or curiosity.

Complementary to the printed book, there is a web page[2] that includes all the data, the code, and the examples presented in this book. As a reader you can always use it to download and test the examples, to comment or ask questions, or just to browse for updates on the subjects related to *Spatial Analysis in Geology Using R*.

[2] https://github.com/pnogas67/SpatialAnalysisinGeologyUsingR

Acknowledgements

I would like to thank the following people for their help in the preparation of this book:

Dalia Cristovão for the love and the thorough and critical review of the text with invaluable patience, insights, and comments.

Marcelo Silva, José Roseiro and Miguel Maia for reading some chapters and discussing with me much of the material used.

The students of GIS and Digital Cartography at Évora University and Porto Universities (Portugal), Eduardo Mondlane University (Mozambique), who shared with me so much about their spatial analysis doubts and questions, hence contributing to a better text.

To the Editor, Rob Calver, who believed in this project and patiently led me to completion, and the reviewers who provided helpful feedback on the book.

The R community, especially the authors of the open-source software packages that we used in the book.

About the Author

Pedro M. Nogueira received his undergraduate and PhD degrees in Geology from Porto University. He also completed the curricular part of an MSc in Engineering Informatics at Évora University. From 1997 to 2022, he was Assistant Professor at Évora University and currently holds the position of Associate Professor. He has more than 30 years of experience as a field geologist and mineral resources specialist. Pedro has worked in diverse countries such as East Timor, Mozambique, and the Dominican Republic. He has published more than 30 peer-reviewed papers in international journals and has served as an editor of several books and book chapters.

Pedro is a full member of the Earth Sciences Institute and has participated or served as a project leader or researcher in more than 15 funded projects, several of which are internationally run. His current research interests are centred around the study of gold, copper, iron, and rare earth mineral deposits, interweaved with geochemistry, geophysics, remote sensing, and spatial data analysis – let's call it spatial inteligence.

About the Author

Part I

Introduction to R

1

R as a Geologist's Tool

R is a powerful, open-source programming language[1] and software environment for data analysis and statistical computing. It was first developed in 1993 by Ross Ihaka and Robert Gentleman at the University of Auckland, New Zealand. Over the years, it has grown into a widely used and well-regarded language for data analysis, with a large and active community of users and contributors. It provides a flexible and efficient environment for data analysis, statistical modelling, and graphical representation of data, and it is particularly well-suited for data analysis because of its many built-in functions and libraries for data manipulation, statistical modelling, and visualisation. This makes it ideal for researchers, data scientists, statisticians, and for those who need to analyse and understand large and complex datasets.

R is often compared with other programming languages, such as Fortran and Python, or computer applications such as MATLAB (high-level programming language and environment for numerical computation, visualisation, and algorithm development), SAS (Statistical Analysis System, a software suite used for advanced analytics, business intelligence, data management, and predictive modelling), or SPSS (Statistical Package for the Social Sciences, a software tool used for statistical analysis, data mining, and decision support). Each of these has its own strengths and weaknesses, and the choice of which to use depends on the specific needs of the analysis. For example, Python is often favoured for its simplicity, versatility, and large community of users, while MATLAB is popular for its strong support for matrix operations and scientific computing. SAS and SPSS, on the other hand, are proprietary software packages that offer a wide range of statistical functions and graphical representation options.

In comparison, R is open-source and free, making it an accessible option that offers a large and well-supported set of packages and functions, making it a go-by tool to perform numerous data analysis and statistical modelling tasks. Recent advances powered it with a strong visualisation component, with several built-in and third-party packages available for creating high-quality plots and maps, which is quite in hand for spatial analysts and researchers who deal with spatial data, such as geographers or geologists.

[1] For more details, see https://en.wikipedia.org/wiki/R_(programming_language) and R Core Team (2023).

DOI: 10.1201/9781032651880-2

Another advantage of the R programming language is its flexibility and extensibility, as users can write custom functions, create new packages, and import data from multiple sources. This allows an easy integration with existing workflows and tools and to customise the data analysis processes according to the necessities of each specific project.

Despite these strengths, it does have some disadvantages compared to other programming languages. For example, it's not as fast as some other languages for large-scale computations, and its syntax can be more complex and challenging. However, these disadvantages are outweighed by the many benefits that include its openness, versatility, and strong community of users and contributors.

Geological studies involve the analysis of complex datasets to understand the earth's structure, composition, processes, and history. By leveraging R's versatility, one can harness advanced statistical techniques, create interactive visualisations, and outline workflows that enhance our research.

It is clear by just looking at recent scientific publications that many of the published works could benefit from the advanced capabilities that the use of a programming language with a powerful analytical focus such as R can offer to geologists. For highlighting this idea herein is presented a short list of examples that support our view on this matter.

Some of the benefits of using R for geological studies include the following:

1. Data handling and manipulation: R offers various packages like 'readr', 'dplyr', or 'tidyverse' to read, clean, transform, and manage large and complex geological datasets. Geoscientists can quickly filter, sort, and aggregate data, enabling efficient and accurate analysis. Example: In Chapter 8, a study of stream sediments data from Northern Territory mining companies is presented. This data undergoes a process of filtration and cleaning to remove instances where values were not detected. The purpose of this filtration and cleaning is to prepare the data for analysis, specifically to examine the outcomes of mineral exploration campaigns. This type of work analysing stream sediment data is fundamental not only in mineral exploration projects such as those demonstrated by Lipp et al. (2023) but also in studying the role in contaminated soils (e.g. Taylor, 2007). Many authors are presenting new mapping techniques using stream sediment campaigns where these approaches are meaningful, with publishable results (e.g. Salomão et al., 2020).
2. Geospatial analysis: R's spatial analysis packages, such as 'sf', 'sp', 'raster', and 'terra', facilitate the processing and analysis of spatial data, including vector and raster formats. Often, after bibliographical recollection of old maps or during fieldwork, it is necessary to perform operations like coordinate transformation, spatial interpolation, and spatial statistics.

Example: Mapping the distribution of geological formations or understanding geochemical patterns is a daily task for geologists. Beniest & Schellart (2020) combined information from several sources, including geological, geophysical, and bathymetric, to provide a final georeferenced map, thus demonstrating that data publicly available today allows us to create offshore geological maps in remote, inaccessible offshore domains. In Chapter 13, we present an example on how to relate geochemical information with geological units and structures.

3. Advanced statistical analysis: R provides a wide range of statistical techniques, from basic descriptive statistics to advanced multivariate analyses. With packages like 'stats', or 'caret', one can model complex relationships, test hypotheses, or classify geological units based on machine learning approaches.

   Example: In recent years compositional data analysis has changed the panorama of geochemical data analysis. Authors such as Filzmoser et al. (2018) and Grunsky et al. (2024) suggested a workflow for working with compositional data that includes several steps that span from unsupervised learning, including principal component analysis (PCA) to reduce dimensionality and identify key factors of supervised learning where methods such as random forests can help to understand the relations between geochemical variables.

4. Visualisation: Effective visualisation is essential for communicating geological findings. R provides a range of packages like 'ggplot2', 'leaflet', and 'rayshader' to create high-quality static, interactive, and 3D visualisations.

   Example: Creating interactive maps displaying geological features, such as fault lines or mineral occurrences, to facilitate exploration and decision-making. Martínez-Graña et al. (2013) presented a possible virtual tour in geological heritage sites based on an interactive visualisation using Google Earth. Along this book, several examples of interactive visualisations using the 'leaflet' package are presented in various chapters.

5. Reproducibility and automation: R's script-based approach ensures reproducibility and enables geoscientists to automate repetitive tasks, enhancing efficiency and reducing the potential for errors.

   Example: Developing automated scripts for routine processing and analysis of survey data, such as gravity or magnetic measurements, is a common task in geophysical campaigns. One example of this is present in Cruz et al. (2023) where we conducted several gravimetric survey campaigns to obtain gravimetric data to model the installation of a plutonic complex. The repetitive job of treating the data and compatibilising the data from the different campaigns is a task that was easily accomplished by using a R script.

6. Extensive support and resources: R boasts a large, active community of users and developers, ensuring continuous improvements and access to a wealth of resources, such as tutorials, forums, and blogs.

   Example: Geoscientists can leverage existing R packages and resources specifically tailored for geoscience applications. Many other packages for specific applications are continuously being developed for ever new challenges in science. One such example is the study of provenance of sediment samples based on chemical affinities presented recently by Vermeesch (2019). Another example is the package presented by Janoušek et al. (2006) for the interpretation of igneous rocks geochemical data, the GCDkit package.

From this unordered list of examples, it becomes clear about the usefulness and applicability of using a programming language like R as a tool to help solving many of the geological problems, either you are a field geologist, an exploration geologist, a geomorphologist, a structural geologist, a geophysicist, or even an experimental petrologist.

References

Beniest, A., & Schellart, W. P. (2020). A geological map of the Scotia Sea area constrained by bathymetry, geological data, geophysical data and seismic tomography models from the deep mantle. Earth-Science Reviews, 210, 103391.

Cruz, C., Nogueira, P., Máximo, J., Noronha, F., & Sant'Ovaia, H. (2023). New insights for an emplacement model for the Santa Eulália Plutonic Complex (SW of Iberian Peninsula). Journal of the Geological Society, https://doi.org/10.1144/jgs2022-131.

Filzmoser, P., Hron, K., & Templ, M. (2018). Applied compositional data analysis: With Worked Examples in R. Springer Series in Statistics. Switzerland. https://doi.org/10.1007/978-3-319-96422-5

Grunsky, E., Greenacre, M., & Kjarsgaard, B. (2024). GeoCoDA: Recognizing and validating structural processes in geochemical data. A workflow on compositional data analysis in lithogeochemistry. Applied Computing and Geosciences. https://doi.org/10.1016/j.acags.2023.100149

Janoušek, V., Farrow, C. M. & Erban, V. 2006. Interpretation of whole-rock geochemical data in igneous geochemistry: introducing Geochemical Data Toolkit (GCDkit). Journal of Petrology, 47(6), 1255–1259.

Lipp, A. G., de Caritat, P., & Roberts, G. G. (2023). Geochemical mapping by unmixing alluvial sediments: An example from northern Australia. Journal of Geochemical Exploration, 248, 107174.

Martínez-Graña, A. M., Goy, J. L., & Cimarra, C. A. (2013). A virtual tour of geological heritage: Valourising geodiversity using Google Earth and QR code. Computers & Geosciences, 61, 83–93.

R Core Team. (2023). R: A language and environment for statistical computing. R Foundation for Statistical Computing, Vienna, Austria. https://www.R-project.org/.

Salomão, G. N., Dall'Agnol, R., Sahoo, P. K., Angélica, R. S., de Medeiros Filho, C. A., Júnior, J. D. S. F., … & de Siqueira, J. O. (2020). Geochemical mapping in stream sediments of the Carajás Mineral Province: Background values for the Itacaiúnas River watershed, Brazil. Applied Geochemistry, 118, 104608.

Taylor, M. P. (2007). Distribution and storage of sediment-associated heavy metals downstream of the remediated Rum Jungle Mine on the East Branch of the Finniss River, Northern Territory, Australia. Journal of Geochemical Exploration, 92(1), 55–72.

Vermeesch, P. (2019). Exploratory analysis of provenance data using R and the provenance package. Minerals, 9, 193.

2

Getting Started

2.1 Downloading and Installing R and RStudio

R is primarily a command-line language, which often necessitates extensive coding and debugging. Therefore, it is commonly utilised in conjunction with RStudio to streamline the development process. Combined they are a powerful and efficient tool for data analysis and statistical computing. To get started using these tools, you'll need to download and install both R and RStudio on your computer.

The first step is to download R. You can download R from the official website of the R project, which is located at https://cran.r-project.org/. On the website, you'll see links to download R for Windows, Mac, or Linux, depending on your operating system. Once you've clicked the appropriate link, you'll be taken to a page with download options for the latest version of R.

Once you've downloaded the R installation package, you'll need to install it on your machine. The installation process is straightforward and similar to the installation of other software. On Windows, you'll double-click the downloaded '.exe' file to start the installation process. On Mac, double-click the downloaded '.dmg' file to start the installation process. On Linux, you'll extract the downloaded '.tar.gz' extension file and run the R installation script.

The next step is to download RStudio. RStudio is an integrated development environment (IDE) for R that provides a user-friendly interface. You can download RStudio from the official RStudio website, which is located at https://rstudio.com/.[1] On the website, you'll see a link to download RStudio for Windows, Mac, or Linux, depending on your operating system. Once you've clicked the link, you'll be taken to a page with download options for the latest version of RStudio. If you are at beginner level, download the RStudio Desktop version. There are several other versions including an advanced one named RStudio Pro, and another one named RStudio Server version is for advanced users who want to create a server using RStudio.

[1] Currently you are redirected to the new https://posit.co/ site.

DOI: 10.1201/9781032651880-3

Once you've downloaded the RStudio installation package, you'll need to install it on your machine. The installation process for RStudio is similar to the installation of other software.

After the installation is complete, you'll have both R and RStudio installed on your machine and ready to use. To start using RStudio, you'll simply launch the RStudio application. From within RStudio, you'll have access to all the features of R as well as many additional features and tools that make working with R easier and more efficient.

2.2 Understanding the RStudio Environment and Workspace

RStudio is an IDE that provides a more user-friendly interface for working and makes it easier to organise and perform data analysis and statistical computing tasks. In this section, we'll take a look at the RStudio environment and workspace and discuss how to get the most out of RStudio. Deeper details on using and fine-tuning RStudio can be found in the web page of RStudio[2] or in Verzani (2011).

The RStudio environment consists of several panes organised by the user that can contain the console, the source editor, the environment, the history, the plot, the files, etc. (Figure 2.1). Each of these panes serves a specific purpose and provides different functions for working with R.

The Console panel is where you can enter R commands and immediately see the output. This is where you'll spend the majority of your time when using R. The source editor, i.e. the Scripts panel area, is where R scripts are created and edited. A script is a plain text file that contains R code; the scripts are used to automate tasks, run code in batch mode, and share code with others. The Environment panel displays the objects that are currently in the workspace (i.e. the computer memory), including data frames, vectors, and functions. The History panel displays a record of the commands that you've entered in the console. The Plot panel shows the plot outputs from your code, the Help panel displays the information about a function or package, the Files panel displays a list of files and directories in your workspace, whereas the Viewer panel displays interactive information.

In addition to the panels, RStudio provides several tools that make programming and conducting several tasks easier. For example, RStudio includes a code editor with syntax highlighting, code completion, and other features that facilitate writing and editing the code. A plotting tool allows users to create, customise, and export graphs and plots. RStudio also provides a package manager to install and manage R packages more easily. R

[2] https://www.r-studio.com/downloads/Recovery_Manual.pdf or https://rstudio-education.github.io/hopr/starting.html

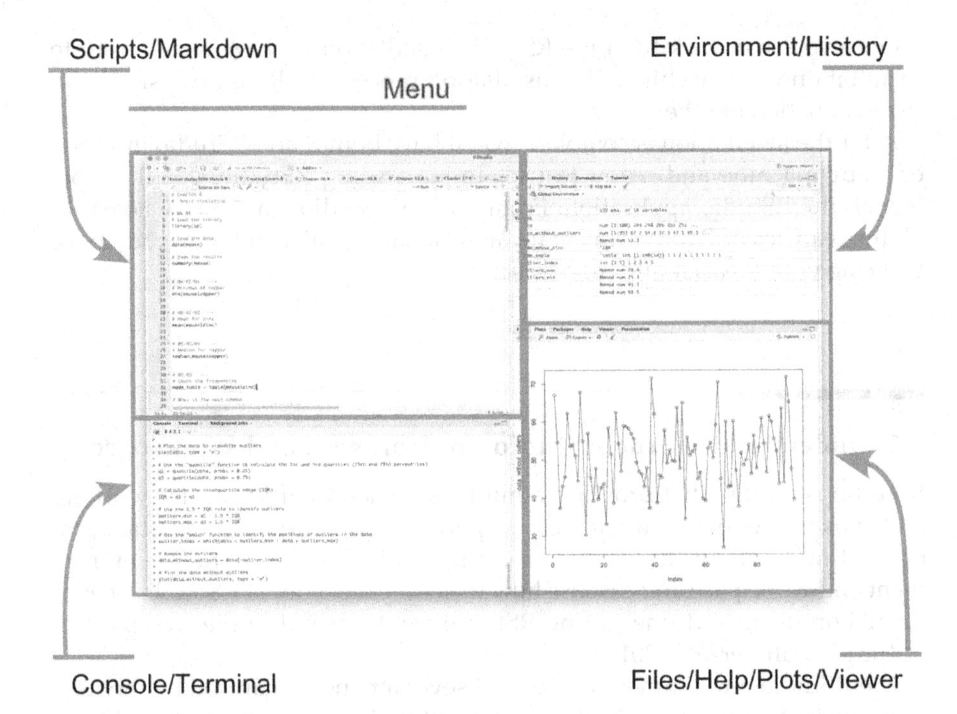

FIGURE 2.1
The RStudio environment. The different panes are outlined.

packages are collections of functions, data, and documentation that can be used to extend its functionalities (see Section 2.6 for more).

RStudio uses the concept of "Projects" to manage work. A project in RStudio is essentially a working directory associated with a specific R endeavour, which can include the R scripts, data files, and other related files. When a new project is created, RStudio sets the project's directory as the working directory.

2.3 Basic Syntax and Data Structures in R

One of the reasons for the popularity of R is its simple and intuitive syntax. In this section, we'll take a look at its basic syntax and data structures. Let the journey commence.

The R syntax is similar to other programming languages and consists of keywords, operators, functions, and variables. Keywords are reserved words that have a specific meaning, such as 'if' and 'for'. Operators are used to perform operations on variables, such as addition, subtraction, or comparison. Functions are predefined procedures that perform a specific task, and they are activated by their name and configured by passing arguments or parameters to them. Variables are used to store data, and they are assigned values using the equal sign '=' or the attribution sign '<-'. In the examples of this book, we will always use the '=' sign, but in literature and internet examples, you might find the '<-' sign often.

Several data structures are available that can be used to store data. The most commonly used data structures in R include scalars, vectors, matrices, arrays, data frames, and lists (Figure 2.2).

Scalars can have different types, including numeric, integer, logical, character, complex, and raw. A scalar is a single value that is not part of a vector or matrix. It can be a single numeric value, a single character, a single logical value, etc. To create a scalar variable named 'x', we use the code:

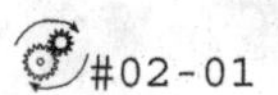
#02-01

```
x = 42
```

Vectors are one-dimensional arrays of data, and they can be numeric, character, or logical (among other types). To create a numeric vector in R, one can

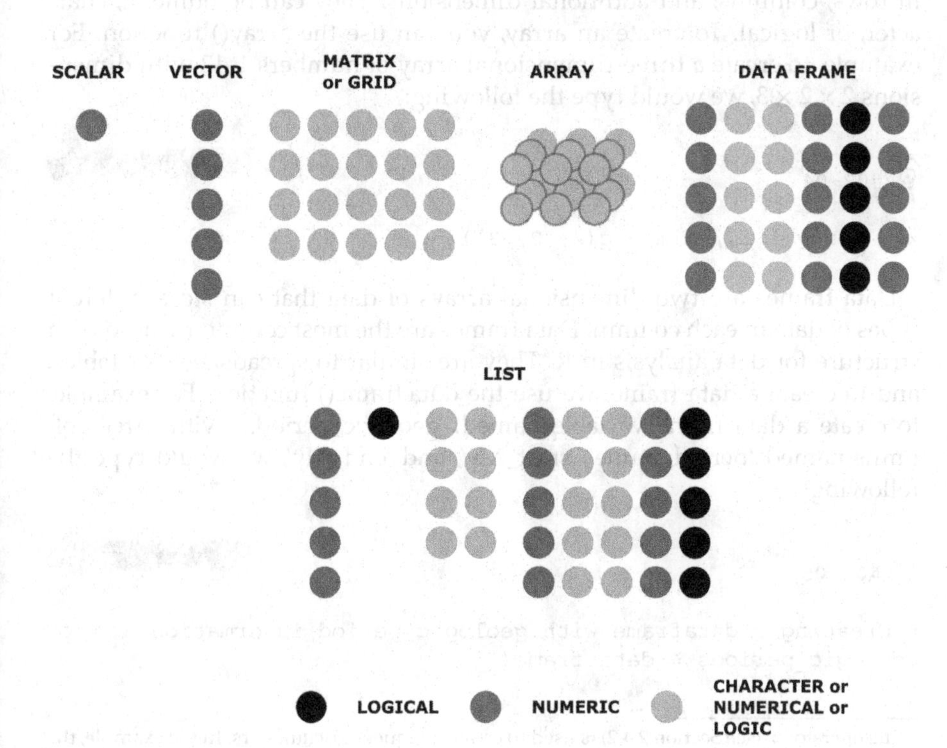

FIGURE 2.2
Basic types of data structures in R.

use the c() function, which stands for combine. For example, to create a vector of numbers, named 'x', we would type the following:

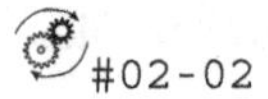#02-02

```
x = c(1, 2, 3, 4, 5)
```

Matrices are specifically two-dimensional arrays that can also be numeric, character, or logical (among other types). To create a matrix, you can use the matrix() function. For example, to create a matrix with the numbers 1–6,[3] named also 'x', we would type the following:

#02-03

```
x = matrix(1:6, ncol = 2, nrow = 3)
```

An array is a data structure used to store homogeneous data elements of the same data type. It is a multi-dimensional collection of elements organised in rows, columns, and additional dimensions. They can be numeric, character, or logical. To create an array, you can use the array() function. For example, to create a three-dimensional array of numbers 1–12 with dimensions 2 × 2 × 3, we would type the following:

#02-04

```
x = array(1:12, dim = c(2, 2, 3))
```

Data frames are two-dimensional arrays of data that can store different types of data in each column. Data frames are the most commonly used data structure for data analysis in R. They are similar to spreadsheets or tables, and to create a data frame, we use the data.frame() function. For example, to create a data frame variable named, 'geologic_periods', with three columns named 'period_name', 'start_My', and 'end_My', we would type the following[4]:

#02-05

```
# Creating a dataframe with geologic period information
geologic_periods = data.frame(
```

[3] The operator ':' (see Section 2.4.2) is used to create a sequence of numbers. In the example, the operation 1:6 creates a list with the integer numbers 1–6.

[4] Notice the '#' symbol in this code snippet that indicates R that the text in this line is a comment; therefore, it is not executable.

```
  period_name = c("Cambrian", "Devonian", "Permian"),
  start_My = c(541, 419, 298),
  end_My = c(485, 359, 252)
)
```

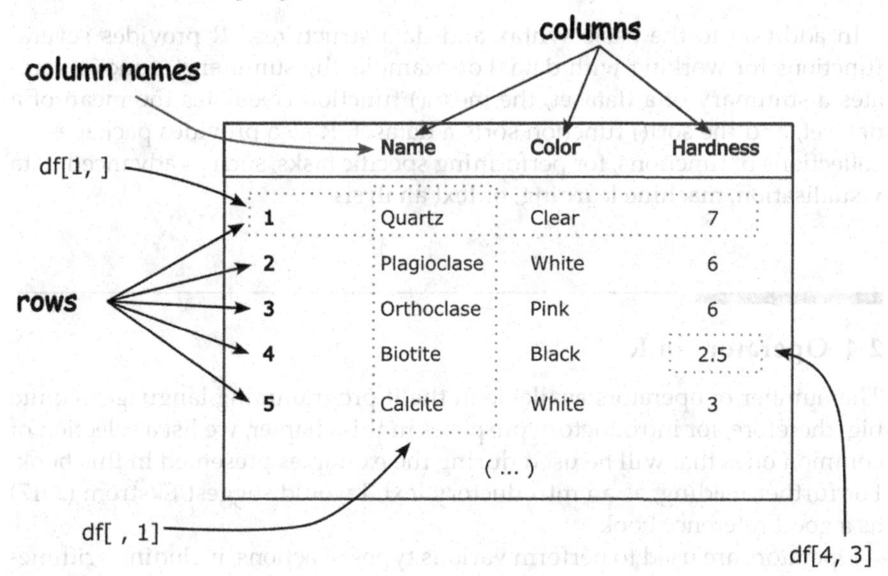

FIGURE 2.3
Illustration of the structure of a data frame.

Figure 2.3 displays an illustration of the structure of a data frame named 'df'. This data frame has three columns named 'Name', 'Color', and 'Hardness' and five rows. Notice that the row numbers of a data frame in R start with number 1.

Lists are collections of data, and they can contain any type of data, including vectors, matrices, arrays, data frames, and even other lists. To create a list, you can use the list() function. For example, to create a list named 'geology_data', with two data frames, one containing mineral names and corresponding hardness's and the other one with rock names and its corresponding seismic velocity, we would type the following:

#02-06

```
# Creating a list with two data frames
geology_data = list(
  minerals = data.frame(
      mineral_name = c("Quartz", "Feldspar", "Calcite"),
      hardness = c(7, 6, 3)
```

```
    ),
    rocks = data.frame(
        rock_name = c("Granite", "Sandstone", "Limestone"),
        velocity = c(6000, 4000, 5000)
    )
)
```

In addition to the basic syntax and data structures,[5] R provides several functions for working with data. For example, the summary() function creates a summary of a dataset, the mean() function calculates the mean of a dataset, and the sort() function sorts a dataset. R also provides packages, or collections of functions, for performing specific tasks, such as advanced data visualisation, machine learning, or text analysis.

2.4 Operators in R

The number of operators available in the R programming language is quite big; therefore, for introductory purposes in this chapter, we list a selection of common ones that will be used during the examples presented in this book. For further reading, as an introductory text, I would suggest Ekstrom (2017) as a good reference book.

Operators are used to perform various types of actions, including arithmetic, logical, comparison, or assignment operations.

2.4.1 Basic Operators

Basic operators are the ones that perform simple arithmetic, logical, comparison, and assignment operations. A few selected operators are presented in Table 2.1.

It's important to note that the precedence of operators determines the order in which they are evaluated in expressions. Typically, arithmetic operations are evaluated before comparison operations, which in turn are evaluated before logical operations. Parentheses have the highest precedence and can be used to explicitly control the evaluation order. In complex or unclear situations, it is advisable to use parentheses to clarify the execution order of operations.

2.4.2 Other Operators

Apart from the basic operators, there are a plethora of other operators that perform specific functions and that are often used in R programming. Most

[5] In fact, these functions are provided by the library named 'base' R that provides the fundamental set of functions and data structures that are automatically available when you start using R. It includes a wide range of essential tools for data manipulation, statistical analysis, plotting, and programming.

TABLE 2.1

Selected Operators in R

Arithmetic Operators	
+	Addition
-	Subtraction
*	Multiplication
/	Division
^	Exponentiation
( )	Parenthesis
%/%	Integer division
%%	Modulus (remainder)
Logical Operators	
&	Logical AND
|	Logical OR
!	Logical NOT
Comparison Operators	
==	Equal to
!=	Not equal to
<	Less than
>	Greater than
<=	Less than or equal to
>=	Greater than or equal to
Assignment Operators	
<-	Assignment operator
=	Another form of assignment operator

of these operators will be applied in examples in this book, and when its use is not clear, the necessary explanation is provided.

A selection of specific operators is listed in Table 2.2.

These operators are commonly used for data manipulation, analysis, and visualisation. By understanding its functionalities and the corresponding syntax, it is possible to design more efficient and effective R code.

2.4.3 A Special Operator: The Pipe

Among the operators with specific functions previously listed, there is one that deserves a special note, the pipe operator as presented in (Wickham et al., 2023a; Wickham et al., 2023b). Sometimes, when browsing code examples and repositories such as stack overflow,[6] one might find cryptic coding,

[6] https://stackoverflow.com/ is a popular question-and-answer website specifically designed for programming and software development. It serves as a community-driven platform where individuals can ask questions related to coding, software development, and other technical topics.

TABLE 2.2
More R Operators for Specific Functions

Element-wise Operators	
:	Sequence operator (creates a sequence of numbers)
c()	Concatenation operator (concatenates objects into a vector)
Matrix Operators	
%*%	Matrix multiplication
t()	Matrix transpose
Apply Family Operators	
apply()	Applies a function to the margins of an array
lapply()	Applies a function to each element of a list and returns a list
sapply()	Simplifies the result of lapply() into an array or matrix
tapply()	Applies a function over subsets of a vector
mapply()	Applies a function to the first elements of several lists, and then to the second elements, and so on
Subsetting Operators	
[ ]	Subsetting operator (selects elements from an object)
[[ ]]	Double bracket operator (selects elements from a list)
$	Dollar sign operator (selects elements from a list or data frame by name)
filter()	Filters rows from a data frame based on a condition
slice()	Selects rows from a data frame based on a condition
select()	Selects columns from a data frame
arrange()	Sorts rows in a data frame
Grouping Operators	
group_by()	Groups data into subsets based on one or more variables
summarize()	Summarises grouped data into a single row

supposedly in R language. One common example is the generalised use along the code of the '%>%' operator, also known as the 'pipe' operator. This operator is a key feature of the 'dplyr' and 'magrittr' packages. It allows to chain multiple operations together in a single line of code, making the code more concise, and hopefully more readable.

The '%>%' operator takes the output from the expression on its left and passes it as the first argument to the function on its right. This makes it easy to perform multiple operations on a data frame without having to repeatedly type the data frame name.

For example, consider the code below, where the filter() function is used to keep only the rows in the 'df' data frame where the 'hardness' column is greater than 5. The result of the filter() function, the 'filtered_data' data frame, is then passed to the select() function, which is used to select only the 'mineral_names' and 'color' columns from the filtered data frame.

```
filtered_data = filter(df, hardness > 5)
selected_data = select(filtered_data, mineral_names, color)
```

The same code is greatly simplified using the pipe operator.

```
df %>% filter(hardness > 5) %>% select(mineral_names, color)
```

The use of the '%>%' operator makes it easy to see the sequence of operations performed on the data frame and makes the code more readable and easier to maintain.

2.5 Flow Control

Flow control functions are functions that manage the execution of the code. These are common in many programming languages and for those who are not familiar with these, I suggest the book 'The art of R programming' by Matloff (2011), where more details on these and other operators are explained.

These functions help to make decisions and repeat sets of actions based on conditions. In this point, we will discuss the most commonly used flow control functions, namely if()-else statements, for() loops, and while() loops.

2.5.1 If()-Else Statements

If()-else statements are one of the most basic flow control functions. They allow to execute a certain block of code, that is between braces '{} ', if a certain condition is met, and execute another block of code, the one that is between braces after the else statement, if the initial condition is not met. The syntax for an if()-else statement is as follows:

```
if (condition) {
  # code to be executed if condition is TRUE
} else {
  # code to be executed if condition is FALSE
}
```

Here's an example of using an if()-else statement written in R:

```
#02-07

x = 5

if (x > 0) {
  print("x is positive")
} else {
  print("x is not positive")
}
```

This code will print, in the console panel, the text 'x is positive'. The first line of code assigns the value of 5 to the 'x' variable, and therefore, the condition 'x > 0' is true. The result in the console panel is:

```
[1] "x is positive"
```

2.5.2 For() Loops

For() loops are used to repeat a certain block of code, that is between braces, for a specified number of times. The syntax for a for() loop is as follows:

```
for (variable in sequence) {
  # code to be executed for each iteration
}
```

Suppose that we want to loop through the sequence of numbers 1–5 and print each one. The 1:5 operation retrieves the sequence of numbers. Here's an example of the code using a for() loop:

```
#02-08

for (i in 1:5) {
  print(i)
}
```

This code will print the numbers 1–5 in the console panel, one number per iteration of the loop. The result presented in the console panel is as follows:

```
[1] 1
[1] 2
[1] 3
```

```
[1] 4
[1] 5
```

2.5.3 While() Loops

While() loops are similar to for() loops, but they repeat a certain block of code while a certain condition is met. The syntax for a while() loop in R is as follows:

```
while (condition) {
  # code to be executed while condition is TRUE
}
```

Here's an example of using a while() loop for printing the numbers 1–5. The 'x' variable is used as a counting variable and also to print its value. Inside the loop, the 'x' variable is increased; otherwise, the loop would be infinitely printing the number 1.

```
#02-09

x = 1

while (x <= 5) {
  print(x)
  x = x + 1
}
```

The result is printed in the console panel as in the for() loop example.

Whether you're making decisions with if()-else statements, repeating actions with for() loops, or repeating actions while a certain condition is true with while() loops, these functions provide the ability to write more complex and powerful code.

2.6 Packages and Libraries

In R, the terms 'package' and 'library' are often used interchangeably, but they technically refer to two different things. A 'package' is a collection of functions, data sets, and documentation that are bundled together for distribution and use by others. Packages are distributed via repositories such as CRAN (Comprehensive R Archive Network) and can be easily installed and loaded using the install.packages() and library() functions, respectively. A 'library' is a location in the file system of the computer where packages are stored. When a package is installed, the functions and data sets are copied to the library directory, and the package can be loaded into the programming environment using the library() function.

In practice, when you install a package, it is stored in a library, which you can then load and use. In this book for simplicity of language, I will use the terms interchangeably.

Packages are key components of the R programming language, offering a vast array of tools and resources to help users effectively and efficiently solve the different computational and statistical problems. They are designed to extend the functionality of some of the pre-installed packages that come with the R installation, providing additional tools and capabilities that can be easily accessed and used.

There are several pre-installed packages that provide the basic functionalities for data manipulation, visualisation, statistical analysis, machine learning, and more. These packages are part of the initial distribution of R, so it is not necessary to install them separately. Here's a list of the essential pre-installed packages:

'base': It is the fundamental package that automatically loads when R is started. It includes the essential functions and data types needed for basic R programming.

'methods': It provides support for object-oriented programming in R and contains functions related to generic functions, method dispatch, and class definitions.

'utils': It contains utility functions that aid in various tasks, such as installing and managing packages, reading and writing data, handling files and directories, and more.

'grDevices': It includes functions for setting up graphical devices (such as X11, PDF, and PNG), managing graphical parameters (e.g. colours, fonts, and line types), and handling graphical output. This package is crucial for generating and customising plots and graphics.

'stats': It includes functions for basic statistical measures, probability distributions, hypothesis testing, linear and nonlinear modelling, and more.

'graphics': It provides various functions for creating plots and graphical visualisations. It includes functions for scatter plots, bar plots, histograms, box plots, and more, making it easy to visualise data.

'datasets': It contains various datasets that are commonly used for teaching and practising data analysis in R. It includes datasets like 'iris', 'mtcars', and 'swiss'.

Some examples of the most popular packages that are nor pre-installed include 'ggplot2' for data visualisation, 'dplyr' for data manipulation, or 'caret' for machine learning.

The R community is also known for its open-source spirit and collaboration, and many users contribute to the development of new packages and tools as well as sharing their work and experiences with others. This has led to a thriving and supportive community of users and developers and has helped to ensure that R remains one of the most widely used programming languages in data science and statistical analysis.

To install a new package, one must use the install.packages() function in the R console. Alternatively, the RStudio has behind the menu 'TOOLS' an option for installing packages that are available in repositories like the CRAN (see Section 2.7).

For example, to install the 'dplyr' package, type in the console the following code:

```
#02-10

install.packages("dplyr")
```

Once the package is installed, one can load it into the current R session using the library() function. For example:

```
#02-11

library(dplyr)
```

This will allow the use of the functions and features provided by this package. The packages only need to be loaded once per R session and can be used as many times as necessary.

Often, in this book, the examples are in sequences of code that go deeper in the analysis. In that case, we just refer to load the necessary libraries at the beginning of each code sequence. When trying to use a function that is not previously installed, R will produce an error message in the console. Usually, these messages are coloured in red.

In addition to packages and libraries, there is a wealth of online resources for users to access, including online forums, discussion groups, and mailing lists where users can connect, ask questions, and share knowledge. There are also a number of tutorials, books, and online courses available for users who are just getting started or who want to deepen their understanding of the language and its capabilities. For more of these resources, please refer to Section 2.11.

2.7 An Overview of CRAN (Comprehensive R Archive Network)

CRAN,[7] acronym for the Comprehensive R Archive Network, is a collection of over 19878 packages (July 2023 and growing) and a central repository to find, download, and install the latest packages and updates. It is a key component of the R ecosystem, providing users with a vast array of tools,

[7] Accessible in https://cran.r-project.org/

libraries, and resources that can be easily accessed and utilised to enhance the programming experience.

CRAN was established in 1997 and since then has become one of the largest and most comprehensive repositories of open-source packages. The packages are designed to extend the functionality of the pre-installed ones and can be used for a myriad of applications, including data visualisation, machine learning, spatial data analysis, or statistical modelling.

One of the key benefits of CRAN is its ease of use. Users can access the repository directly from within the R environment and can easily install and manage packages using the install.packages() function. This function can also be used to upgrade existing packages, ensuring that users always have access to the latest version of the packages they use.

In addition to ease of use, CRAN also provides a level of quality assurance for its packages. All packages are reviewed and tested by the CRAN team to ensure that they are well-documented, reliable, and free from errors or bugs. This ensures that users can trust the packages they download and use and that they can rely on them for their work.

Another benefit of CRAN is the sheer breadth and depth of the packages it contains. Whether you are interested in data visualisation, spatial analysis, machine learning, or complex statistical modelling, you are likely to find a package in CRAN that meets your needs. The packages in CRAN are also well-maintained, with regular updates and bug fixes, ensuring that they continue to be relevant and useful over time.

In addition to its packages, CRAN also provides a wealth of resources and support for its users. There are online forums and discussion groups where users can connect with and get help with their programming questions.

One of the most important things to understand about CRAN is that it is an open-source community and that its development and growth are driven by contributions from the R user community. Many users contribute to the development of new packages as well as sharing their work and experiences with others.

2.8 Popular Packages for Data Analysis and Visualisation

One of the most required and used functionalities of the R programming language is its capabilities to perform data analysis and visualisation. The reason for its popularity is the vast array of packages available. In this book, we will deal with some of the most useful packages in what concerns data management and analysis, namely:

'dplyr': It is used for data manipulation and cleaning. It is founded in a grammar for handling data (Wickham et al. 2023b) and profits from a concise syntax for filtering, transforming, and aggregating data. 'dplyr' is a must-have package for anyone working with large datasets.

'tidyr': It is specially designed for reshaping data; it provides functions for transforming data from wide to long formats and vice versa (Wickham et al. 2023c). This is a useful package for anyone working with data in a tabular format.

'ggplot2': It is the most popular package for creating data visualisations, such as bar graphs, scatter plots, and histograms (Barber, 2023). It provides a high-level interface for creating complex visualisations with ease.

These are the cherry-picked packages that we consider as fundamental. In the next chapters, these, and many others, will be introduced and explained as necessary. One must emphasise that each specific subject has its own group of packages. For the case of spatial analysis, there is a set of specialised packages that will be detailed in the following chapters.

There are many other packages available in CRAN, and new packages are added regularly. To stay up to date with the latest packages, it is important to regularly check CRAN for new packages and updates to existing packages.

2.9 R Packages for Geologists

Apart from the most commonly used R packages, there are specialised ones for geology matters. Here are some popular ones created and used by geologists, petrologists, and mineralogists:

GCDkit[8]: It is a renowned package created by Janousek and collaborators that is used for geochemical data analysis, specifically for geochemistry, mineralogy, and petrology (Janousek et al., 2016).

ggtern[9]: It is a powerful tool for creating ternary plots in R, which are often used in geology for visualising the composition of mineral systems (Hamilton & Ferry, 2018). It is an extension of the 'ggplot2' package.

geophys[10]: It is a package that provides functions for geophysical modelling and inversion, including gravity and magnetic modelling (Lees, 2018).

provenance[11]: It is a package in R that provides datasets and R scripts for conducting provenance analysis and other geological research

[8] Can be accessed in: https://www.gcdkit.org/
[9] Can be accessed in: https://cran.r-project.org/web/packages/ggtern/index.html
[10] Can be accessed in: https://www.rdocumentation.org/packages/geophys/versions/1.4-1
[11] Can be accessed in: https://www.ucl.ac.uk/~ucfbpve/provenance/

tasks. It aims to develop the visualisation and interpretation of 'Big Data' in the context of sedimentary provenance analysis (Vermeesch et al., 2016).

whitebox[12]: It is a frontend package for 'Whitebox Tools' library that provides advanced tools for geospatial analysis including geomorphological, hydrological and remote sensing analysis (Lindsay, 2016; Wu & Brown, 2022).

These packages provide a range of functions for working with geologic, geomorphologic, petrologic, mineralogic, and geophysical data, including retrieval, manipulation, and visualisation.

2.10 Versions and Updates

The R programming language and RStudio are frequently updated with new releases, each bringing enhancements, bug fixes, and additional functionalities. These updates reflect the active development and community-driven nature of R.

When working with R, it is crucial to carefully choose the version that best suits your needs, considering the specific libraries and tasks involved. Different versions of R may have varying compatibility with certain packages, and using an appropriate version ensures smooth execution and access to the desired features. It is recommended to refer to library documentation or community resources to determine the recommended version for compatibility.

However, it is worth noting that in certain cases, using older versions may be necessary. For instance, if specific packages or functions require compatibility with a particular R version, reverting to an older release becomes essential.

The R community is mindful of this need and provides options to access and install older versions of R. Again, the CRAN is the primary repository for R packages and provides access to previous versions of R. It offers several versions for users to choose from, ensuring flexibility in selecting the most suitable version for their needs.

2.11 Other Resources for Learning R

When interested in deepening the R skills, there are many resources available to help one get started. Some of the most helpful resources for learning R include the following:

[12] Can be accessed in: https://www.whiteboxgeo.com/manual/wbt_book/r_interface.html

Books: There are more than a handful of books available that cover different aspects of R, when considered adequate; the bibliographic or web references are listed at the end of each chapter.

Online tutorials: There are many online tutorials available that cover different aspects of R. For example, the DataCamp website has a comprehensive library of tutorials, ranging from introductory courses[13] to advanced topics. Another great resource is Codecademy's R Programming course, which covers the basics of R programming.

Forums: There are many forums available where you can ask questions and get help with R. For example, Stack Overflow is 'the' resource for getting help with most of the programming languages, including R. The R-help mailing list is another resource, where you can ask questions and receive answers from experienced R users.

MOOCs: MOOCs, or massive open online courses, are a great way to learn R. As an example, Coursera or Udemy academies offer multiple MOOCs, from basic[14] to advanced topics.[15]

Websites: There are many websites that provide resources for learning R. For example, the RStudio website has a comprehensive library of tutorials, resources, and support. The CRAN website is another great resource, providing access to the latest version of R as well as documentation and tutorials. Additionally, R-Bloggers emerges as a standout community-driven platform, aggregating blog posts from around the web on various R-related topics. This site is particularly useful for keeping abreast of the latest trends, techniques, and packages within the R ecosystem, thanks to its wide range of articles and tutorials contributed by a diverse community of R users, ranging from data scientists to statisticians and researchers.

YouTube: There are many YouTube channels dedicated to R and data science. For example, StatQuest[16] is a great channel for learning about statistics and data science.

2.12 Concluding Remarks

With so many options, it's easy to find the right resources to help you get started and advance your skills. Whether you're a beginner or an experienced R user, these resources can help you learn and grow as a programmer.

[13] See for example: https://www.datacamp.com/courses/free-introduction-to-r
[14] See for example: https://www.coursera.org/projects/getting-started-with-r
[15] See for example: https://www.udemy.com/course/intermediate-spatial-data-analysis-with-r-qgis-more/
[16] The channel is at: https://www.youtube.com/@statquest

Anyway, this book will guide you from the most common R basics to more advanced tools, especially the ones needed for spatial analysis with practical applications in geology.

References

Barber, C. (2023). Visualizing data with ggplot2 in R. Available in: https://library.rice.edu/sites/default/files/documents/ggplot_resource.pdf

Ekstrom, C. T. (2017). R Primer. CRC Press. Taylor & Francis Group, Boca Raton, FL.

Hamilton, N. E. & Ferry, M. (2018). ggtern: Ternary diagrams using ggplot2. Journal of Statistical Software, Code Snippets, 87(3), 1–17. https://doi.org/10.18637/jss.v087.c03

Janousek, V., Moyen, J. F., Martin, H., Erban, V., & Farrow, C. (2016). Geochemical Modelling of Igneous Processes – Principles and Recipes in R Language. Bringing the Power of R to a Geochemical Community. Springer-Verlag, Berlin, Heidelberg, pp. 1–346.

Lees, J. (2018). geophys: Geophysics, Continuum Mechanics, Gravity Modeling. R package version 1.4-1. Available at CRAN: https://CRAN.R-project.org/package=geophys

Lindsay, J. B. (2016). Whitebox GAT: A case study in geomorphometric analysis. Computers & Geosciences, 95, 75–84. http://dx.doi.org/10.1016/j.cageo.2016.07.003

Matloff, N. (2011). The Art of R Programming: A Tour of Statistical Software Design. No Starch Press, San Francisco, CA.

Vermeesch, P., Resentini, A., & Garzanti, E. (2016). An R package for statistical provenance analysis. Sedimentary Geology, 336, 14–25. https://doi.org/10.1016/j.sedgeo.2016.01.009

Verzani, J. (2011). Getting Started with RStudio. O'Reilly Media, Inc, Sebastopol, CA.

Wickham, H., Çetinkaya-Rundel, M., & Grolemund, G. (2023a). R for data science. O'Reilly Media, Inc, Sebastopol, CA.

Wickham, H., François, R., Henry, L., Müller, K., & Vaughan, D. (2023b). dplyr: A Grammar of Data Manipulation. R package version 1.1.4. https://CRAN.R-project.org/package=dplyr

Wickham, H., Vaughan. D., & Girlich, M. (2023c). _tidyr: Tidy Messy Data_. R package version 1.3.0, <https://CRAN.R-project.org/package=tidyr>.

Wu, Q., & Brown, A. (2022). whitebox: 'WhiteboxTools' R frontend. R package version 2.2.0. https://CRAN.R-project.org/package=whitebox

3

Reading and Writing Data

Reading and writing data from and to various file formats is the foremost task in any data science operation. R has a pleiad of functions and options to accomplish these tasks that go from the 'base' R package to advanced or supplier specific packages. Among these, particularly relevant to the open science paradigm is the possibility of fetching open data that is deposited in web servers. In this section, we'll explore some of the different methods.

3.1 Data in CSV Files

Comma-Separated Values, i.e. CSV, files are a simple and widely used format for storing and sharing tabular data. This format consists of rows and columns of data, with each value separated by commas.[1] The files are plain text, making them easy to read, edit, and transfer between different software applications. When it comes to collecting geological data, CSV files are commonly used to manage, analyse, and share information, might the data be provided from field campaigns or from many types of laboratorial equipment (e.g. GPS, digital compass in smartphone, field magnetometer, or SEM-EDS [scanning electron microscope with energy dispersive spectrometry]). For geological applications, its simple format allows an easy data entry, organisation, and analysis.

The CSV files have a structure that consists of i) rows, where each one represents a record or data point, and ii) columns, separated by commas, represent fields or attributes of the record. The first row often serves as a header, providing the names of the columns.

3.1.1 Reading CSV Files

The read.csv()[2] and read.csv2()[3] functions are used to read data from a CSV file. For example, if we have a CSV file named 'data.csv' in the working folder,

[1] In fact, it can be customised to different separators, such as semicolons, tab characters, or other user defined characters.

[2] The read.csv() function is part of the 'utils' package which is a fundamental package that provides various utility functions and tools for performing common tasks. It is a pre-installed package (see Section 2.6) that is automatically available when we install and start R.

[3] The read.csv2() function is designed for reading CSV files that use semicolons as column separators and commas as decimal points in numeric values. It is particularly useful for data files created in European countries where comma is commonly used as a decimal point.

DOI: 10.1201/9781032651880-4

we can read it into the variable named 'data', located in the working environment, with the following code:

```
data = read.csv("data.csv")
```

3.1.2 Writing CSV Files

To write data to a CSV file, we can use the write.csv()[4] function from the 'utils' package. For example, to write the content of the 'data' variable from the previous example to a CSV file named 'output.csv', we can use the following code:

```
write.csv(data, "output.csv")
```

3.2 Retrieving Data from a Website

Many organisations, government departments, and companies make their data available for public use, typically in a compressed file format like ZIP. If the data is available as a file or folder link, it can be downloaded and used for teaching, research, or other types of projects.

In this example, we will retrieve data from the site: https://forest-gis.com/2012/01/portugal-shapefiles-gerais-do-pais.html/

This site contains a link to the Portuguese roads in the link: https://biogeo.ucdavis.edu/data/diva/rds/PRT_rds.zip

For this exercise, one must previously create the necessary folders in the computer. For this case, it is necessary to create the sub-folders named 'Spatial/Data/' and 'Spatial/Data/Roads Pt/' under the working folder location.

Notice also that the code uses two libraries, i.e. the 'sf' and 'terra' packages. If it is the first time that you are using this packages in your R installation, don't forget to previously run the install.packages("sf") and install.packages("terra") commands, otherwise you will receive a friendly message in the console:

```
Error in library(sf): there is no package called 'sf'
```

[4] As for read.csv2(), there is a write.csv2() function. Same comments apply.

Please note this is valid for all the following code snippets, where it's expected that you have already installed the required packages.

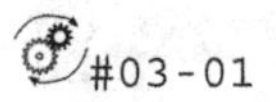#03-01

```
# Load the libraries
library(sf)
library(terra)

# Define the working folder
setwd("Spatial/Data/")

#Define the link to the data
url ="https://biogeo.ucdavis.edu/data/diva/rds/PRT_rds.zip"

# Download the data
download.file(url, destfile = "Roads_Pt.zip")

# unzip the file
unzip(zipfile = "Roads_Pt.zip", exdir = 'Roads Pt')
```

Let's analyse the different steps that compose this code snippet.

setwd() is a function from the 'base' R package to define the working folder, i.e. the place where R will look for files or folders in the computer disk. It is the root of the folders and sub-folders that will be used. 'url' is a string/text variable with the URL[5] address for fetching the data. download.file() is a function from the 'utils' package that is used to download files from the internet. The 'url' parameter is the address of the file to download, while the 'destfile' parameter specifies the filename and location where to save the file. In some cases, the arguments or parameters to a function need to be indicated (e.g. destfile = 'Roads_PT.zip'); other times, the function can 'guess' what argument is being passed in the case the 'url' argument could be indicated as 'url = url', but the function 'understands' the data without an explicit indication that the 'url' argument is the 'url' variable. unzip() is a function from the 'utils' package that is used to extract the contents of a ZIP archive. The 'zipfile' parameter is the name of the ZIP file, while 'exdir' specifies the directory where the contents of the archive will be extracted.

With this code, the working directory is defined as 'Spatial/Data/'. The file will be downloaded to that folder with the name 'Roads_Pt.zip'. After that, the file is unzipped to the folder 'Spatial/Data/Roads Pt'.

---

[5] A URL (Uniform Resource Locator) is a simple way to describe the address or location of a resource, such as a web page, file, or document, on the internet. It's like the 'street address' of a website or an online file allowing to find and access it by typing the URL into a web browser. URLs typically start with 'http://' or 'https://' for websites, followed by the domain name (e.g. www.example.com) and any additional path or file information that leads you to the specific resource you want to view or download.

To verify the results from this operation, use the command dir() from the 'base' R package. This command lists the contents of the corresponding folder.

#03-02

```
dir("Roads Pt/")
```

The result is as follows:

```
[1] "PRT_roads.dbf" "PRT_roads.prj" "PRT_roads.shp" "PRT_roads.
shx"
```

The resulting output in the console shows that there are four files created in the folder which are as follows: 'PRT_roads.dbf', 'PRT_roads.prj', 'PRT_roads.shp', and 'PRT_roads.shx'. These files are the basic ones from an ESRI (Environmental Systems Research Institute) shapefile format[6] (see also Chapter 10).

For reading the files, it is necessary to use the read_sf() function from the 'sf' package. Other functions can be used to acknowledge the data. In the example below, the functions class() from the 'base' R package, st_bbox() from the 'sf' package, crs() from the 'terra' package, and head() and plot() functions, that are extensions[7] of the 'base' package, are used to retrieve information about the 'roads.pt' variable and to plot it.

#03-03/01

```
# Read the data
roads_pt = read_sf("Roads Pt/PRT_roads.shp")

# Verify its class
class(roads_pt)
```

The result is as follows:

```
[1] "sf"              "tbl_df"          "tbl"              "data.frame"
```

It displays a vector of four strings: "sf", "tbl_df", "tbl", and "data.frame". These strings represent the different classes or types of objects that the 'roads_pt' variable belongs to.

"sf" indicates that 'roads.pt' is an object of the 'sf' class. This class is used to store and manipulate spatial data, such as points, lines, or polygons. "tbl_df" and "tbl" both indicate that 'roads_pt' is also a 'tibble' or 'table' object. Tibbles

[6] ESRI. (1998). ESRI Shapefile Technical Description, or see more in: www.esri.com/content/dam/esrisites/sitecore-archive/Files/Pdfs/library/whitepapers/pdfs/shapefile.pdf

[7] Sometimes a package can 'extend' previously existing functions, adding new functionalities or data types to its basic usage. That's what is called to 'extend' a function.

are an alternative and enhanced version of data frames, providing some additional features and improvements for working with tabular data. "data.frame" indicates that 'roads_pt' can also be treated as a traditional data frame object, which is a common way to store and manipulate structured data.

This result is useful for understanding which are the classes or types of objects that 'roads_pt' belongs to, and it is relevant for the analysis or operations that can be performed on this variable.

To retrieve the bounding coordinates of the data, use the st_bbox() function from the 'sf' package.

```
#03-03/02

st_bbox(roads_pt)
```

The result is as follows:

```
     xmin      ymin      xmax      ymax
-31.257694  32.636250  -6.254389  42.143025
```

To know the coordinate reference system[8] (CRS), one must use the crs() function from the 'terra' package.

```
#03-03/03

crs(roads_pt)
```

The result is as follows:

```
[1] "GEOGCRS[\"WGS 84\",\n      DATUM[\"World Geodetic System 1984\",\n
      ELLIPSOID[\"WGS 84\",6378137,298.257223563,\n LENGTHUNIT[\"metre\",1]]],\n
      PRIMEM[\"Greenwich\",0,\n
      ANGLEUNIT[\"degree\",0.0174532925199433]],\n CS[ellipsoidal,2],\n
AXIS[\"geodetic latitude (Lat)\",north,\n ORDER[1],\n
ANGLEUNIT[\"degree\",0.0174532925199433]],\n    AXIS[\"geodetic longitude
(Lon)\",east,\n  ORDER[2],\n  ANGLEUNIT[\"degree\",0.0174532925199433]],\n
ID[\"EPSG\",4326]]"
```

The 'roads_pt' data is in the WGS84 datum format in a latitude and longitude coordinate form.

---

[8] A coordinate reference system (CRS) is a way to describe the location of objects or points on the earth's surface using a set of coordinates. It's like a system of rules that help us understand where something is located on a map or in a geographic space. A CRS defines a consistent method of measuring distances, angles, and directions, allowing us to accurately represent and analyse spatial data. By using a CRS, we can communicate and share geographic information effectively, ensuring that everyone understands the precise location of different features on the earth's surface.

To visualise a sample of the data contained in the 'roads_pt' variable, the head() function can be used, as in the example:

#03-03/04 

```
head(roads_pt)
```

The result is as follows:

```
Simple feature collection with 6 features and 5 fields
Geometry type: MULTILINESTRING
Dimension:     XY
Bounding box:  xmin: -8.865778 ymin: 41.69567 xmax: -8.196134
ymax: 42.13969
Geodetic CRS:  WGS 84
# A tibble: 6 × 6
  MED_DESCRI  RTT_DESCRI     F_CODE_DES ISO   ISOCOUNTRY
       geometry
  <chr>          <chr>          <chr>  <chr> <chr>
       <MULTILINESTRING [°]>
1 Without Median Secondary Route Road      PRT    PORTUGAL
((-8.228927 42.13301, -8.227361 42.13269, -8.216251 42.13042,
-8.20625 4...
2 Without Median Secondary Route Road      PRT    PORTUGAL
((-8.589778 42.04839, -8.587277 42.04894, -8.585333 42.04847,
-8.582306 ...
3 Without Median Secondary Route Road      PRT    PORTUGAL
((-8.589778 42.04839, -8.611834 42.04353, -8.617444 42.03992,
-8.622861 ...
4 Without Median Secondary Route Road      PRT    PORTUGAL
((-8.633639 42.02883, -8.634389 42.02991, -8.638234 42.03542))
5 Without Median Primary Route   Road      PRT    PORTUGAL
((-8.633639 42.02883, -8.626111 42.01647, -8.626638 42.01003,
-8.632722 ...
6 Without Median Primary Route   Road      PRT    PORTUGAL
((-8.651361 41.98283, -8.665944 41.9805, -8.68675 41.96994,
-8.701834 41...
```

The data includes six variables named 'MED_DESCRI', 'RTT DESCRI', 'F_CODE_DES', 'ISO', 'ISOCOUNTRY', and 'geometry'.

The plot() function draws the data. In this case, we plot the 'geometry' variable[9] that contains the geometric information of the data, i.e. the coordinates.

#03-03/05 

```
plot(roads_pt$geometry)
```

[9] To identify a variable from a data frame, the $symbol is used following the data frame name.

FIGURE 3.1
Simple plot of the road's shapefile.

The resulting plot displays the roads downloaded from mainland Portugal and its Atlantic Islands (Figure 3.1).

3.3 Retrieving Data from a Web Server

Data can also be retrieved from a web server[10] that allows queries to its database. This code example is for retrieving information of the earthquakes that occurred in January 2023 and that are available from the USGS (United States Geological Survey) server. To query the server database, the read.csv() function from the 'utils' package is used. The code snippet retrieves the information; selects the columns "time", "mag", "place", and "type" using brackets to subset the data (see more in Section 4.2 in Chapter 4); and displays the first elements using the head() function from the 'utils' package (see more in Section 5.3 in Chapter 5).

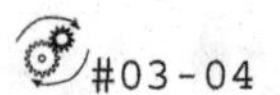#03-04

```
# Define the URL for the earthquake data
url = "https://earthquake.usgs.gov/fdsnws/event/1/query?format=
csv&starttime=2023-01-01&endtime=2023-01-31&minmagnitude=5"

# Load the earthquake data into a data frame
earthquakes = read.csv(url)
```

[10] We define a web server as a website that provides data based on a query. Contrary to the website that is a web location where the data is only deposited and can be downloaded in an 'as is' basis, in a web server the query can be parameterised to retrieve subsets of the data.

```
# Subset the relevant columns from the data frame
earthquakes = earthquakes[, c("time", "mag", "place", "type")]

# Summarise the earthquake data
head(earthquakes)
```

The URL parameters passed to the `earthquake.usgs.gov' server specify how to retrieve earthquake data from the USGS database.

The URL parameters used in the example code are as follows:

format=csv: It specifies that the data should be returned in CSV format.

starttime=2023-01-01: It refers to the start date for the search, in this case January 1, 2023.

endtime=2023-01-31: It refers to the end date for the search, in this case January 31, 2023.

minmagnitude=5: It indicates the minimum magnitude of earthquakes to include in the search results. In this case, only earthquakes with a magnitude of 5 or greater will be included.

By passing these URL parameters to the 'earthquake.usgs.gov' site, the server is able to filter and retrieve only the relevant earthquake data for the specified date range and magnitude range. This makes it easier to work with the data and to analyse the results.

The 'earthquake.usgs.gov' server supports a number of parameters that can be used to customise the data retrieval process. Some of the most commonly used parameters include the following:

- format: It specifies the format of the data to be returned. Available options include csv, geojson, kml, quakeml, and text.
- starttime and endtime: These specify the start and end dates for the search, respectively. The dates should be in the format YYYY-MM-DD.
- minmagnitude and maxmagnitude: These specify the minimum and maximum magnitudes of earthquakes to include in the search results, respectively.
- latitude and longitude: These specify the latitude and longitude of the centre of the search area, respectively.
- maxradius: It specifies the maximum radius of the search area, in kilometres.
- orderby: It specifies the order in which the results should be returned. Available options include time, magnitude, and depth.
- limit: It specifies the maximum number of results to return.
- offset: It specifies the number of results to skip before returning the results.

By passing these and other parameters to the server, one can customise the data retrieval process and retrieve only the data that is relevant to the task purposes. For a complete list of parameters and their descriptions, see the USGS API (Application Programming Interface) documentation.[11]

The results from this query are visualised using the head() function that retrieves the first *n* elements of a vector, matrix, table, data frame, or function. In case of omission of *n*, it retrieves the first six elements.

```
                     time mag                               place      type
1 2023-01-30T19:55:25.696Z 5.3   103 km S of Zurrieq, Malta earthquake
2 2023-01-30T18:45:37.506Z 5.0   105 km ENE of Port Blair, India earthquake
3 2023-01-30T17:07:44.514Z 5.0   152 km S of Hotan, China earthquake
4 2023-01-30T10:25:12.698Z 5.1   126 km SSW of Ishigaki, Japan earthquake
5 2023-01-29T23:49:38.212Z 5.7        southern Xinjiang, China earthquake
6 2023-01-29T19:57:14.754Z 5.0    171 km NW of Tobelo, Indonesia earthquake
```

3.4 Retrieving Data from Excel Files

For reading some types of files, it is necessary to use special packages. Excel is one of these cases. There are several packages for reading the Excel files, and for this example, we will use the 'readxl' package that supports both the legacy '.xls' format and the modern xml-based '.xlsx' format. 'readxl' is part of a bigger collection of R packages used for data science named 'tidyverse'.[12]

3.4.1 Reading Excel Files

The read_excel() function from the 'readxl' package is used to read data from Excel files. In the following example, an Excel file named 'data.xlsx' that is in the working directory is read and its contents are stored in the variable named 'data' in the working environment.

```
# Load the library
library(readxl)

data = read_excel("data.xlsx")
```

3.4.2 Writing Excel Files

To write data to an Excel file in R, we use the 'writexl' package. This package has only one function, i.e., 'write_xlsx' that allows the creation of an

[11] https://earthquake.usgs.gov/fdsnws/event/1/

[12] For more information on 'tidyverse' collection, visit their website at https://www.tidyverse.org/

Excel format file in the disk. For instance, to write the data from the variable named 'data' from the previous example to an Excel file named 'output.xlsx', we use the following code:

```
# Load the library
library(writexl)

write_xlsx(data, "output.xlsx")
```

Note that there are several other packages that can be used to read and write Excel files, besides 'readxl' and 'writexl'. A few examples are as follows:

'openxlsx'[13]: It was created by Schauberger & Walker (2023), which provides a high-level interface for reading, writing, and manipulating Excel files. It supports a wide range of Excel features, including formatting, charts, and worksheets, and is particularly well-suited for working with large or complex Excel files.

'xlsx'[14]: It was developed by Dragulescu & Arendt (2020), which provides a low-level interface for reading and writing Excel files. It supports a more limited range of Excel features than 'openxlsx', but it is faster and more memory-efficient, making it a good choice for working with large datasets. Ensure you have a jdk (Java Development Kit, version higher than or equal to 1.5) installed for your operating system.

'XLConnect'[15]: It was provided by Mirai Solutions GmbH (2023) which has a comprehensive set of functions for reading, writing, and manipulating Excel files. It supports a wide range of Excel features, including formatting, charts, and worksheets, and is particularly well-suited for working with Excel files that contain complex data structures or macros.

The best package for a task will depend on the complexity of the data, the size of the datasets, and the specific requirements for reading, writing, and manipulating the files.

---

[13] For reference, see https://ycphs.github.io/openxlsx/

[14] For reference, see http://www.sthda.com/english/wiki/r-xlsx-package-a-quick-start-guide-to-manipulate-excel-files-in-r

[15] For reference, see http://www.sthda.com/english/articles/2-r/4-xlconnect-read-write-and-manipulate-microsoft-excel-files-from-within-r/

3.5 Other Types of Data

In addition to the packages and examples mentioned above, there are a number of other libraries for importing and exporting data from other sources. For example, the 'rJava' library provides an interface for working with Java, while the 'rpy2' package provides an interface for working with Python. The 'RJSONIO' package provides functions for working with JSON data, and the 'RCurl' package provides functions for working with data from the web.

3.6 Concluding Remarks

This chapter highlights the paramount importance of proficiently reading and writing data, where the existing R functions are simple and clear, particularly within the context of geological applications. The capacity of R to seamlessly handle diverse data formats, coupled with its extensive functionality, empowers the users to navigate through intricate datasets with ease.

In the journey ahead, the proficiency gained in reading and writing data will undoubtedly serve as a solid foundation for more advanced geological analyses and contribute to the continued advancement of our understanding of earth's processes.

References

Dragulescu, A., & Arendt, C. (2020). xlsx: Read, write, format Excel 2007 and Excel 97/2000/XP/2003 Files. R package version 0.6.5. https://CRAN.R-project.org/package=xlsx

Mirai Solutions GmbH. (2023). XLConnect: Excel connector for R. R package version 1.0.8. https://CRAN.R-project.org/package=XLConnect

Schauberger, P., & Walker, A. (2023). openxlsx: Read, write and edit xlsx files. R package version 4.2.5.2. https://CRAN.R-project.org/package=openxlsx

4

Data Cleaning and Manipulation

Data cleaning and manipulation is the first step in data analysis. Often neglected or considered of minor relevance, this step is usually one of the most time-consuming tasks in data science and if not done properly the popular saying 'garbage in, garbage out' applies. In an article in Forbes magazine,[1] Gil Press states that *'Data scientists spend 60% of their time on cleaning and organising data. Collecting data sets comes second at 19% of their time, meaning data scientists spend around 80% of their time on preparing and managing data for analysis'.*

The process involves subsetting, eliminating, transforming, and reshaping data in order to make it suitable for further analysis. Among the available tools for these tasks, there are multiple 'base' R functions and dedicated packages.

Firstly, the data should be stored as variables in the R environment, often as a data frame. As outlined previously, these are essentially tables with rows and columns (see Figure 2.3). A data frame can be created from scratch, or imported from a file, web server, or a database.

The second step in data manipulation is to examine the structure of the data and identify any missing values, outliers, or inconsistencies. This can be done using functions such as str(), summary(), and head().

4.1 Know Your Data

Even before doing any operations in the data, it is important to understand what are its contents and structure. To examine the data, the pre-installed functions from the 'base' and 'utils' packages can be used. Here, we reference the most commonly used ones.

class(): This function returns the class or data type of an object, such as 'numeric', 'character', 'factor', 'data.frame', 'matrix', and 'array'.

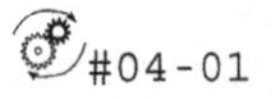#04-01

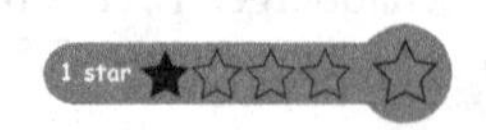


```
# Create a vector named 'x'
x = c(1, 2, 3)
```

[1] https://www.forbes.com/sites/gilpress/2016/03/23/data-preparation-most-time-consuming-least-enjoyable-data-science-task-survey-says/

DOI: 10.1201/9781032651880-5

```
# Verify its class
class(x)
```

The console output is as follows:

```
[1] "numeric"
```

str(): This function returns a concise summary of the structure of an object, including the class, length, and a sample of the data. In the example below, a data frame with two columns/variables is created. The column names are 'mineral_name' and 'hardness'. The str() function retrieves the structure of the 'df' data frame.

#04-02

```
# Create a data frame named 'df'
df = data.frame(mineral_name = c("quartz","K feldspar"),
hardness = c (7, 6))

# Verify its structure
str(df)
```

The console output is as follows:

```
'data.frame':    2 obs. of  2 variables:
$ mineral_name: chr  "quartz" "K feldspar"
$ hardness : num  7 6
```

Notice that the data frame contains two rows (obs. – observations) and two columns (variables). Remember that the variables in a data frame are referred using a dollar sign ($), e.g. the 'hardness' column is referred as 'df$hardness'.

typeof(): This function returns the type of an object, such as 'double', 'integer', 'logical', and 'character'.

#04-03

```
# Create a vector named 'x'
x = c(1, 2, 3)

# Verify its type
typeof(x)
```

The console output is as follows:

```
[1] "double"
```

length(): This function returns the length of an object, i.e. the number of elements in a vector and the number of rows in a matrix or data frame.

```
# Create a vector named 'x'
x = c(1, 2, 3)

# Verify its length
length(x)
```

The console output is as follows:

```
[1] 3
```

attributes(): This function returns the attributes of an object, such as the dimension of an array, the levels of a factor, or the elements in a data frame. Note that the row names are not defined for the 'df' data frame; hence, they are considered by its position: 1, 2, etc.

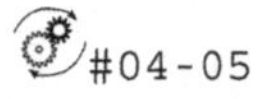

```
# Create a data frame named 'df'
df = data.frame(mineral_name = c("quartz","K feldspar"),
hardness = c (7, 6))

# Verify its attributes
attributes(df)
```

The console output is as follows:

```
$names
[1] "mineral_name"        "hardness"

$class
[1] "data.frame"

$row.names
[1] 1 2
```

By harnessing these functions, one can quickly determine the data type, structure, and attributes of an object. This will be useful when working with new data or when debugging the R code.

4.2 Subsetting Data

Subsetting data involves extracting a specific portion of a larger dataset based on certain conditions or criteria. This process allows us to focus on the relevant part of the data for analysis, visualisation, or further exploration.

Subsetting is also an important task to isolate and work with smaller portions of data, allowing better processing time and memory management of the computer resources.

In geological campaigns, subsetting data is done in many circumstances and occasions:

Region of Interest Analysis: Often departing from large spatial datasets, such as geological maps or satellite images, subsetting allows us to extract data from a specific region of interest, enabling more efficient analysis and characterisation of a study area.

Filtering Data: Many geological campaigns need to filter out data points that meet certain geological criteria. For instance, one can subset earthquake data to focus on events of a particular magnitude range or depth, for evaluating seismic hazard.

Temporal Analysis: Subsetting time-series data is helpful to analyse geological processes and phenomena over specific time intervals, such as identifying trends in groundwater levels or glacier movement.

Resource Exploration: In mineral resources exploration, subsetting can help in targeting specific areas with potential mineral deposits by extracting relevant data such as type or mode of mineral occurrences, lithological or structural units, or geochemical data.

Environmental Studies: When necessary to focus the study on specific environmental parameters, such as soil properties, pollution levels, or water quality, in specific geographical regions, subsetting data is paramount.

Spatial Analysis of a Geological Feature: Subsetting spatial data based on geometric shapes (e.g. polygons) can be used for geospatial analysis, such as geochemical characteristics of a rock formation, land-use planning, geomorphological studies, or habitat mapping.

R's subsetting capabilities, coupled with its 'base' package or with specific data manipulation packages like 'dplyr' and 'tidyr', provide geologists with versatile tools to extract, analyse, and visualise subsets of data that are crucial for making informed geological interpretations, predictions, and decisions.

4.2.1 Subset Numeric Data

The simplest way of subsetting data is using the 'base' R functions. For the next example, we use the functions set.seed() and rnorm(). The rnorm() function from the 'stats' package is used to generate random numbers from a normal distribution, whereas the set.seed() function from the 'base' R is used to set the starting point for the random number generator. This may sound a bit technical, but essentially, it allows us to control the randomness of the random number generation, making the results reproducible.

The most common approach to subset data is to use square brackets '[ ]' that subset data structures such as vectors, matrices, or data frames. The basic syntax for subsetting a data structure is 'data_structure[condition]', where 'condition' is an expression[2] that returns a logical vector indicating which elements of the data structure should be returned. For example, if we have a vector 'x', we can use the following syntax to return all elements of 'x' that are greater than 3:

```
x[x > 3]
```

When working with data frames, we can use square brackets to subset rows. For example, if we have a data frame 'df', we can use the following syntax to return all rows of 'df' where the value of the 'column_name' variable is greater than 3:

```
df[df$column_name > 3, ]
```

This concept is better exemplified by a real example of how to subset data. The code below creates a data frame with x, y, and z variables that have 100 elements with a random normal distribution, with mean = 0 and standard deviation = 1, which are the default values for the rnorm() function.

```
#04-06

# Generate some artificial data
set.seed(1)
data = data.frame(x = rnorm(100), y = rnorm(100), z = rnorm(100))

# Subset the data to x > 0
data_subset = data[data$x > 0, ]
```

In this example, the subsetting is made directly by selecting values in the data frame that meet the condition 'data$x > 0'. To view the results, the summary() function is used.

Notice that the conditional expression is followed by a comma, i.e. to instruct R to select 'all the columns' in which the rows match the condition.

[2] This is the condition or criteria that you want to apply to filter or subset the data. It can be a logical statement, comparison, or any other valid expression in R.

The designation for a data frame is dataframe [row, column] ; therefore, if the 'column' element is missing, it means 'all the columns'.

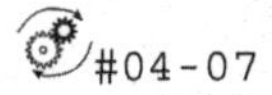

#04-07

```
# Summarise all the data
summary(data)

# Summarise the subset of the data
summary(data_subset)
```

And the resulting outputs are as follows:

```
# summarise all the data
       x                   y                   z
 Min.   :-2.2147   Min.   :-1.91436   Min.   :-2.888921
 1st Qu.:-0.4942   1st Qu.:-0.65105   1st Qu.:-0.455149
 Median : 0.1139   Median :-0.17722   Median :-0.001606
 Mean   : 0.1089   Mean   :-0.03781   Mean   : 0.029673
 3rd Qu.: 0.6915   3rd Qu.: 0.50090   3rd Qu.: 0.698491
 Max.   : 2.4016   Max.   : 2.30798   Max.   : 2.649167

# summarise the subset of the data

       x                   y                    z
 Min.   :0.001105   Min.   :-1.66497   Min.   :-2.285236
 1st Qu.:0.374432   1st Qu.:-0.49535   1st Qu.:-0.637936
 Median :0.615276   Median :-0.04581   Median :-0.001606
 Mean   :0.764452   Mean   : 0.05321   Mean   :-0.012721
 3rd Qu.:1.090794   3rd Qu.: 0.51431   3rd Qu.: 0.660860
 Max.   :2.401618   Max.   : 2.30798   Max.   : 1.778429
```

Notice that the 'x' variable in the original 'data' variable ranges from –2.2147 to 2.4016, whereas in the variable 'data_subset', that was subsetted, the 'x' variable ranges from 0.001105 to 2.401618.

Using brackets provides precise control and clarity, allowing the extraction of specific elements or subsets from the dataset for focused analysis and manipulation.

Apart from using brackets to subset data, there are more complex functions from data analysis packages. Among the many packages available, 'dplyr' and 'tidyr' (Wickham et al., 2023a; Wickham et al., 2023b) are the most commonly used.

Below is an example of subsetting data using the 'dplyr'[3] package and the function filter().

[3] Just a reminder that you must use the install.packages ("dplyr") if you haven't done it before.

#04-08

```
# Load the dplyr package
library(dplyr)

# Generate some artificial data
set.seed(1)
data = data.frame(x = rnorm(100), y = rnorm(100), z = rnorm(100))

# Subset the data to x > 0
data_subset = filter(data, x > 0)
```

In this example, the operation is the same as in the snippet #04-06 and the results are similar, but it is achieved by using the filter() function from the 'dplyr' package. Notice that in this case, we use the filter() function with two arguments, the 'data' that is the variable to be evaluated and 'x > 0' that is the condition to be met.

Naturally, the applications can be more complex depending on the conditions used and the data available. The following example is such a case, where we subset the data to keep only observations where 'x' is greater than 0 and 'y' is less than –0.5, and 'group' is either 'A' or 'B'. The '&' symbol is used in the expression to combine conditions (AND logical operation), and the '|' symbol is used to specify multiple possible values (OR logical operation) for a categorical variable. For creating the 'group' variable, the sample() function is used. This function from the 'base' package generates a random sample of 100 elements from the vector c("A", "B", "C"). The replace = TRUE argument allows for sampling with replacement, meaning that the same element can be chosen more than once.

#04-09

```
# Generate some artificial data
set.seed(1)
data = data.frame(x = rnorm(100), y = rnorm(100), z = rnorm(100),
    group = sample(c("A", "B", "C"), 100, replace = TRUE))

# Subset the data to keep only observations where x is greater
# than 0 and y is less than -0.5, and group is "A" or "B"
data_subset = filter(data, x > 0 & y < -0.5 & (group == "A" |
group == "B"))
```

4.2.2 Subset Text Data

Selecting elements from a dataset based on a text string is a common requirement in some cases. For this example, a data frame is created containing 13 rock names, where 4 of them start with the text 'gran'.

Subsetting text can be made using the 'base' R as well as the 'dplyr' package. For using the 'base' R, it is suggested to use the grep() function that searches for matches of an argument pattern within each element of a character vector. The grep() function is used in combination with the regular expression pattern[4] '^Gran' to select only the rock names that start with 'gran'.

#04-10 

```
# Create a sample data frame with 13 rock names
rock_df = data.frame(rocks = c("Granite", "Gneiss", "Limestone",
"Sandstone", "Marble", "Granodiorite", "Schist", "Gabbro",
"Granite Porphyry", "Dolomite", "Basalt", "Granite Gneiss"))

# Select only the rock names that start with "gran" using base R
base_select = rock_df[grep("^Gran", rock_df$rocks), ]

# View the result
base_select
```

The result is the four rocks that start with 'Gran'.

```
[1] Granite    Granodiorite    Granite Porphyry    Granite Gneiss
```

Another example with the 'dplyr' and 'stringr' packages is presented hereafter.

#04-11 

```
# Load the dplyr and stringr packages
library(dplyr)
library(stringr)

# Select only the rock names that start with "gran"
gran_select = filter(rock_df, str_detect(rocks, "^Gran"))

# View the result
gran_select
```

[4] Regular expressions are powerful patterns used to search and manipulate text in R. When combined with the grep() function, regular expressions provide a way to match and extract specific patterns within character vectors or strings. By incorporating regular expressions as the pattern argument in grep(), one can define complex search patterns with great flexibility. Regular expressions allow us to specify a combination of characters, special characters, metacharacters, and quantifiers to define the pattern that we want to match. For example, one can use regular expressions to find all words starting with a specific letter, extract phone numbers in a certain format, or identify email addresses within a text.
For more information, see https://cran.r-project.org/web/packages/stringr/vignettes/regular-expressions.html

The results are as follows:

```
            rocks
1         Granite
2    Granodiorite
3 Granite Porphyry
4   Granite Gneiss
```

The filter() function, from the 'dplyr' package, combined with the str_detect() function from the 'stringr' package is used to achieve the same result. The resulting output is a new data frame 'gran_select' with the selected rock names. When we have a vector of strings and want to find out which elements contain a particular pattern, the str_detect() function returns a logical vector, where each element corresponds to a string in the input vector, indicating whether the specified pattern is found (TRUE) or not (FALSE) in that string.

Note that the '^' character in the regular expression pattern denotes the start of a string, and the str_detect() function in 'stringr' is equivalent to grep() in 'base' R for pattern matching. This expression is case sensitive. To make the regular expression case-insensitive, one can add the 'ignore_case = TRUE' argument.

4.3 Missing Values

Once the data has been examined, a second step in data management is to handle any missing values. Missing values, in R, are represented by NA for not available or NULL for not defined. Some statistical or mathematical functions can be performed with missing values; others cannot be applied with this type of data.

The simplest approach to handling missing values is to remove the entire row or column that contains any missing value. However, this can result in loss of considerable amounts of data and information. A more sophisticated approach is to impute the missing values, by replacing them with some meaningful value, e.g. a central tendency value, such as the mean, median, or mode of the variable. There is also the possibility of using more advanced techniques such as regression or multiple imputation.

In geological applications, namely in geochemical and mineral exploration studies, there are other strategies that can be employed. An example is to first evaluate what is the Limit of Detection (LOD) of the method or equipment that is being used for analysis. A possible approach is to replace NAs by the LOD or, in a more conservative way, replace by LOD/sqrt(2). Advanced packages such as 'zCompositions'[5] allow more advanced data imputations.

[5] https://cran.r-project.org/web/packages/zCompositions/zCompositions.pdf

A detailed discussion of the imputation and handling missing values can be found in Palarea-Albaladejo & Martín-Fernández (2015).

4.3.1 Handling Missing Values

The pre-installed R packages have several functions to deal with missing values. Remember that the best strategy depends on the nature of the data, the type of analysis to be performed, and the assumptions made. One must always consider the implications of the chosen strategy on the integrity of the proposed analysis.

The following list represents some of the possible strategies:

Identify Missing Values: Before handling missing values, it's important to identify them, often represented as NA. Functions like is.na() or is.null() can be used to detect missing values in the data.

Exclude Missing Values: If the selected analysis allows, one can exclude rows with missing values using the function na.omit(). Alternatively, the complete.cases() function can be utilised to identify rows without any missing values, allowing for manual filtering of the dataset to only include complete cases, i.e. rows without NAs.

Imputation: Imputing missing values involves filling in the missing data with estimated or calculated values. The functions like mean() or median() can assist in imputation. One can replace missing values with mean or median values of the non-missing values in the same variable.

Let's look at examples.

4.3.1.1 The is.na() Function

This function is used to identify the missing values in a data frame. It returns a logical vector indicating whether each element of a data frame is missing or not.

#04-12

```
# Create the data
x = c(1, 2, NA, 4, 5)

# View the result
is.na(x)
```

And the result is as follows:

```
[1] FALSE FALSE  TRUE FALSE FALSE
```

4.3.1.2 The na.omit() Function

This function removes all the rows with missing values from a data frame.

#04-13 1 star

```
# Create the data
y = data.frame(col1 = c(11, 12, NA, 14, 15), col2 = c(NA, 6, 7,
8, 9))

# View the result
na.omit(y)
```

And the result is as follows:

```
   col1 col2
2    12    6
4    14    8
5    15    9
```

4.3.1.3 The na.rm Argument

Many functions have an 'na.rm' argument that allows to remove missing values when calculating the result.

#04-14 1 star

```
# Create the data
x = c(1, 2, NA, 4, 5)

# View the result
mean(x, na.rm = TRUE)
```

And the result is as follows:

```
[1] 3
```

Without the na.rm argument, the result would be NA.

4.3.2 Replacing NAs

It's important to note that imputation of missing values is a subjective process, and it's essential to understand the assumptions behind the method used. In some cases, it may be more appropriate to drop the missing values instead of imputing them. In other cases, more advanced imputation methods, such as multiple imputation, may be necessary.

Let's see how to make an imputation in an example data frame 'df'. In the example, the choice is to impute the NA values with a chosen specific value (let's say 0).

#04-15 1 star

```
# Create the data frame
df = data.frame(x = c(1, 2, NA, 4, 5), y = c(NA, 2, 3, 4, NA))

# Impute NA values with 0
df[is.na(df)] = 0

# View the results
df
```

The result is as follows:

```
  x y
1 1 0
2 2 2
3 0 3
4 4 4
5 5 0
```

If the choice is to replace each NA in a column by the mean column value, this is a two-step process. First calculate the mean for each column and secondly input the NAs by this mean value.

#04-16 1 star

```
# Create the data frame
df = data.frame(x = c(1, 2, NA, 4, 5), y = c(NA, 2, 3, 4, NA))

# Impute NA values with mean
df$x[is.na(df$x)] = mean(df$x, na.rm = TRUE)
df$y[is.na(df$y)] = mean(df$y, na.rm = TRUE)

# View the results
df
```

Notice that the mean, in both cases, is 3. The result is as follows:

```
  x   y
1  1 3.0
2  2 2.0
3  3 3.0
4  4 4.0
5  5 3.0
```

4.4 Transforming Variables in R

The task of data cleaning also involves transforming variables. This can include converting continuous variables to categorical variables (e.g. turning geological age into Periods or Eras), or vice versa. In R, this can be done using functions such as cut() and as.numeric() from 'base' R. It may also involve creating new variables from existing ones.

4.4.1 Converting Character to Numeric

In this example, a vector of character variables is converted to its numeric version using the as.numeric() function.

#04-17

```
# Create a vector of character values
char_vals = c("10", "15", "20", "25", "30")

# Convert the values to numeric using the as.numeric()
function
num_vals = as.numeric(char_vals)

# View the result
num_vals
```

And the result is as follows:

```
[1] 10 15 20 25 30
```

The group of as.*() functions allows several transformations of the variables. In this case, it is very simple to make the conversion of a character variable (notice the double quotes to identify character/text value) to a numeric one. Other examples of this group of functions are as follows:

as.character(): Converts an object to character (string) type.

as.numeric(): Converts an object to numeric (numeric or floating-point) type.

as.integer(): Converts an object to integer type.

as.logical(): Converts an object to logical (Boolean) type.

as.factor(): Converts an object to a factor (categorical) type.

as.Date(): Converts an object to a date type (if possible).

as.POSIXlt(): Converts an object to a POSIXlt type (date and time with attributes).

4.4.2 Converting Continuous to Categorical

In next example, with the purpose of converting a continuous variable into a categorical one, a data frame with 12 minerals and its hardness is created. The cut() function is used to create the categorical variable with the following values: soft, medium, and hard.

#04-18

```
# Create example data
minerals = data.frame(name = c("Apatite", "Calcite",
"Feldspar", "Fluorite", "Gypsum", "Halite", "Muscovite",
"Orthoclase", "Quartz", "Topaz", "Tourmaline", "Zircon"),
hardness = c(5, 3, 6, 4, 2, 2.5, 3, 6, 7, 8, 7.5, 7.5))

# Convert hardness to categorical variable
minerals$hardness_cat = cut(minerals$hardness, breaks =
c(-Inf, 3, 7, Inf), labels = c("soft", "medium", "hard"))

# View the result
minerals
```

The new categorical variable is called 'hardness_cat' and categorises the minerals as 'soft', 'medium', or 'hard' based on their hardness values. The cut() function is used with the argument 'breaks' to define the intervals (i.e.] -inf, 3],] 3, 7] and] 7, +Inf[) and the argument 'labels' to define the text values for the intervals. Note that 'Inf' stands for infinite.

The result is as follows:

```
   name       hardness hardness_cat
1  Apatite    5.0      medium
2  Calcite    3.0      soft
3  Feldspar   6.0      medium
4  Fluorite   4.0      medium
5  Gypsum     2.0      soft
6  Halite     2.5      soft
7  Muscovite  3.0      soft
8  Orthoclase 6.0      medium
9  Quartz     7.0      medium
10 Topaz      8.0      hard
11 Tourmaline 7.5      hard
12 Zircon     7.5      hard
```

4.4.3 Converting Categorical to Continuous

To demonstrate the opposite procedure, we present an example containing a categorical variable with 12 common rock names, some repeated, and the code transforms the 'rock_names' variable into a numerical one, i.e. a factor.

#04-19

```
# Create example data
rock_names = c("Granite", "Basalt", "Sandstone", "Limestone",
"Granite", "Marble", "Shale", "Gneiss", "Schist", "Quartzite",
"Basalt", "Sandstone")

# Convert the rock_names variable to a numerical one
rock_numbers = as.numeric(as.factor(rock_names))

# View the results
rock_numbers
```

And the output is as follows:

```
[1] 3 1 7 4 3 5 9 2 8 6 1 7
```

In this example, the as.factor() function is used to convert the 'rock_names' variable into a factor vector, and the as.numeric() function is used to convert the factor vector into a numerical vector. The result of these operations is stored in the 'rock_numbers' variable. Observe that repeating rock names have the same number, i.e. factor.

Notice that data type errors are one of the most common sources of headaches in R. It's important to check the data type before and after transforming variables to ensure that the transformation was successful and that the values are stored in the expected data type.

4.4.4 Appending Rows

Appending rows to a data frame means adding one or more new rows (i.e. observations) to an existing data frame. There are several ways to append rows to a data frame, including the rbind() function from the 'base' R package or using square brackets '[ ]'.

Here is a demonstration of appending rows to a data frame with eight rows containing information (columns) about the mineral names, mineral colours, and mineral hardness.

#04-20/01

```
# Create a data frame
minerals = data.frame(mineral_names = c("Quartz",
"Plagioclase", "Orthoclase", "Biotite", "Muscovite", "Calcite",
"Dolomite", "Magnetite"), mineral_color = c("Clear", "White",
"Pink", "Black", "Green", "White", "White", "Black"), mineral_
hardness = c(7, 6, 6, 2.5, 2.5, 3, 4, 5.5))

# View the result
minerals
```

The resulting data frame is as follows:

```
  mineral_names mineral_color mineral_hardness
1     Quartz              Clear                7.0
2     Plagioclase         White                6.0
3     Orthoclase          Pink                 6.0
4     Biotite             Black                2.5
5     Muscovite           Green                2.5
6     Calcite             White                3.0
7     Dolomite            White                4.0
8     Magnetite           Black                5.5
```

To append new data (i.e. rows), the rbind() function is used.

#04-20/02

```
# Append two more minerals to the data frame
new_minerals = data.frame(mineral_names = c("Hematite",
"Galena"), mineral_color = c("Red", "Silver"), mineral_
hardness = c(5.5, 2.5))

# Row bind
minerals_complete = rbind(minerals, new_minerals)

# View the results
new_minerals

minerals_complete
```

The 'new_minerals' variable is as follows:

```
    mineral_names mineral_color mineral_hardness
1     Hematite      Red           5.5
2     Galena        Silver        2.5
```

and the new data frame, named 'minerals_complete', is as follows:

```
   mineral_names mineral_color mineral_hardness
1     Quartz            Clear              7.0
2     Plagioclase       White              6.0
3     Orthoclase        Pink               6.0
4     Biotite           Black              2.5
5     Muscovite         Green              2.5
6     Calcite           White              3.0
7     Dolomite          White              4.0
8     Magnetite         Black              5.5
9     Hematite          Red                5.5
10    Galena            Silver             2.5
```

Here is another example now using square brackets.

#04-21

```
# Create a data frame
minerals = data.frame(mineral_names = c("Quartz",
"Plagioclase", "Orthoclase", "Biotite", "Muscovite", "Calcite",
"Dolomite", "Magnetite"), mineral_color = c("Clear", "White",
"Pink", "Black", "Green", "White", "White", "Black"), mineral_
hardness = c(7, 6, 6, 2.5, 2.5, 3, 4, 5.5))

# Create a new data frame
new_minerals = data.frame(mineral_names = c("Hematite",
"Galena"), mineral_color = c("Red", "Silver"), mineral_
hardness = c(5.5, 2.5))

# Append two more minerals to the data frame
n1 = nrow(minerals) + 1
n2 = n1 + nrow(new_minerals) - 1
minerals[n1:n2,] = new_minerals
```

The nrow() function retrieves the number of rows from a data frame (see more in Section 5.3 in Chapter 5).

4.4.5 Appending Columns

Similar to the previous example, appending a new column to a data frame can be done using the cbind() function. To append a new column, simply specify the data frame and the new values to add. Let's see how to do it using the data frame 'minerals' that contains information on eight mineral names, their colours, and hardness.

#04-22/01

```
# Create a data frame
minerals = data.frame(mineral_names = c("Quartz",
"Plagioclase", "Orthoclase", "Biotite", "Muscovite",
"Hornblende", "Garnet", "Feldspar"), mineral_color =
c("Clear", "White", "Pink", "Black", "Green", "Black", "Red",
"White"), mineral_hardness = c(7, 6, 6, 2.5, 2.5, 7, 6.5, 6))

# View the result
minerals
```

The resulting data frame is as follows:

```
     mineral_names     mineral_color     mineral_hardness
1    Quartz            Clear             7.0
```

```
2      Plagioclase       White           6.0
3      Orthoclase        Pink            6.0
4      Biotite           Black           2.5
5      Muscovite         Green           2.5
6      Hornblende        Black           7.0
7      Garnet            Red             6.5
8      Feldspar          White           6.0
```

To append another column, the cbind() function is used.

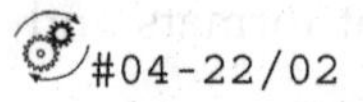#04-22/02

```
# Create a new column of mineral density
mineral_density = c(2.65, 2.71, 2.54, 2.88, 2.83, 3.5, 4.2, 2.56)

# Append the new column to the existing data frame
minerals = cbind(minerals, mineral_density)

# View the result
minerals
```

And the resulting data frame is as follows:

```
  mineral_names mineral_color mineral_hardness mineral_density
1 Quartz        Clear          7.0              2.65
2 Plagioclase   White          6.0              2.71
3 Orthoclase    Pink           6.0              2.54
4 Biotite       Black          2.5              2.88
5 Muscovite     Green          2.5              2.83
6 Hornblende    Black          7.0              3.50
7 Garnet        Red            6.5              4.20
8 Feldspar      White          6.0              2.56
```

The resulting data frame named 'minerals' will now contain the original information on the mineral names, colour, and hardness, as well as the new column with the mineral density.

As often in R, there is more than one way of performing a task. The same can be done using brackets or the dollar sign. This has already been presented informally in the #04-18 snippet.

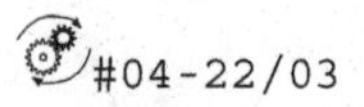#04-22/03

```
# Append the new column to the existing data frame V2
minerals[,"mineral_density_v2"] = mineral_density

# Append the new column to the existing data frame V3
minerals$mineral_density_v3 = mineral_density
```

4.5 Reshaping Data

In addition to the basic data manipulation and cleaning functions, there are several types of changes that are often necessary to be performed. These changes might be necessary to present the data in a certain format or to pass information to a function in a pre-determined way. In the following chapters, some visualisation packages for graphs or maps will be used, and each of these packages will expect the data in a certain format. That is why it is relevant to be able to transform data between different formats and shapes.

4.5.1 Transposing a Data Frame

In order to show how to transpose a data frame, we begin with a list of ten mineral samples and some of its properties.

The transpose function t() is simple, and in this example, the table is transposed from a row = sample, column = variable to a row = variable, column = sample.

04-23

```
# Create a data frame with 10 samples
df = data.frame(Sample_Number = 1:10, Mineral_name =
c("Quartz", "Plagioclase", "Orthoclase", "Biotite",
"Muscovite", "K-feldspar", "Hornblende", "Garnet", "Pyroxene",
"Olivine"),Hardness = c(7, 6, 6, 2.5, 2.5, 6, 5, 7, 6, 7),
color = c("Clear", "White", "Pink", "Black", "Green", "Pink",
"Black", "Red", "Green", "Yellow"), density = c(2.65, 2.71,
2.54, 2.87, 2.83, 2.56, 3.5, 4.3, 3.3, 4.4))

# Transpose the data frame
df_transposed = t(df)

# View the result
df

df_transposed
```

The original data frame is as follows:

```
    Sample_Number Mineral_name Hardness color density
1               1       Quartz      7.0 Clear 2.65
2               2  Plagioclase      6.0 White 2.71
3               3   Orthoclase      6.0  Pink 2.54
4               4      Biotite      2.5 Black 2.87
```

```
5          5      Muscovite  2.5  Green 2.83
6          6    K-feldspar   6.0   Pink 2.56
7          7    Hornblende   5.0  Black 3.50
8          8       Garnet    7.0    Red 4.30
9          9      Pyroxene   6.0  Green 3.30
10        10      Olivine    7.0 Yellow 4.40
```

The resulting transpose data frame is as follows:

```
              [,1]         [,2]           [,3]           [,4]         [,5]
Sample_Number " 1"         " 2"           " 3"           " 4"         " 5"
Mineral_name  "Quartz"     "Plagioclase"  "Orthoclase"   "Biotite"    "Muscovite"
Hardness      "7.0"        "6.0"          "6.0"          "2.5"        "2.5"
color         "Clear"      "White"        "Pink"         "Black"      "Green"
density       "2.65"       "2.71"         "2.54"         "2.87"       "2.83"
              [,6]         [,7]           [,8]           [,9]         [,10]
Sample_Number " 6"         " 7"           " 8"           " 9"         "10"
Mineral_name  "K-feldspar" "Hornblende"   "Garnet"       "Pyroxene"   "Olivine"
Hardness      "6.0"        "5.0"          "7.0"          "6.0"        "7.0"
color         "Pink"       "Black"        "Red"          "Green"      "Yellow"
density       "2.56"       "3.50"         "4.30"         "3.30"       "4.40"
```

* Note that the font size is reduced to better observe the results.

The transpose is also a useful operation in matrix algebra. In that case, the function works exactly in the same way and the results are as expected in a matrix transposition.

4.5.2 Long to Wide and Vice Versa

Reshaping data is crucial for data presentation and analysis, as it ensures that the data is in the appropriate format for the intended operations. There are several packages that make the process of reshaping a data frame easier and more efficient. As stated previously, 'dplyr' and 'tidyr' packages are the 'go to' ones, as they are designed to facilitate data manipulation and cleaning and reshaping data, e.g., by converting wide data to long data, or vice versa.

Let's look at examples of these reshaping functions.

Suppose you have a data frame with 'mineral_names', 'hardness', and 'color' variables from five different minerals.

#04-24/01

```
# Create a data frame
df = data.frame(mineral_names = c("Quartz", "Plagioclase",
"Orthoclase", "Biotite", "Muscovite"),hardness = c(7, 6, 6,
2.5, 2.5), color = c("Clear", "White", "Pink", "Black",
"Green"))

# View the result
df
```

The data frame is as follows:

```
  mineral_names hardness color
1 Quartz             7.0  Clear
2 Plagioclase        6.0  White
3 Orthoclase         6.0  Pink
4 Biotite            2.5  Black
5 Muscovite          2.5  Green
```

4.5.2.1 The pivot_longer() Function

Using the pivot_longer() function from the 'tidyr' package will transform the data.

#04-24/02

```
# Load the tidyr package
library(tidyr)

# Convert the "hardness" column to character type
df$hardness = as.character(df$hardness)

# Convert the data frame to long format
df_long = pivot_longer(df, names_to = "Variable", cols = 2:3,
values_to = "Value")

# View the result
df_long
```

And the result is as follows:

```
# A tibble: 10 × 3
   mineral_names Variable Value
   <chr>         <chr>    <chr>
 1 Quartz        hardness 7
 2 Quartz        color    Clear
 3 Plagioclase   hardness 6
 4 Plagioclase   color    White
 5 Orthoclase    hardness 6
 6 Orthoclase    color    Pink
 7 Biotite       hardness 2.5
 8 Biotite       color    Black
 9 Muscovite     hardness 2.5
10 Muscovite     color    Green
```

This reshaping creates an intermediate variable named 'Variable' that identifies what the 'Value' variable contains, in this case "hardness" or "color".

4.5.2.2 The pivot_wider() Function

The pivot_wider() function provides the inverse transformation. To revert the 'df_long' variable, we apply the following code:

#04-24/03

```
# Transform the data frame to wide format
df_wide = pivot_wider(df_long, names_from = Variable, values_
from = Value )

# View the result
df_wide
```

The result is as follows:

```
# A tibble: 5 × 3
  mineral_names hardness color
  <chr>         <chr>    <chr>
1 Quartz        7        Clear
2 Plagioclase   6        White
3 Orthoclase    6        Pink
4 Biotite       2.5      Black
5 Muscovite     2.5      Green
```

4.6 Concluding Remarks

In geological applications, data cleaning and preparation play a pivotal role in ensuring the accuracy and reliability of analyses conducted. The geological datasets often involve diverse sources, including field measurements, laboratory experiments, and remote sensing data. Ensuring the quality of these datasets and its integrity is crucial, as geological analyses heavily rely on accurate and consistent information.

Data cleaning involves identifying and rectifying errors, such as missing values, and inconsistencies, which can distort geological interpretations. Through effective data cleaning, one can enhance the integrity of the datasets, leading to more robust models and geospatial analyses.

R provides a powerful environment for data cleaning in geological applications, offering a range of functions and packages that facilitate efficient data manipulation, transformation, and validation. Before using any new function or package, it is important to examine the data type that it expects and how it handles missing and NA values. If necessary, use the functions to transform and reshape the variables. Finally, remember to check for outliers in order to ensure that the data is of high quality and suitable for analysis.

References

Palarea-Albaladejo, J., & Martín-Fernández, J. A. (2015). zCompositions — R package for multivariate imputation of left-censored data under a compositional approach. *Chemometrics and Intelligent Laboratory Systems*, 143, 85–96. https://doi.org/10.1016/j.chemolab.2015.02.019

Wickham, H., François, R., Henry, L., Müller, K., & Vaughan, D. (2023a). dplyr: A Grammar of Data Manipulation. R package version 1.1.4. https://CRAN.R-project.org/package=dplyr

Wickham, H., Vaughan, D., & Girlich, M. (2023b). tidyr: Tidy Messy Data_. R package version 1.3.0. https://CRAN.R-project.org/package=tidyr

5

Functions for Everyday Data Analytics

This chapter focuses on functions frequently used in data handling or spatial analysis, particularly within the context of geological data analysis. Unlike the basic functions covered in earlier chapters, this section introduces additional, commonly used functions, selected based on the authors' extensive experience. Although a cherry-picking approach, it is intended to demonstrate and explain these functions to provide readers with practical insights into their application in geological studies. Several authors provide different sets of functions – one that is recommended for developing exploratory data analysis is Friendly (2008).

5.1 Arrays and Vectors

These are functions that handle arrays and vector-like datasets and complement the abilities to manipulate the data, thus contributing to the data preparation and handling stages, frequently used in conjunction with other functions to shape data into a suitable format for the analytical tasks.

5.1.1 The sort() Function

This function, from the 'base' R package, sorts the elements in a vector in ascending or descending order. For example, consider the following example:

```
#05-01/01

# Create the data
x = c(3, 1, 4, 2)

# View the result
sort(x)
```

The result is as follows:

```
[1] 1 2 3 4
```

DOI: 10.1201/9781032651880-6

You can sort elements in descending order by using the 'decreasing' argument defined as 'TRUE'.

#05-01/02

```
# Sort descending
sort(x, decreasing = TRUE)
```

The result is as follows:

```
[1] 4 3 2 1
```

5.1.2 The unique() Function

The unique() function is used to extract the unique values from a vector or an array. It helps in identifying and removing duplicates from the data. For example, consider the following example:

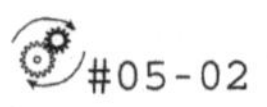
#05-02

```
# Create the data
x = c(3, 1, 4, 2, 1, 4)

# View the result
unique(x)
```

The result is as follows:

```
[1] 3 1 4 2
```

5.2 Strings and Texts

5.2.1 The paste() and paste0() Functions

The paste() and paste0() functions are included in the 'base' R package and concatenate elements from a vector into a single string. For example, consider the following vector with three strings that we want to join. The use of the paste() function allows this operation.

#05-03

```
# Create the data
x = c("Granite", "Gabro", "Syenite")

# View the result
paste(x, collapse = " & ")
```

The result is as follows:

```
[1] "Granite & Gabro & Syenite"
```

The 'collapse' argument indicates what string is used to join the elements.

The paste0() function is a shorthand version of paste() where the separator is an empty string by default. It simply concatenates the strings without any separation.

5.2.2 The substr() Function

This function, from the 'base' R package, extracts a portion of a string. For example, consider that you want to retrieve the first letters of a string, then using the substr() function allows such an operation.

#05-04

```
# Create the data
x = "Granodiorite"

# View the result
substr(x, 1, 5)
```

The first argument number (1) indicates the first character and the second argument number (5) refers to the number of characters to consider.

The result is as follows:

```
[1] "Grano"
```

5.2.3 The gsub() Function

This function, from the 'base' package, replaces all occurrences of a pattern in a string. In the example below, the space character is replaced by the underscore character.

#05-05

```
# Create the data
x = "Wehrlite Peridotite Troctolite"

# Substitute a sequence of character
gsub(" ", "_", x)
```

The result is as follows:

```
[1] "Wehrlite_Peridotite_Troctolite"
```

5.3 Data Frame-Related Functions

Data frames are the central variable type for data analysis as they are similar and behave like a table. Many functions and packages have data frames as input variables. Below are the most common examples for retrieving data frame insights.

5.3.1 The nrow() Function

This 'base' R function returns the number of rows in a data frame. For example, consider the following data frame with two columns and three rows.

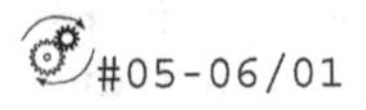

```
#05-06/01

# Create the data
x = data.frame(a = c(1, 2, 3), b = c(4, 5, 6))

# Number of rows
nrow(x)
```

The result is as follows:

```
[1] 3
```

5.3.2 The ncol() Function

This function returns the number of columns in a data frame. Consider the following:

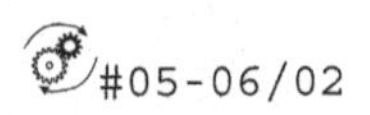

```
#05-06/02

# View the result
ncol(x)
```

The result is as follows:

```
[1] 2
```

5.3.3 The head() Function

This function is from the utils package and returns the first 'n' rows of a data frame. For the same data frame as above, the head() function can be used to show the first two rows.

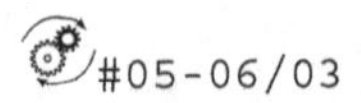

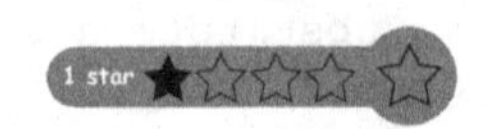

```
#05-06/03

# View the header part
head(x, 2)
```

The result is as follows:

```
  a b
1 1 4
2 2 5
```

5.3.4 The tail() Function

As above, the function returns the last 'n' rows of a data frame. Consider the following:

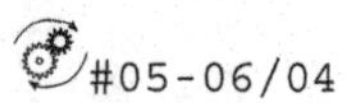#05-06/04

```
# View the tail part
tail(x, 2)
```

The result is as follows:

```
  a b
2 2 5
3 3 6
```

5.3.5 The colnames() and rownames() Functions

These are other 'base' R functions that retrieve the names of the columns and rows, respectively, in a data frame. Let's consider the following data frame:

#05-07/01

```
# Create the data
df = data.frame(mineral_name = c("Quartz", "Plagioclase",
"Orthoclase"), hardness = c(7, 6, 6), color = c("Clear",
"White", "Pink"))
```

The colnames() function can be used to retrieve the names of the columns in this data frame.

#05-07/02

```
# View the names of the columns
colnames(df)
```

The result is as follows:

```
[1] "mineral_name" "hardness" "color"
```

Similarly, the rownames() function can be used to retrieve the names of the rows from the data frame.

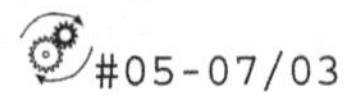

```
# View the names of the rows
rownames(df)
```

The result is as follows:

```
[1] "1" "2" "3"
```

Note that by default, the row names in a data frame are simply the row numbers, but these can be changed by specifying a different vector of row names using the rownames() function.

```
# Define the rownames
rownames(df) = c ("Sample A", "Sample B", "Sample C")

# View the resulting data frame
head(df)
```

The result is as follows:

```
          mineral_name hardness color
Sample A  Quartz       7        Clear
Sample B  Plagioclase  6        White
Sample C  Orthoclase   6        Pink
```

5.4 The apply() Functions

The apply() functions are a group of 'base' R functions where each one has its own purpose and behaviour. Briefly, the main apply() functions are as follows:

apply(): This function is used to apply a function to the rows or columns of a matrix or array. The first argument to apply is the matrix or array where the function will act; the second argument is either 1 or 2, specifying whether to apply the function to the rows (1) or columns (2); and the third argument is the function to consider. apply() returns a vector or an array, depending on the type of input data and the function performed.

lapply(): This function is used to apply a function to each element of a list. The first argument to lapply() is the list, and the second argument is the function to use. lapply() returns a list.

sapply(): This function is similar to lapply(), but it automatically simplifies the result to a vector, matrix, or array, if possible. The first argument to sapply() is the list, and the second argument is the function to apply.

tapply(): This function is used to apply a function to subsets of a vector, based on the values of one or more factors. The first argument to tapply() is the vector, the second argument is the factor or list of factors, and the third argument is the function to apply.

Each of these apply functions has its own strengths and weaknesses, and choosing the right one depends on the specific task you are trying to accomplish. The more general apply() function is a versatile function that allows you to perform a task repeatedly over the margins (rows or columns) of an array, e.g., over the rows or columns of a matrix. It's a way of applying a function to the elements of an object, such as a matrix or data frame.

To demonstrate how to use the apply() function, let's consider a data frame of minerals with columns for mineral name, mineral colour, mineral hardness, and mineral density. We will calculate the distance of each mineral's density to the mean density using the apply() function.

#05-08

```
# Create the data frame
minerals = data.frame(
  name = c("Quartz", "Plagioclase", "Orthoclase", "Biotite",
"Muscovite", "Galena", "Sphalerite", "Chalcopyrite"),
  color = c("Clear", "White", "Pink", "Black", "Green",
"Gray", "Yellow", "Green"),
  hardness = c(7, 6, 6, 2.5, 2.5, 2.5, 4, 3.5),
  density = c(2.65, 2.71, 2.54, 2.88, 2.83, 7.5, 4.1, 4.3))

# Calculate the mean density
mean_density = mean(minerals$density)

# Calculate the distance of the density to the mean
mineral_density_distance =
   apply(minerals, 1, function(x) abs(as.numeric(x[4]) -
mean_density))

# Add the resulting column to the minerals data frame
minerals$density_dist = mineral_density_distance

# Show the new data frame
minerals
```

The result is a data frame with a new variable that states the distance from the mean density (3.69).

```
  name         color  hardness density density_dist
1 Quartz       Clear  7.0        2.65     1.03875
2 Plagioclase White   6.0        2.71     0.97875
3 Orthoclase   Pink   6.0        2.54     1.14875
4 Biotite      Black  2.5        2.88     0.80875
5 Muscovite    Green  2.5        2.83     0.85875
6 Galena       Gray   2.5        7.50     3.81125
7 Sphalerite  Yellow 4.0        4.10     0.41125
8 Chalcopyrite Green 3.5        4.30     0.61125
```

In this example, the apply() function is applied to the 'minerals' data frame, over the first margin (remember the parameter 1 means apply to the rows), and the function used to perform the calculation is 'abs(x[4] - mean_density', which calculates the absolute value of the difference between each mineral's density ('x[4]' – the fourth column of the data frame), and the mean density variable ('mean_density'), previously calculated. The resulting vector of distances is then added as a new column to the minerals data frame (minerals$Mineral_Density_Distance = mineral_density_distance).

It is also possible to use the apply() function (in this case, the lapply()) to create categorical variables. In the following example, a variable named 'dist_cat' is created that states that the density of a given mineral in the 'minerals' data frame is 'near' when the density difference is less than 1 or 'far' when the density difference is higher or equal to 1 from the mean density of the considered minerals.

#05-09

```
# Use lapply to calculate a categorical variable.
minerals$dist_cat = lapply(minerals$density, function(x) {
  if (abs(as.numeric(x) - mean_density) < 1) {
      return("near")
  } else {
      return("far")
  }
})

# Show the results
minerals
```

The results are as follows:

```
  name         color  hardness density density_dist dist_cat
1 Quartz       Clear  7.0        2.65     1.03875      far
2 Plagioclase White   6.0        2.71     0.97875      near
```

```
3 Orthoclase   Pink   6.0      2.54    1.14875      far
4 Biotite      Black  2.5      2.88    0.80875      near
5 Muscovite    Green  2.5      2.83    0.85875      near
6 Galena       Gray   2.5      7.50    3.81125      far
7 Sphalerite   Yellow 4.0      4.10    0.41125      near
8 Chalcopyrite Green  3.5      4.30    0.61125      near
```

Notice that in this case, the function uses an if() conditional test to create the 'dist_cat' variable, depending on the value of 'abs(as.numeric(x) - mean_density)' being higher or lower than 1.

In this example, a function is created by declaring it using the function() function. The next section explains a little more about functions.

5.5 Creating Functions

Creating user defined functions is a key aspect of any programming language. Functions allow the user to repeat sets of operations in the data that can be evoked by the function name. The function() function is used to create user-defined functions. The basic syntax for creating a function is as follows:

```
my_function = function(arg1, arg2, ...) {
  # Function body: Code that defines the function's behavior
  # Use arguments (arg1, arg2, ...) in the function body
  # Perform operations, calculations, or any desired tasks
  # Return a value if necessary
}
```

The breakdown of the elements is as follows:

- my_function: It is the name given to the function. One can choose any valid variable name, but it's a good practice to give a descriptive name that reflects the purpose of the function.
- function(arg1, arg2, ...): It is the declaration of the function where the arguments necessary are specified (arg1, arg2, ...). Arguments are the values or data to be passed to the function when it is called. One can have multiple arguments separated by commas.
- Function body: It is enclosed in curly braces {}. This is where it is defined what the function does. One can perform calculations, execute statements, and manipulate data within the function body.
- Return statement (optional): If function needs to output a result, one can use the return() statement to specify the value to be returned. Not all functions need to return a value.

Here is a simple example of a function named 'add_numbers' that adds two numbers.

#05-10 3 stars

```
# Define a function that adds the numbers of a vector
add_numbers = function(x, y) {
  addition = x + y
  return(addition)
}


# Call the function
sum_result = add_numbers(3, 5)

# View the result
print(sum_result)
```

The output result is as follows:

```
[1] 8
```

5.6 Concluding Remarks

In a slightly overstatement, one might contend that the potentialities for data manipulation in R are virtually boundless. At the very least, these prospects are extensive and transcend the limitations of a conventional textbook. Fortunately, the available resources in published documents, such as this book, published papers, or on the internet, when driven by problem solving searches provide the answer to most of the quests.

The complementary-to-basic functions presented in this chapter are what 15 years of teaching GIS and spatial analysis to geologists and geographers advises me to emphasise. The apply() functions are the most challenging for newcomers to computer programming in R, but an approach driven by examples will almost certainly be successful.

References

Friendly, M. (2008). *Exploratory Data Analysis with R*. CRC Press, Boca Raton, FL.

6

Basic Statistics

It is recognised that R has a vast collection of specialised tools, functions, and packages dedicated to the various statistical techniques, being composed of an extensive ecosystem that empowers statisticians to perform complex analyses and implement sophisticated models efficiently. Notwithstanding this complexity, this chapter is primarily designed to present the readers with the basic tools to deal with what is commonly designated as exploratory data analysis (EDA).

The EDA refers to the process of visually and quantitatively exploring and summarising data to understand its underlying patterns, relationships, and characteristics. The knowledge presented here can be complemented with more advanced statistics found in subject-specific texts such as the ones presented by Williams (2017) or Pearson (2018) or in one of the many massive open online courses (MOOCs) found in the web, such as the introductory course[1] of 'Exploratory data analysis with R' from Coursera.

6.1 A Primer in Exploratory Data Analysis

EDA is used in almost all areas of research, including all of the earth sciences. For these first examples, we will use the 'meuse' dataset (Pebesma, 2022) that is a widely used dataset for exemplifying some techniques applied in environmental statistics and spatial analysis. Originally, in R, it was made available as a part of the 'sp' package[2] and consists of environmental data collected from the Meuse River in the Netherlands, encompassing heavy metal concentrations in soils and other environmental variables. The data includes a set of spatial coordinates as well as some geochemical variables such as the contents in several metals like zinc or cadmium or the land use or the type of soil.

6.1.1 The 'Meuse' Dataset

As stated previously, R packages often provide free usable datasets to exemplify its applications and functionalities. The 'meuse' dataset[3] can be used

[1] https://www.coursera.org/projects/exploratory-data-analysis-in-r
[2] It is also available in a slightly modified version, in the 'gstats' package, see Pebesma (2022).
[3] https://cran.r-project.org/web/packages/gstat/vignettes/gstat.pdf

DOI: 10.1201/9781032651880-7

for exemplifying various purposes, including the analysis of environmental factors and their relationship with heavy metal concentrations, and the visualisation of these relationships on a map. The data can also be used to demonstrate spatial statistical methods, such as intersection, spatial interpolation, or kriging.

The dataset has the following variables:

x: A numeric vector; Easting (m) in Rijksdriehoek (RDH) (Netherlands topographical) map coordinates.

y: A numeric vector; Northing (m) in RDH coordinates.

cadmium: Topsoil cadmium concentration, mg kg^{-1} soil ('ppm'); zero cadmium values in the original dataset have been shifted to 0.2 (half the lowest non-zero value).

copper: Topsoil copper concentration, mg kg^{-1} soil ('ppm').

lead: Topsoil lead concentration, mg kg^{-1} soil ('ppm').

zinc: Topsoil zinc concentration, mg kg^{-1} soil ('ppm').

elev: Relative elevation above local riverbed, m.

dist: Distance to the Meuse river; obtained from the nearest cell in meuse.grid, which in turn was derived by a spatial operation, horizontal precision 20 metres; then normalised to [0,1].

om: Organic matter, kg $(100\ kg)^{-1}$ soil (percent).

ffreq: Flooding frequency class: 1 = once in 2 years; 2 = once in 10 years; 3 = one in 50 years.

soil: Soil type according to the 1:50000 soil map of the Netherlands.

lime: Lime class: 0 = absent and 1 = present by field test with 5% HCl.

landuse: Land use classification.

dist.m: Distance to Meuse river in metres, as obtained during the field survey.

The 'meuse' dataset is frequently employed as a teaching and learning resource in spatial data analysis due to its accessibility and suitability for introductory examples, and yes, we also used it in some of the examples of this book.

6.1.2 Basic Functions

EDA always starts with loading the dataset; in most cases, the data is imported from a file (such as a Comma-Separated Values [CSV] or Excel file). As previously declared, for these first examples, we will use the 'meuse' dataset.

By loading the 'sp' library, the 'meuse' dataset is made available. The data() function from the 'utils' R package can be used to list the existing datasets or is called to make the 'meuse' dataset user accessible. Once the dataset is loaded, a thoughtful commencement is to get a summary of the data. We can do this by using the summary() function.

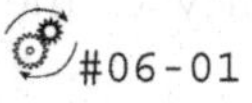#06-01 

```
# Load the library
library(sp)

# Load the data
data(meuse)

# Show the results
summary(meuse)
```

The resulting output is as follows:

```
       x                y             cadmium           copper            lead            zinc
 Min.   :178605  Min.   :329714  Min.   : 0.200  Min.   : 14.00  Min.   : 37.0  Min.   : 113.0
 1st Qu.:179371  1st Qu.:330762  1st Qu.: 0.800  1st Qu.: 23.00  1st Qu.: 72.5  1st Qu.: 198.0
 Median :179991  Median :331633  Median : 2.100  Median : 31.00  Median :123.0  Median : 326.0
 Mean   :180005  Mean   :331635  Mean   : 3.246  Mean   : 40.32  Mean   :153.4  Mean   : 469.7
 3rd Qu.:180630  3rd Qu.:332463  3rd Qu.: 3.850  3rd Qu.: 49.50  3rd Qu.:207.0  3rd Qu.: 674.5
 Max.   :181390  Max.   :333611  Max.   :18.100  Max.   :128.00  Max.   :654.0  Max.   :1839.0

      elev             dist               om           ffreq   soil    lime      landuse
 Min.   : 5.180  Min.   :0.00000  Min.   : 1.000   1:84    1:97    0:111   W      :50
 1st Qu.: 7.546  1st Qu.:0.07569  1st Qu.: 5.300   2:48    2:46    1: 44   Ah     :39
 Median : 8.180  Median :0.21184  Median : 6.900   3:23    3:12            Am     :22
 Mean   : 8.165  Mean   :0.24002  Mean   : 7.478                           Fw     :10
 3rd Qu.: 8.955  3rd Qu.:0.36407  3rd Qu.: 9.000                           Ab     : 8
 Max.   :10.520  Max.   :0.88039  Max.   :17.000                           (Other):25
                                  NA's   :2                                 NA's   : 1
     dist.m
 Min.   : 10.0
 1st Qu.: 80.0
 Median : 270.0
 Mean   : 290.3
 3rd Qu.: 450.0
```

The summary() function, from the 'base' R package, returns a set of summary statistics for the data, including the mean, median, minimum, maximum, and quartiles. Nevertheless, in the R constellation of functions and packages, there are other ways to get not only this but also more specific information about the data. For example, we can use the min() function, from the 'base' R, to get the minimum value of a specific column. In the example below, the minimum value of the 'copper' variable is retrieved.

#06-02/01 

```
# Minimum of copper
min(meuse$copper)
```

and the output is as follows:

```
[1] 14
```

6.1.2.1 Central Tendency

Central tendency is the set of measures that describe the centre of the data distribution. The most commonly used measures of central tendency are the mean, median, and mode.

The mean is the average of the data and is calculated as the sum of the values divided by the number of values. In R, we can calculate the mean using the mean() function. If the data contains not available (NA), one must use the 'na.rm = TRUE' parameter to calculate the mean without considering the NA. In the example below, R calculates the mean of the 'zinc' variable.

```
#06-02/02

# Mean for zinc
mean(meuse$zinc)
```

And the output is as follows:

```
[1] 469.7161
```

The median is the middle value of the data when the data is ordered. In R, we can calculate the median using the median() function that is part of the 'stats' package.

```
#06-02/03

# Median for copper
median(meuse$copper)
```

And the output is as follows:

```
[1] 31
```

The mode corresponds to the most frequently occurring value in the data. To calculate the mode, the combination of the table() and names() functions, from the 'base' package, can be used.[4] The table() function is used to count the frequency of each unique value in a vector, and the names() function is used to extract the names of the most frequent values. The which.max() function, from the base R package, determines the location, i.e. index of the first minimum or maximum of a numeric or logical vector.

In this example, the mode of the 'zinc' variable is calculated.

[4] Note that there is a mode() function in R that is part of the 'base' R which is not used to calculate the mode central tendency value.

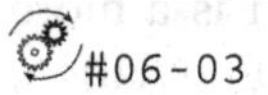 

```
# Count the frequencies
mode_table = table(meuse$zinc)

# What is the most common
mode_meuse_zinc = names(which.max(mode_table))

# Show the results
mode_meuse_zinc
```

And the output is as follows:

```
[1] "180"
```

6.1.2.2 Dispersion

Dispersion is the measure that describes the spread of the data. The most commonly used measures of dispersion are the range, variance, and standard deviation.

The range is defined as the minimum and maximum values of the data. To retrieve these values, the range() function from the 'base' packages can be used. In the following example, we retrieve the range of the elevation data from the 'meuse' dataset.

```
# Range of the elevation data
range(meuse$elev)
```

And the output is as follows:

```
[1]  5.18 10.52
```

The variance is a measure of the spread of the data around the mean. To calculate the variance, use the var() function from the 'stats' package.

In this example, the variance for the 'cadmium' is calculated.

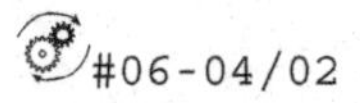 

```
# Variance of the cadmium
var(meuse$cadmium)
```

And the output is as follows:

```
[1] 12.41678
```

The standard deviation is the square root of the variance and is a more commonly used measure of dispersion. To calculate the standard deviation, use the sd() function, also from the 'stats' package.

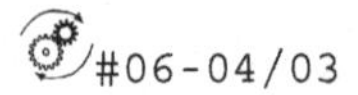

```
# Copper standard deviation
sd(meuse$copper)
```

And the output is as follows:

```
[1] 23.68044
```

6.2 Handling Outliers

Once the data has been cleaned and transformed (see Chapter 4), it is important to check for outliers. Outliers are values that are significantly different from the majority of the data and can have a major impact on the results of an analysis. Outliers can be visually detected using box plots or scatter plots or by calculating summary statistics such as the mean and standard deviation. If outliers are detected, they can be removed or transformed in order to make the data suitable for analysis.

Let's look at an example of creating and plotting outliers. Don't panic, this is a long example, but it is quite simple in its fundamentals.

In the first part of this example, we create a sample dataset, named 'data', and plot it to visualise the outliers. The dataset is created using the rnorm() function, to obtain a random normal distribution, with 1000 values comprising a mean of 50 and a standard deviation of 10. To the first five values, the value 200 is added assuring that these became outliers. The plot() function from the 'base' package produces a simple line and point graph (for further details, refer to Chapter 7).

#06-05/01

```
# Data set with some outliers
set.seed(123)
data = rnorm(100, mean = 50, sd = 10)

# Create the outliers
data[1:5] = data[1:5] + 200

# Plot the data to visualise outliers
plot(data, type = "o")
```

Figure 6.1 shows the result from the plot.

In the second part of the code, the quantile() function from the 'stats' package is used to calculate the first and third quartiles (25th and 75th percentiles) of the data with the 'probs = 0.25' and 'probs = 0.75' parameters. The result is stored in the variables 'q1' and 'q3', respectively.

The interquartile range (IQR) is calculated as the difference between the third and first quartiles and stored in the variable 'IQR'; for more on calculating outliers, please refer to Zwillinger & Kokoska (2000). The 1.5 * IQR rule is used to identify outliers. Any data points that are less than q1 – 1.5 * IQR or greater than q3 + 1.5 * IQR are considered outliers. The minimum and maximum values that define the range of non-outlier values are stored in the variables 'outliers_min' and 'outliers_max', respectively.

The which() function from 'base' R identifies the positions of the outliers in the data. Any data points that meet the condition 'data < outliers_min' or 'data > outliers_max' will be assigned the value of TRUE, and the index of those data points are stored in the variable 'outlier_index'.

After, the outliers are removed from the original dataset by indexing the dataset with the negative values of the 'outlier_index' variable, and the resulting dataset with outliers removed is stored in the variable 'data_without_outliers'.

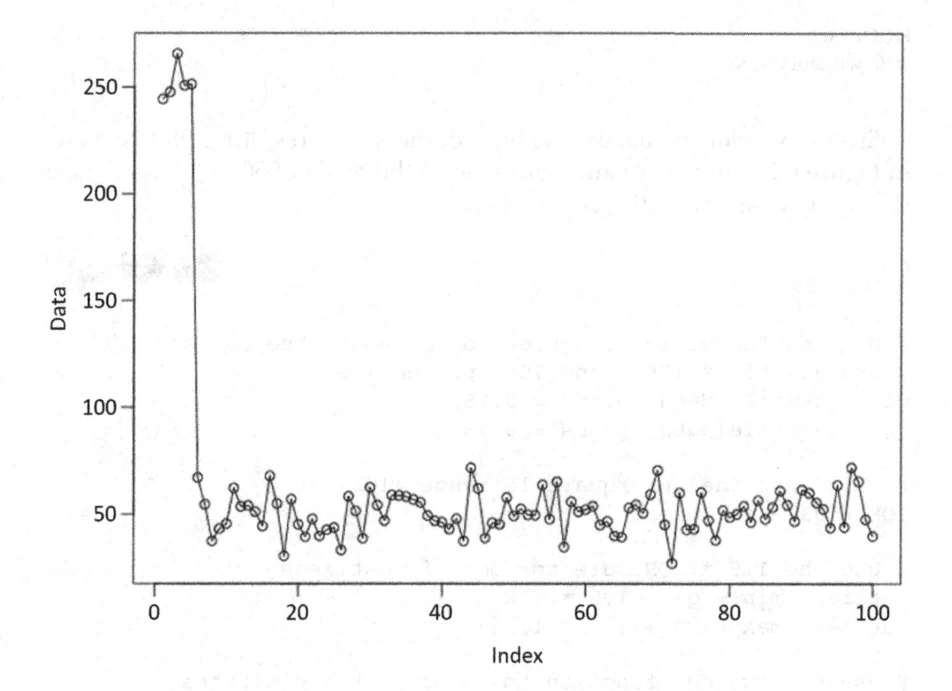

FIGURE 6.1
A plot of the data with a random normal distribution with five outliers.

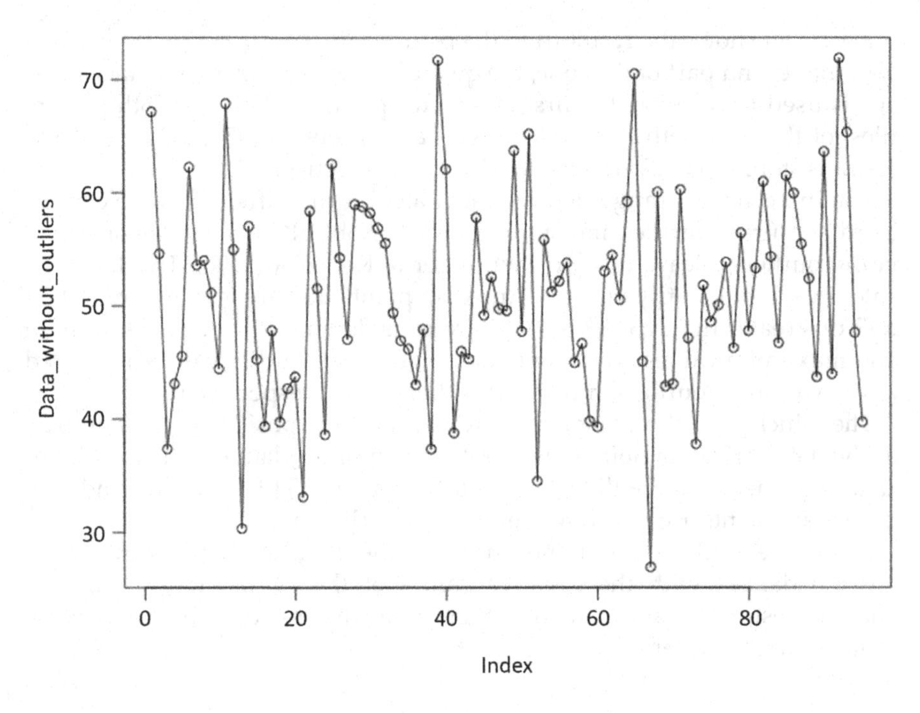

FIGURE 6.2
Data without outliers.

Finally, we plot the dataset without outliers. The resulting plot is shown in Figure 6.2, where the values are around the mean of 50 and with a mean standard deviation of 10.

#06-05/02

3 stars

```
# Use the "quantile" function to calculate the 1st and
# 3rd quartiles (25th and 75th percentiles)
q1 = quantile(data, probs = 0.25)
q3 = quantile(data, probs = 0.75)

# Calculate the interquartile range (IQR)
IQR = q3 - q1

# Use the 1.5 * IQR rule to identify outliers
outliers_min = q1 - 1.5 * IQR
outliers_max = q3 + 1.5 * IQR

# Use the "which" function to identify the positions of
# outliers in the data
outlier_index = which(data < outliers_min | data >
outliers_max)
```

```
# Remove the outliers
data_without_outliers = data[-outlier_index]

# Plot the data without outliers
plot(data_without_outliers, type = "o")
```

Notice the three stars difficulty of this example. Some of the functions used will be detailed in the next chapters.

6.3 Correlating Variables

The cor() function, from the 'stats' package, is used to calculate the correlation matrix between the variables in a data frame. Consider the 'minerals' data frame as in the following example in which we want to verify if there is a correlation between hardness and density of the minerals. The cor() function can be used to calculate the correlation between the variables 'mineral_hardness' and 'mineral density' that are numeric variables. Here is the code example of how you can use the cor() function with the 'minerals' data frame.

#06-06 

```
# Create the minerals data frame
minerals = data.frame(mineral_names = c("Quartz",
"Plagioclase", "Orthoclase", "Biotite", "Muscovite"), hardness
= c(7, 6, 6, 2.5, 2.5), color = c("Clear", "White", "Pink",
"Black", "Green"), density = c(2.65, 2.72, 2.6, 2.7, 2.8))

# Calculate the correlation matrix between the variables
cor_matrix = cor(as.numeric(minerals[,2]), as.numeric
(minerals[,4]))

# Show the results
cor_matrix
```

The result is as follows:

```
[1] -0.6761714
```

It is important to note that the cor() function only calculates the correlation between numeric variables, so if you have categorical variables, you may need to convert them to numeric variables first, using the as.numeric() function.

The cor() function will return a correlation matrix or value with the values between –1 and 1, where –1 indicates a strong negative correlation, 0 indicates no correlation, and 1 indicates a strong positive correlation. In this example, the value of –0.676 indicates that the relationship between the variables 'hardness' and 'density' is moderately negative.

For those new to mineralogy, it's important to understand that hardness and density are two separate physical properties that reveal different features of a mineral. Although they don't directly influence each other, patterns of association may emerge due to the mineral's composition and the arrangement of its atoms. Additionally, it's crucial to remember that just because two properties are associated doesn't imply that one causes the other, i.e. correlation does not mean causation.

Interpreting correlation results implies examining the strength and direction of the relationship between two variables. The strength of the relationship is measured by the correlation coefficient, which ranges from –1 to 1. It is not straightforward to create classes for correlation values, but in general, correlation coefficients between 0.7 and 1.0 are considered strong, while coefficients between 0.3 and 0.7 are considered moderate, and coefficients below 0.3 are considered weak. However, the exact cut-off values for strong and weak correlations can vary depending on the field of study and in the particular application.

It is important to reinforce the idea that correlation does not imply causation. A relationship between two variables may be due to a third variable or a spurious relationship. Further analysis is advisable before considering correlation between variables as a causality demonstration.

6.4 Inference Statistics

Inferential statistics is a branch of statistics that involves drawing conclusions about populations or groups based on data obtained from samples. This is in contrast to EDA, which summarises and describes the features of a dataset, but does not make any inferences about the population, unless the data being analysed is the population. For further reading on inference methodologies, please consult Witte & Witte (2021) or Agresti et al. (2023).

As expected, in R there are a variety of packages and functions available for conducting inferential statistical analysis, including t-tests, ANOVA, regression, among others.

6.4.1 The T-test

A t-test is a statistical hypothesis test used to determine whether there is a significant difference between the mean of two groups. It is commonly used when the variable of interest is numerical and the groups being compared are independent of each other. For a one-sample test, we can use the t.test() function, from the 'stats' package, with the following syntax:

```
t.test(x, mu = desired_mean, alternative = "two.sided", var.
equal = TRUE)
```

where 'x' is the vector of observations and 'mu' is the desired mean. The alternative argument specifies the alternative hypothesis and can take on the values 'greater', 'less', or 'two.sided'. The 'var.equal' argument is a logical value that indicates whether the variances of the two groups are equal.

For example, let's say we have a dataset of mineral density values for a new mineral, and we want to test whether a sample with a density of 6.5 can be considered to belong to the mineral. We can perform a one-sample t-test as follows:

#06-07/01

```
# Vector of density
density_values = c(6.8, 6.7, 6.6, 6.5, 6.3)

# Perform the t-test to a mean of 6.5
t.test(density_values, mu = 6.5, alternative = "two.sided",
var.equal = TRUE)
```

The output will give us the t-value and p-value, which we can use to make inferences about the null hypothesis. If the p-value is small, it indicates that the observed differences between the sample mean and the desired mean are statistically significant, and we can reject the null hypothesis.

```
      One Sample t-test

data:  density_values
t = 0.92998, df = 4, p-value = 0.405
alternative hypothesis: true mean is not equal to 6.5
95 percent confidence interval:
 6.341161 6.818839
sample estimates:
mean of x
      6.58
```

The results of the one-sample t-test suggest that we fail to reject the null hypothesis that the mean density of the mineral sample is equal to 6.5. This is because the p-value (0.405) is greater than the significance level of 0.05, which means that there is not enough evidence to suggest that the mean density of the mineral sample is significantly different from 6.5.

The estimate of the mean density of the mineral sample is 6.58. The 95% confidence interval for the mean density of the mineral sample is between 6.34 and 6.82. This means that we are 95% confident that the true mean density of the mineral sample falls within this range. Let's now test the density value of 3.2 and verify what is the response of the t-test.

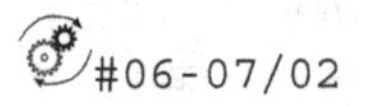
#06-07/02

```
# Perform the t-test to a mean of 3.2
t.test(density_values, mu = 3.2, alternative = "two.sided",
var.equal = TRUE)
```

The result is as follows:

```
        One Sample t-test

data:  density_values
t = 39.292, df = 4, p-value = 2.507e-06
alternative hypothesis: true mean is not equal to 3.2
95 percent confidence interval:
 6.341161 6.818839
sample estimates:
mean of x
      6.58
```

In this case, the population mean was tested to be 3.2. The t-value is 39.292 and the p-value is 2.507×10^{-6}, which is very small. This suggests that there is a significant difference between the mean of the sample (6.58) and the population mean (3.2).

6.4.2 Welch Two-Sample T-test

Another example of inference is the two-sample tests. These are statistical tests used to determine if there is a significant difference between two independent groups (e.g. Witte & Witte, 2021). In the context of mineralogy examples, we can use two-sample tests to compare the properties (such as hardness or density) of two different minerals.

For instance, let's say we have two groups of minerals: 'group1' and 'group2'. Each group contains eight observations of density values. We can perform a t-test to determine if there is a significant difference between the means of the density values for each group. The code for this test is as follows:

#06-08

```
# Example of data
group1 = c(6.0, 6.1, 6.2, 6.3, 6.4, 6.5, 6.6, 6.7)
group2 = c(7.0, 7.1, 7.2, 7.3, 7.4, 7.5, 7.6, 7.7)

# Perform the t-test to compare the two groups
t.test(group1, group2)
```

As in the previous example, this t-test give us information on the t-value, degrees of freedom, p-value, and the confidence interval. Based on these results, we can determine if there is a significant difference between the means of the density values for each group. The results are as follows:

```
        Welch Two Sample t-test

data:  group1 and group2
t = -8.165, df = 14, p-value = 1.079e-06
alternative hypothesis: true difference in means is not equal to 0
95 percent confidence interval:
 -1.2626816 -0.7373184
sample estimates:
mean of x mean of y
      6.35      7.35
```

The results of the Welch Two-Sample t-test indicate that there is a significant difference in the means of the two groups (group 1 and group 2) with a p-value of 1.079×10^{-6}, which is less than 0.05 (the commonly used significance level). The t-value of –8.165 also supports this conclusion, as it falls far in the negative tail of the t-distribution.

The 95% confidence interval for the difference in means is between –1.2626816 and –0.7373184, indicating that the true difference in means lies within this range with 95% confidence. This implies that we are 95% confident that the true difference in means of the two groups is between –1.26 and –0.74.

The sample estimate for the mean of group 1 is 6.35 and the sample estimate for the mean of group 2 is 7.35, indicating that, on average, group 2 has a higher mean density value than group 1.

6.4.3 ANOVA

Another inferential statistical method is ANOVA (analysis of variance), which is used to determine if there is a significant difference between the means of three or more groups. In R, the aov() function from the 'stats' package is used to conduct an ANOVA. If we have three groups of data, 'group1', 'group2', and 'group3', and we want to determine if there is a significant difference in their means, the ANOVA could be used. The following code exemplifies its use.

```
# Run the test
aov_result = aov(value ~ group, data = data)

# Show the results
summary(aov_result)
```

Here's an example using the 'meuse' dataset:

#06-09

```
# Divide the data into three groups based on the elevation
# (6.5 and 8 metres)
group1 = meuse[meuse$elev > 8, ]
group2 = meuse[meuse$elev <= 8 & meuse$elev > 6.5, ]
group3 = meuse[meuse$elev <= 6.5, ]

# Create a new data frame with the zinc concentration and group
data = data.frame(
  value = c(group1$zinc, group2$zinc, group3$zinc),
  group = c(rep("group1", nrow(group1)), rep("group2",
nrow(group2)), rep("group3", nrow(group3)))
)

# Perform an ANOVA on the zinc concentration
aov_result = aov(value ~ group, data = data)

# Show the results
summary(aov_result)
```

In this case, the 'meuse' dataset is used as in Section 6.4.2. example to create three groups based on the elevation (elev). A new data frame is created with the variables 'zinc' concentration and 'group', and the aov() function is used to perform an ANOVA on the zinc concentration. The aov() function takes two arguments: the response variable 'value' and the explanatory variable 'group'. The result of the ANOVA is stored in the 'aov_result' variable. Finally, the summary() function is used to summarise the results of the ANOVA. The summary includes the F-statistic, the degrees of freedom, the p-value, and the mean squares for the three groups.

The p-value can be used to determine if the difference in the mean zinc concentration between the groups is statistically significant. A p-value less than 0.05 indicates that the difference is statistically significant, and we can reject the null hypothesis that the means are equal.

The obtained results are as follows:

```
               Df   Sum Sq Mean Sq F value   Pr(>F)
group           2  7021254 3510627   38.87 2.33e-14 ***
Residuals     152 13729193   90324
---
Signif. codes:  0 '***' 0.001 '**' 0.01 '*' 0.05 '.' 0.1 ' ' 1
```

The result of the ANOVA provides the following information:

- Df: This is the degrees of freedom for each term in the model. In this case, there are 2 degrees of freedom for the group and 152 degrees of freedom for the residuals.

- Sum Sq: This is the sum of squares for each term in the model, which is a measure of the variability in the data explained by that term. In this case, the group explains 7021254 units of variability, while the residuals explain 13729193 units of variability.
- Mean Sq: This is the mean squares for each term in the model, which is the sum of squares divided by the degrees of freedom.
- F value: This is the F-statistic, which is the ratio of the mean squares for the group and the residuals. The F-statistic is used to determine if the difference in means between the groups is statistically significant.
- Pr(>F): This is the p-value, which is the probability of observing an F-statistic as extreme or more extreme than the one calculated, given that the null hypothesis (that the means of the groups are equal) is true. A small p-value (less than 0.05) indicates that the difference between the means is statistically significant, and we can reject the null hypothesis.

Based on this result, we can conclude that the difference in the mean zinc concentration between the three groups is statistically significant, as indicated by the low p-value (2.33×10^{-14}). This is also supported by the large F-value (38.87), which indicates that the group explains a significant amount of variability in the data. The '***' next to the p-value indicates that the p-value is less than 0.001, which is considered highly significant.

It is easily concluded that both tests indicate that the zinc value is different for the three groups, making this division highly significant, i.e. the expected zinc value is different depending on if the elevation is lower than 6.5 metres, between 6.5 and 8 metres or higher than 8 metres.

6.4.4 Regression Analysis

Regression analysis is another common inferential statistical method (Agresti et al. 2023). It is used to model the relationship between a dependent variable and one or more independent variables. The lm() function, from the 'stats' package, can be used to fit a linear regression model (Lilja, 2016). For example, if we have a dependent variable 'y' and an independent variable 'x', and we want to fit a linear regression model, we could run the following code to verify the model.

```
# Fit a linear regression model
reg_fit = lm(y ~ x, data = data)

# Summarise the results of the regression
summary(reg_fit)
```

Returning to the 'meuse' dataset, here's an example on how to calculate the regression between the zinc concentration and the elevation.

#06-10/01

```
# New data frame with zinc concentration and elevation
data = data.frame(x = meuse$zinc, y = meuse$elev)

# Fit a linear regression model
reg_fit = lm(y ~ x, data = data)

# Summarise the results of the regression
summary(reg_fit)
```

In this example, a new data frame is created with the zinc concentration 'x' and elevation 'y'. Next, the lm() function is used to fit a linear regression model to the data. The lm() function takes two arguments: the response variable 'y' and the explanatory variable 'x'. The result of the regression is stored in the 'reg_fit' variable. Notice that the 'y ~ x' formula specifies the model, where 'y' represents the response variable and 'x' represents the predictor variable. That is, 'y' is dependent on 'x'.

Finally, the summary() function is used to retrieve the results of the regression. The summary includes information such as the coefficients, standard errors, t-statistics, p-values, and the R-squared value (Lilja, 2016).

The coefficients can be used to interpret the relationship between the elevation and the zinc concentration. The R-squared value provides a measure of how well the regression line fits the data, with a value closer to 1 indicating a better fit. In this case, the p-value can be used to determine if the relationship between the elevation and the zinc concentration is statistically significant.

The results obtained are as follows:

```
Call:
lm(formula = y ~ x, data = data)

Residuals:
     Min       1Q   Median       3Q      Max
-2.44460 -0.62376  0.07227  0.52779  2.09075

Coefficients:
              Estimate Std. Error t value Pr(>|t|)
(Intercept)  8.9736397  0.1114095  80.546  < 2e-16 ***
x           -0.0017207  0.0001871  -9.196 2.57e-16 ***
---
Signif. codes:  0 '***' 0.001 '**' 0.01 '*' 0.05 '.' 0.1 ' ' 1

Residual standard error: 0.8524 on 153 degrees of freedom
Multiple R-squared:  0.356,    Adjusted R-squared:  0.3518
F-statistic: 84.57 on 1 and 153 DF,  p-value: 2.568e-16
```

The results of the linear regression provide the following information:

Call: This is the call to the lm() function that was used to fit the linear regression model.

Residuals: This provides a summary of the residuals, including the minimum, first quartile, median, third quartile, and maximum values.

Coefficients: This provides the estimated coefficients for the intercept and the explanatory variable x, along with their standard errors, t-statistics, and p-values. The estimated coefficient for x is –2.44460, which indicates that for every unit increase in 'x' (zinc), the expected decrease in 'y' (elevation) is –2.44460. The p-value for x is 2.57×10^{-16}, which is less than 0.05, indicating that the relationship between 'x' and 'y' is statistically significant.

Residual standard error: This is a measure of the average deviation of the residuals from the regression line.

Multiple R-squared: This provides a measure of the proportion of the variability in the response variable y that is explained by the explanatory variable x. In this case, the R-squared value is 0.356, which indicates that 35.6% of the variability in the zinc concentration can be explained by the elevation.

F-statistic: This is used to determine if the relationship between x and y is statistically significant. The F-statistic is calculated as the ratio of the explained variance to the residual variance. The p-value for the F-statistic is 2.568×10^{-16}, which is less than 0.05, indicating that the relationship between 'x' and 'y' is statistically significant.

These results can be visualised and easily perceivable using the plot() function.

```
#06-10/02

# Plot the relationship zinc / elevation
plot(data$x, data$y, xlab = "Zinc Concentration",
ylab = "Elevation",
      main = "Relationship between Zinc Concentration and
Elevation",
      cex.lab = 1.5, cex.axis = 1.5,
      xlim = c(min(data$x), max(data$x)),
      ylim = c(min(data$y), max(data$y)))
abline(reg_fit, col = "red")

# Display the regression formula
text(800, max(data$y), paste("y =", format(reg_fit$coef[1]),
" + ", format(reg_fit$coef[2]), "x"), cex = 1.5, pos = 4)
```

```
# Display the p-value
text(800,10.2, paste("p-value =", format(summary(reg_fit)
$coef[2,4], digits = 3)), cex = 1.5, pos = 4)
```

After the calculations (code snippet #06-10/01), the plot() function is used to create a scatter plot of the relationship between the elevation and zinc concentration. The abline() function, from the 'graphics' package, is then used to add the regression line to the plot. The abline() function takes the result of the regression 'reg_fit' and the colour of the line 'col = "red"'as arguments.

The text() functions are from the 'graphics' package and are used to print the text of the regression formula and the p-value. The format() function, from the 'base' R package, allows the definition of the number of digits and other parameters to obtain a pretty printing.

The plot in Figure 6.3 shows the relationship between the elevation and zinc concentration, with the line indicating the regression. The regression

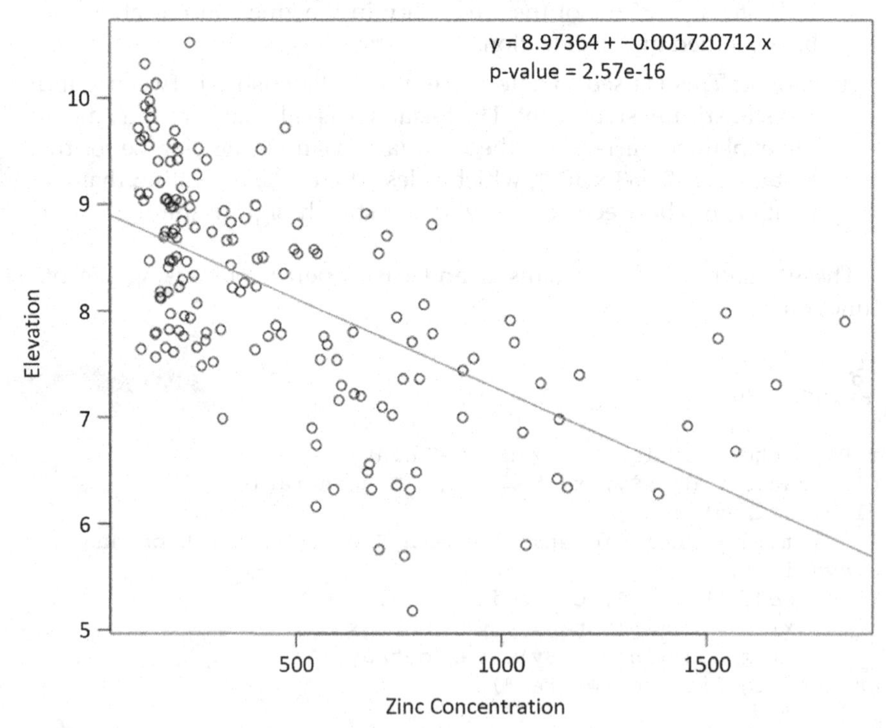

FIGURE 6.3
Plot of the relationship between the elevation and the zinc concentration in the 'meuse' dataset.

line can be used to make predictions about the zinc concentration based on the elevation.

The plot() functions and the best way to create graphics will be addressed in Chapter 7 on data visualisation.

These are just a few examples of the many inferential statistical methods that are available in R. The choice of the methods will depend on the research questions and the nature of the data. It is important to carefully consider the assumptions underlying each method and to check for any violations of these assumptions before conducting the analysis.

6.5 Concluding Remarks

R stands out as a powerful and versatile programming language for basic statistics, as so in the realm of geological applications. Its adaptability is evident in its capacity to handle diverse statistical tasks, making it an invaluable asset for geologists engaged in EDA. R offers a rich environment for conducting EDA in an intuitive and efficient manner. The flexibility of R allows geologists to tailor analyses to the unique characteristics of geological datasets, promoting a deeper understanding of the underlying patterns and trends.

Beyond EDA, R is particularly effective at tackling tasks like detecting outliers, analysing correlations, and conducting statistical inference within geological datasets. This is especially relevant in the case of geological data, where identifying outliers is critical, as anomalous values could either indicate significant geological events or errors in data collection. The available statistical functions enable geologists to identify and interpret outliers effectively. Moreover, the correlation analyses are easy to conduct, allowing to quantify relationships between variables and unveil potential geological dependencies. Statistical inference, including hypothesis testing and confidence interval estimation, is seamlessly conducted in R, empowering users to draw meaningful conclusions from the data with a high degree of statistical rigour.

References

Agresti, A., Franklin, C., & Klingenberg, B. (2023). Statistics: The Art and Science of Learning from Data (Fifth Edition). Pearson, Essex, UK, p. 877.

Lilja, D. (2016). Linear Regression Using R: An Introduction to Data Modeling. University of Minnesota, Minneapolis, MN.

Pearson, R. K. (2018). Exploratory data analysis using R. CRC Press, Boca Raton, FL.

Pebesma, E. (2022). The Meuse Data Set: A Brief Tutorial for the Gstat R Package. ViennaR. https://cran.r-project.org/web/packages/gstat/index.html
Williams, G. J. (2017). The essentials of data science: knowledge discovery using R. CRC Press, Boca Raton, FL.
Witte, R. & Witte, J. S. (2021). Statistics, Eleventh Edition. John Wiley & Sons, Hoboken, NJ, p. 501.
Zwillinger, D., & Kokoska, S. (2000). CRC Standard Probability and Statistics Tables and Formulae. CRC Press, Boca Raton, FL, p. 18. ISBN 1-58488-059-7.

7

Basic Data Visualisation

Visualising data is an essential part of data analysis as it helps to understand and communicate insights about the data in a clear and concise manner. In R, there are a variety of plotting functions available to visualise data, including several types of basic plots such as bar plots, histograms, box plots, and scatter plots. In this chapter, we will deal with the fundamentals of graphics; more advanced examples will be provided in Parts II and III.

The 'graphics' package is part of the 'base' R installation and is a fundamental and versatile package that provides a wide range of tools and functions for creating and customising various types of plots and visualisations. It comes pre-installed and does not require any additional installation steps. It has an extensive set of functions for creating basic plots including features that range from a simple syntax to highly customisable options with compatibility with other packages.

7.1 Bar Plots

Bar plots are used to display the distribution of a categorical variable. In R, bar plots can be created using the barplot() function. The function takes a vector of values as an argument and plots the values as bars. By default, the bars are plotted vertically, but they can be plotted horizontally by specifying the argument "horiz=TRUE".

The 'names.arg' argument is used to specify the names or labels for each bar in the plot, which correspond to different categories or groups in the data. When you provide a vector of names (i.e. categories) to 'names.arg', these names are displayed along the x-axis beneath each bar, making it clear which bar represents which category.

#07-01

```
# Generate sample data
categories = c("A", "B", "C", "D")
values = c(10, 20, 30, 40)

# Create bar plot
barplot(values, names.arg=categories)
```

DOI: 10.1201/9781032651880-8

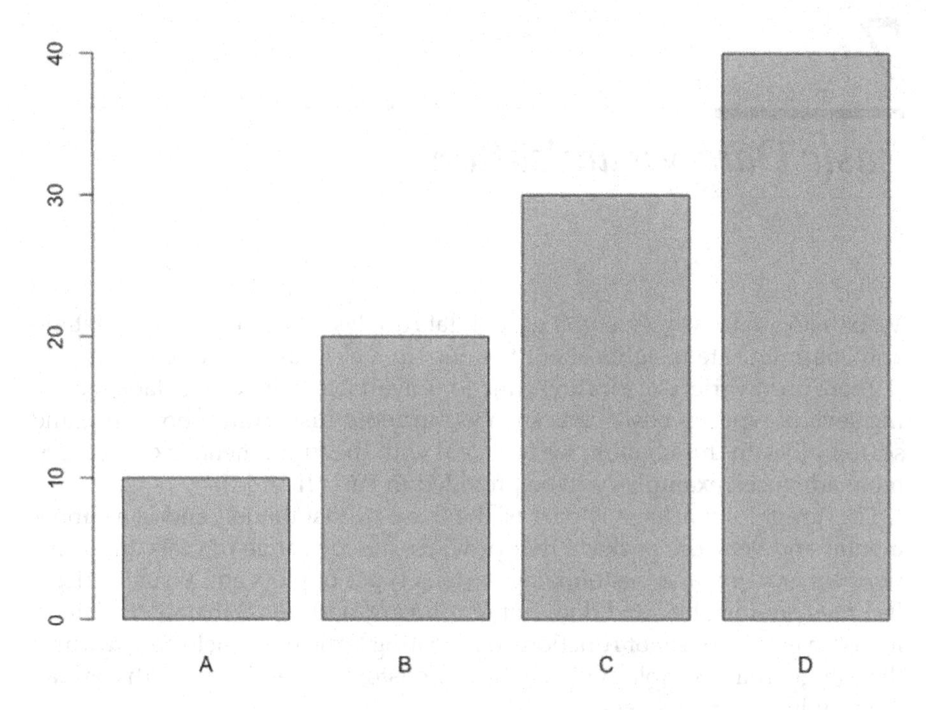

FIGURE 7.1
A simple bar plot.

Figure 7.1 shows the result of this simple barplot().

7.2 Histograms

Histograms are used to visualise the distribution of a continuous variable and can be created using the hist() function. The function takes a vector of values as an argument and plots a histogram of the values. The default number of bins is determined by the 'Sturges' rule (Sturges, 1926), but the number of bins can be specified using the 'breaks' argument (cf. Section 4.4.2, code snippet #04-18 in Chapter 4).

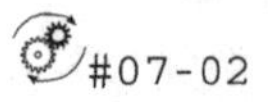

```
# Generate sample data
values = rnorm(100)

# Create histogram
hist(values)
```

Figure 7.2 shows a simple histogram.

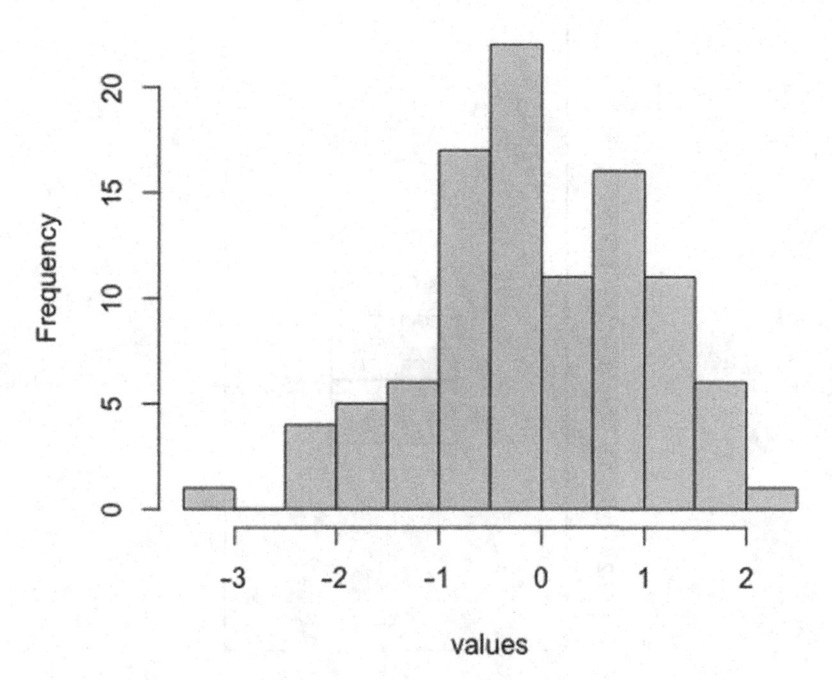

FIGURE 7.2
A simple histogram.

7.3 Box Plots

Box plots are used to display the distribution of a continuous variable and are created using the boxplot() function. The function takes a vector of values as an argument and plots a box plot of the values. The box plot displays the median, quartiles, and outliers of the data.

In the example provided, the box plot is presented in Figure 7.3, and apart from these values, there are two outliers represented as the small open circles.

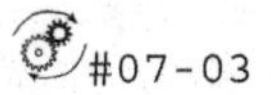

```
# Generate sample data
values = rnorm(100)

# Create box plot
boxplot(values)
```

Figure 7.3. Shows a simple box plot.

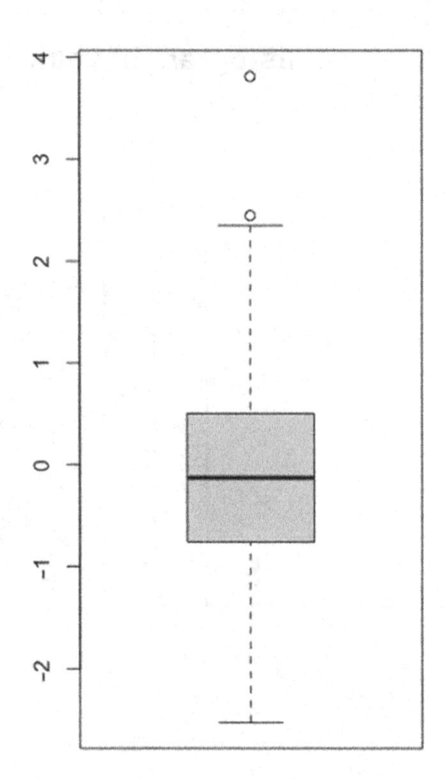

FIGURE 7.3
A simple box plot.

7.4 Scatter Plots

Scatter plots are used to visualise the relationship between two continuous variables and can be created using the plot() function with the argument "type = 'p'" to specify a scatter plot. The function takes two vectors as arguments and plots a x vs y plot of the values.

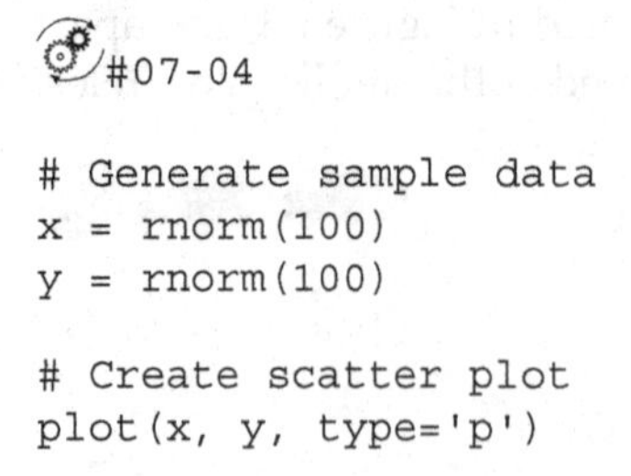

1 star

#07-04

```
# Generate sample data
x = rnorm(100)
y = rnorm(100)

# Create scatter plot
plot(x, y, type='p')
```

Figure 7.4 shows a simple scatter plot.

These basic plots provide a foundation for visualising data. However, there are many other plotting functions available, including advanced plots such

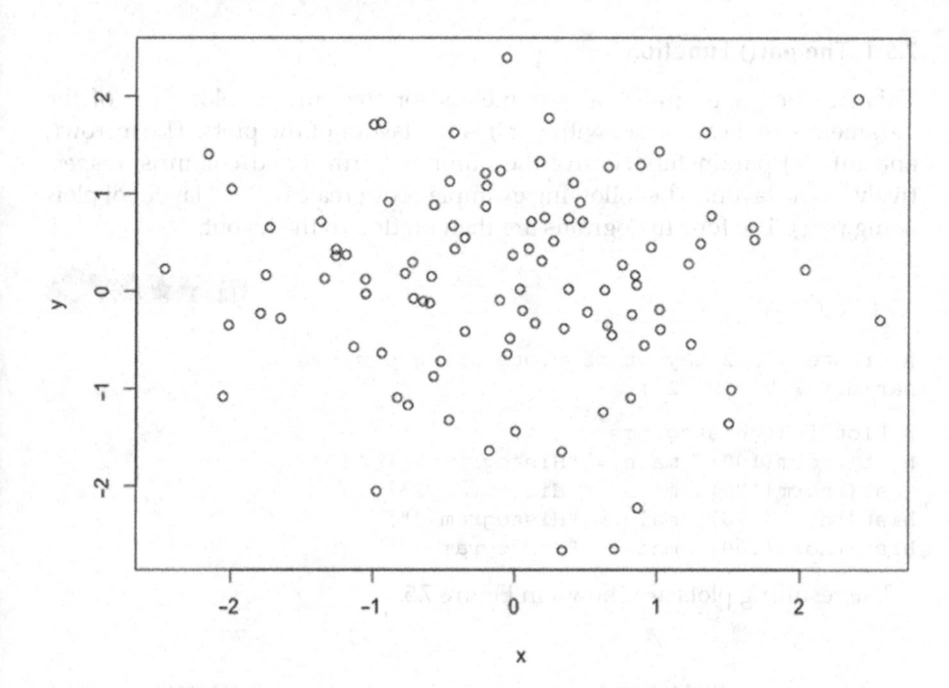

FIGURE 7.4
A simple scatter plot.

as line plots, density plots, and scatter plot matrices. In addition, there are many packages available that enhance this functionality, one such package is 'ggplot2' that we will talk about later (Chapter 11).

Visualising data is an important part of data analysis and science communication. By using basic plots such as bar plots, histograms, box plots, and scatter plots, data analysts can effectively display the distribution of variables and relationships between variables. The importance of using data visualisation to communicate is so much that a major news company such as BBC[1] has its own cookbook for creating visual charts in R.

7.5 Organising the Plots

There are several functions, from the 'graphics' package, that can be used to organise multiple plots in a single figure. These functions include par(), layout(), mfrow(), mfcol(), and split.screen(), among others. Herein it is provided an introduction on how they work and how to obtain publishable results.

[1] see https://bbc.github.io/rcookbook/.

7.5.1 The par() Function

This function sets graphical parameters for the current plot. One of the parameters that can be set with par() is the layout of the plots. The mfrow() and mfcol() parameters control the number of rows and columns, respectively, in the layout. The following example code creates a 2 × 2 layout of plots using par(). The four histograms are then plotted in the layout.

#07-05 2 stars

```
# Create a 2x2 layout of plots using par()
par(mfrow = c(2, 2))

# Plot four histograms
hist(rnorm(100), main = "Histogram 1")
hist(rnorm(100), main = "Histogram 2")
hist(rnorm(100), main = "Histogram 3")
hist(rnorm(100), main = "Histogram 4")
```

The resulting plots are shown in Figure 7.5.

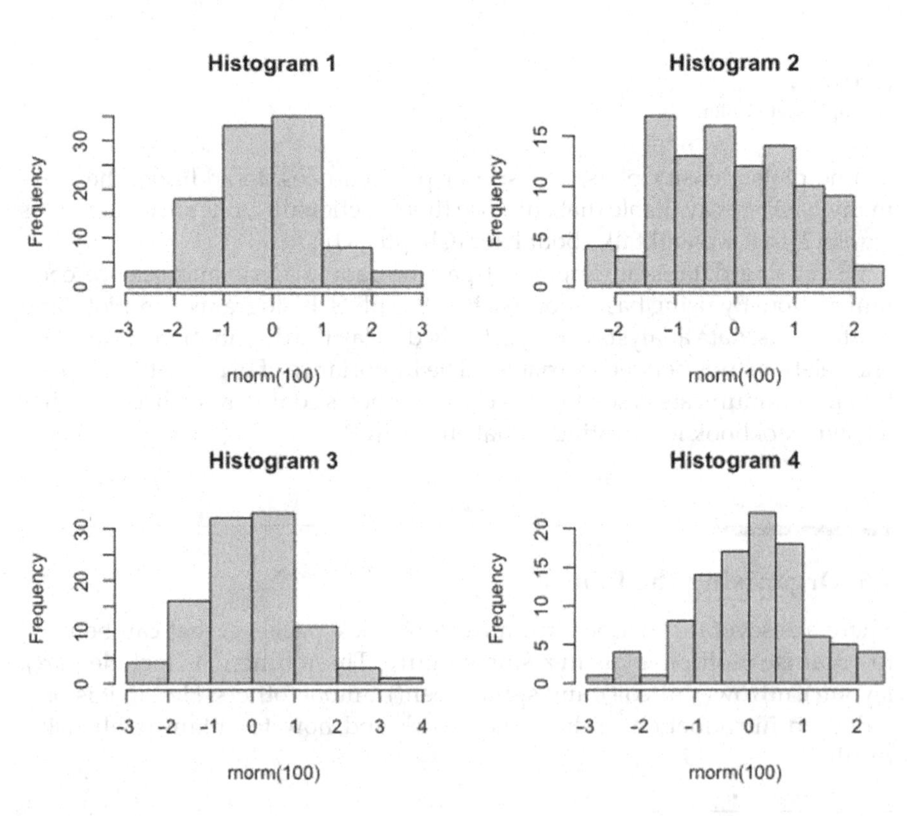

FIGURE 7.5
The result of plotting with the par() function.

7.5.2 The layout() Function

This function allows to create custom layouts for multiple plots. The layout() function takes a matrix as an argument that specifies the placement of the plots. For example:

#07-06 

```
# Create a custom layout of plots using layout()
layout(matrix(c(1, 1, 2, 3), nrow = 2, ncol = 2, byrow = TRUE))

# Plot three histograms
hist(rnorm(100), main = "Histogram 1")
hist(rnorm(100), main = "Histogram 2")
hist(rnorm(100), main = "Histogram 3")
```

The matrix() function creates a matrix that specifies the placement of the plots. The three histograms are then plotted in the layout. The numbers in the matrix indicate the order in which the plots will be filled. In this case, the matrix is set up as a 2×2 grid, and the numbers 1, 1, 2, and 3 specify the order in which the plots will be arranged.

- The first two entries (1, 1) indicate that the first plot will take up the first row of the grid.
- The third entry (2) indicates that the second plot will be in the second row of the grid.
- The fourth entry (3) indicates that the third plot will be in the second column of the grid.

The 'nrow = 2, ncol = 2, byrow = TRUE' parameters define the layout characteristics. In this case, the layout is specified to have two rows (nrow = 2) and two columns (ncol = 2), and the filling of the matrix is done by rows (byrow = TRUE).

Remember that this layout configuration will affect subsequent plots; any subsequent plots will be arranged according to the specifications provided in the layout() function. The numbers in the matrix correspond to the order in which the subsequent plots should fill the grid. For example, the first plot will occupy the top-left position, and the second and third plots will follow the specified order in the grid layout.

The result is shown in Figure 7.6.

7.6 Save the Plots

Plots can be saved to an image file using different functions from different packages. The 'grDevices' package in R is a core package that provides essential functionality for handling graphics devices and controlling various aspects of graphical output. It serves as an interface between R and the

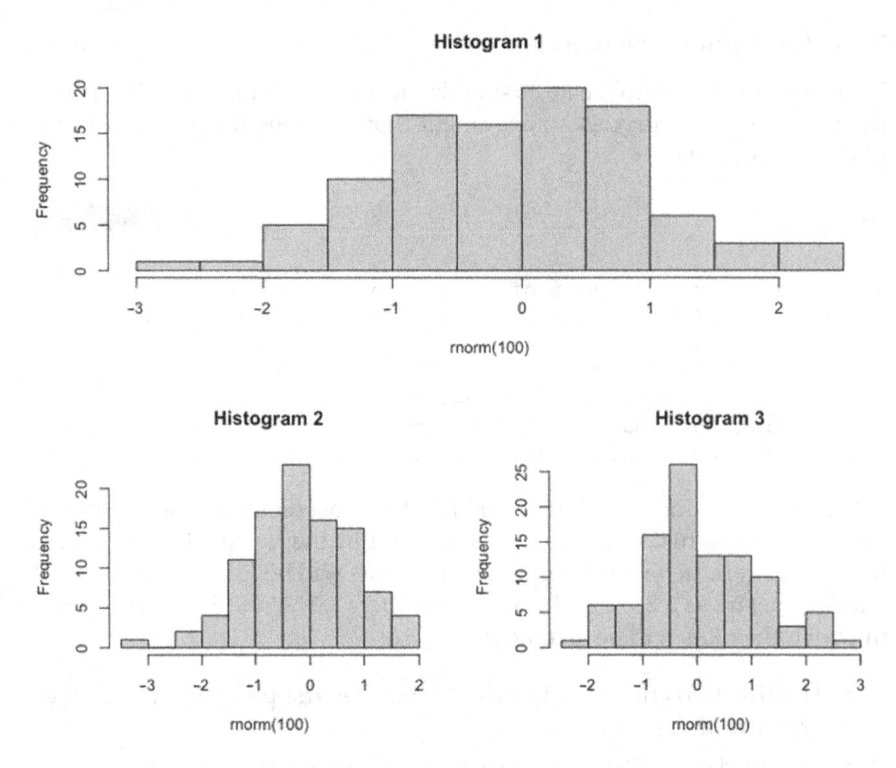

FIGURE 7.6
The use of the layout() function to present three plots.

underlying graphical systems and provides functions for creating, managing, and manipulating graphical devices. One of the primary functions of the package is to communicate with the graphical devices, which are responsible for generating visual output. It provides functions for opening and closing graphics devices, selecting active devices, and controlling device-specific parameters such as size, resolution, and file formats. These functions allow users to generate graphics in different formats, such as 'png', 'pdf', and 'jpeg', or display them on screens or external devices.

Additionally, the 'grDevices' package offers functions for controlling graphical parameters, such as colours, fonts, line types, and point styles. These functions enable users to customise the appearance of plots and ensure consistency across different graphical elements. The package also includes utilities for handling colour palettes, managing transparency, working with different colour models, and converting between colour representations.

The code below is an example of saving a plot to the computer disk.

```
#07-07

# Reset the plot to a 1x1 graph
par(mfrow = c(1, 1))
```

```
# Create the device to save the plot to a PNG file
png("~/my_plot.png")

# Create random data
x = rnorm(50)
y = rnorm(50)

# Plot the result
plot(x, y, main = "My Scatterplot", xlab = "X", ylab = "Y")

# Close the graphical device
dev.off()
```

The png() function is used to open a connection to a PNG file. Next, a scatterplot is created using the plot() function. Finally, we close the png file using the dev.off() function.

This same method can be used to create 'jpeg' or 'tiff' files with the corresponding jpeg() and tiff() functions. Notice that it is necessary to verify the name of the working folder where you want to save the file.

7.7 Concluding Remarks

This chapter has delved into the foundational aspects of data visualisation in R, equipping the reader with essential tools to effectively communicate and analyse data. By exploring a repertoire of basic functions, from simple scatter plots to more intricate graphical representations, the reader becomes empowered to convey data with clarity.

Further examples will navigate the intricate terrain of diverse datasets; the acquired proficiency in data visualisation serves as a cornerstone for insightful interpretations and informed decision-making. Looking ahead, the integration of these basic functions into the geologist's toolkit promises to enhance the communicative power of data visualisation, fostering a deeper understanding of geological phenomena and facilitating more robust scientific inquiry in the field of geoscience.

Reference

Sturges, H. A. (1926). The choice of a class interval. Journal of the American Statistical Association, 21(153). 65–66. https:doi.org/10.1080/01621459.1926.10502161. JSTOR 2965501.

8

Putting It All to Work: Part I

The world is currently witnessing an incessant drive towards making data more readily available and accessible to everyone. As a result, students, researchers, and companies have free access to vast amounts of data, which can be used for learning and training, as well as in research and development projects. This book benefits and embraces these policy initiatives by endeavouring to comprehensively illustrate the functions and methods of spatial analysis and by providing the reader with a multitude of data sources to facilitate learning and research purposes.

For wrapping many of the aforementioned methods and techniques in Part I of the book, an available set of data from stream sediments campaigns in Northern Territory (NT), Australia, will be used to exemplify and deepen the knowledge acquired. The NT Government in its site[1] makes available a large dataset with geological, geochemical, and geophysical information, including its geographic coordinates information.

This chapter is designed as a case study exercise to conduct a detailed assessment of part of the geochemical data, verifying what data is available, creating subsets of data that will be used in future exercises, and providing an exploratory data analysis approach of some selected subsets of the data.

8.1 Reading the Data

The stream sediment repository and the corresponding metadata information can be assessed from the mentioned site. To utilise the data in our project, the first step is to download a 'zip' file with the geochemical data and unzip it to a working folder.

After unzipping the data, by using the folder explorer in the computer one can verify that in the unzipped data folder there is a file named 'GEOCHEM_STREAM_SEDIMENTS.csv' that is the one that contains the information needed for this exercise. The read.csv() function (cf. Section 3.1.1 in Chapter 3) is used to read the data.

[1] https://geoscience.nt.gov.au/gemis/ntgsjspui/community-list

DOI: 10.1201/9781032651880-9

#08-01/01

```
# Northern Territory Government Case Study
# For more information goto
# https://data.nt.gov.au/dataset/northern-territory-geological-
survey-geochemistry-stream-sediment
# Complementary data
# https://geoscience.nt.gov.au/downloads/NTWideDownloads.html

sediment_url = "http://geoscience.nt.gov.au/contents/prod/
Downloads/Geochemistry/GEOCHEM_STREAM_SEDIMENTS_csv.zip"

# Download the file
nt_ss = download.file(sediment_url, destfile = "NT_Stream
Sediments.zip" )

# Unzip the file
unzip(zipfile = "NT_Stream Sediments.zip", exdir = 'NT_SS')

# Read the file
nt_ss = read.csv("NT_SS/GEOCHEM_STREAM_SEDIMENTS.csv")
```

Using the colnames() function, one can retrieve the structure of the variable names included in this large dataset.

#08-01/02

```
# Retrieve the variable names from the data
colnames(nt_ss)
```

The 'nt_ss' variable is populated with 116267 observations of 205 variables (it might be more as the database is constantly growing). To verify the variable names, use the colnames() function. Most of these variables are chemical analysis, but there is a lot of other information, such as the sample reference, the sample ID, and the coordinates in latitude and longitude. However, by looking at the variable names from the 205 variables, one can realise that most of them correspond to chemical analysis in percentage '_PCT', parts per million '_PPM', and even parts per billion '_PPB', represented in the postfixes.

The result is as follows:

```
  [1] "SAMPLEID"        "SAMPLEREF"        "SAMPLE_TYPE"
      "SAMPLE_METHOD"   "DATE_SAMPLED"
  [6] "LONGITUDE"       "LATITUDE"         "LITHOLOGY"
      "MINMESH"         "MAXMESH"
 [11] "SAMPLEWT"        "COMPANY"          "ACCURACY"
      "TITLE"           "MAP_SHEET_100K"
```

```
 [16] "MAP_SHEET_250K" "REPORT_NO"       "OPEN_FILE"
      "JOB_NO"         "COMMENTS"
 [21] "AG_PPB"         "AG_PPM"          "AL_PCT"
      "AL_PPM"         "AS_PPB"
 [26] "AS_PPM"         "AU_PPB"          "AU_PPM"
      "AU_PPT"         "B_PPM"
 [31] "BA_PPM"         "BE_PPM"          "BI_PPM"
      "CA_PCT"         "CA_PPM"
 [36] "CD_PPM"         "CE_PPM"          "CO_PPB"
      "CO_PPM"         "CR_PPM"
...

[166] "P2O5_PCT"       "P2O5_PPM"        "PB_PCT"
      "PB_PPB"         "PB1_PPM"
[171] "PBT_PPB"        "PBT_PPM"         "PD_PCT"
      "PD1_PPB"        "PT_PCT"
...
[
[196] "V_PCT"          "V_PPB"           "V2O5_PPM"
      "W_PPB"          "W1_PPM"
[201] "YB_PPM"         "ZN_PCT"          "ZR_PCT"
      "UNIQ_ID"        "ID"
```

Naturally, not all the variables were analysed for all the samples, and therefore, many NA values are present; therefore, some data cleaning will be necessary.

8.2 Insights from the Data

To learn some of the parameters from the data, the unique() function is used to verify two geologically important variables, the lithology and the company that collected the information, 'LITHOLOGY' and 'COMPANY', respectively.

```
#08-01/03

# Retrieve the LITHOLOGY column data
unique(nt_ss$LITHOLOGY)
```

The result is as follows:

```
 [1] "" "silt" "shale" "dolerite"
 [5] "tuff" "greywacke" "quartz vein" "sand"
 [9] "soil" "laterite" "siltstone" "limestone"
[13] "dolomite/dolostone" "conglomerate" "unclassified sediment"
     "chert"
[17] "basalt" "quartzite" "pisolitic ironstone" "gravel"
[21] "rhyolite" "calcrete" "fault shear zone" "ironstone"
[25] "quartz-feldspar rock" "loam"
```

The lithology dataset has 26 different geological entries, including a blank one and one that is named 'unclassified sediment'.

```
#08-01/04

# Retrieve the COMPANY column data
unique(nt_ss$COMPANY)
```

The result is as follows:

```
  [1] Tidegate Pty Ltd.
  [2] Stockdale Prospecting Ltd.
  [3] Normandy Exploration (Darwin)
...
 [79] Western Desert Resources
 [80]
 [81] Sandfire Resources
...
[228] RIO TINTO EXPLORATION PTY LTD.
[229] Rio Tinto Exploration Pty. Limited
[230] Crossland Mines
...
[299] Western Mining Corporation
[300] Endras Limited
300 Levels:  A.O. (Australia) Pty Ltd. A.O.G Minerals Limited
... Zapopan NL.
```

Regarding the companies, there are 300 (maybe more, see above) that contributed with data, including some data that is blank, i.e. not attributed to any company record [80], and a close inspection also reveals that some companies have duplicate or very similar names, e.g. "Rio Tinto" is present in at least four entries.

To deeper explore the information on this dataset, let's suppose that it is necessary to retrieve the companies that have contributed with more than 1000 samples to the database. This is possible using a for() cycle combined with a conditional expression, in the example using the if() function.

The example is designed to sift through the 'nt_ss' variable to identify companies that have more than 1000 samples recorded. It serves as a practical example of how to perform conditional data filtering and aggregation, particularly useful in data analysis processes where understanding the volume of data per category (in this case, per company) is crucial.

In the first step, the code iterates over each unique company name found in the 'COMPANY' column of the 'nt_ss' variable, using a for() function. For each company, it creates a logical vector 'query' that indicates whether each row (or sample) in the dataset belongs to the current company being considered. This is achieved by comparing the 'COMPANY' column values to 'company_name'. The code then checks if the sum of true values in 'query' (which represents the total number of samples for the company) exceeds 1000. If a company has more than 1000 samples, the code prints a message specifying the company's name along with the exact number of samples it has. This process is repeated for each unique company in the dataset, effectively filtering and reporting on those with a significant number of samples.

#08-02

```
# Verifying the companies with more than 1000 samples
for(company_name in unique(nt_ss$COMPANY)) {
  # Select the company by company name
  query = nt_ss$COMPANY == company_name

  if( sum(query)>1000 ){# Verify if n.  samples > 1000
      # Print the result
      print( paste(company_name,"with", sum(query), "samples
selected"))
  }
}
```

The results display as follows:

```
[1] "Stockdale Prospecting Ltd. with 2661 samples selected"
[1] "Mt Isa Mines Ltd with 21690 samples selected"
[1] "CRA Exploration Pty Ltd. with 5853 samples selected"
...
[1] "Pancontinental Mining Ltd with 1030 samples selected"
[1] "Bureau of Mineral Resources, Geology and Geophysics with
1534 samples selected"
[1] "Australasian Minerals, Inc. with 16072 samples selected"
...
[1] "Western Mining Corporation with 2698 samples selected"
```

Counting the results indicates that the dataset has 20 companies that contributed with more than 1000 samples.

8.3 Subsetting Data

By using the bracketing system, it is possible to create a new variable with the data that is provided by one selected company, e.g. the 'Rio Tinto Exploration Pty Ltd'. Notice that a query variable is created with the expression 'nt_ss$COMPANY == 'Rio Tinto Exploration Pty Ltd''. This query is a vector with logical values (i.e. TRUE or FALSE) indicating which elements from the 'nt_ss$COMPANY' variable match the selection.

The print() function is used to retrieve the obtained results, after which a bracketing subsetting is used to retrieve only the rows that match the query, and the results are stored in the 'nt_ss_01' variable.

#08-03

```
# Example query for selecting the Rio Tinto [78]
query = nt_ss$COMPANY == 'Rio Tinto Exploration Pty Ltd'

# Results obtained
print(paste(sum(query), "samples selected"))

# Make the subsetting using bracketing
nt_ss_01 = nt_ss[query, ]
```

The resulting output is the following:

```
[1] "383 samples selected"
```

Hence, a subset 'nt_ss_01' with 383 samples is created.

It is advisable to remove columns that only have NA values. This is possible with another subsetting operation. For this operation, the which() and colSums() functions from the 'base' package are used. The function which() is used to identify the positions of elements in a data frame that meet certain conditions, and colSums() is used to calculate the sum of values across columns of a matrix or data frame. The is.na() function is called to identify the elements that are NAs.

This returns a variable named 'na_cols' that contains a vector of logical values indicating if an element is an NA (TRUE) or not (FALSE).

#08-04/01

```
# Remove columns that are NA
na_cols = which(base::colSums(is.na(nt_ss_01)) == nrow(nt_ss_01))

# Subset the data frame to exclude the all-NA columns
nt_ss_01_filtered = nt_ss_01[ , -na_cols]
```

For a better understanding of the results, we will display the partial ones from each of the called functions.

The function base::colSums() retrieves a vector of logical TRUE and FALSE values identifying if a column has NA values or not. Notice that the colSums() function from the 'base' R package is explicitly called using the '::' operator. This is done because there are two conflicting functions named colSums(). There is one in the 'base' R package and another one in the 'Matrix' library. Therefore, it is necessary to be explicit in the R command line, which one we are calling.

```
SAMPLEID   SAMPLEREF SAMPLE_TYPE SAMPLE_METHOD DATE_SAMPLED LONGITUDE LATITUDE
FALSE      FALSE     FALSE       FALSE         TRUE         FALSE     FALSE
LITHOLOGY MINMESH    MAXMESH     SAMPLEWT      COMPANY      ACCURACY  TITLE
FALSE      FALSE     TRUE        FALSE         FALSE        FALSE     FALSE
```

In this example, only part of the result, i.e. the columns 'DATE_SAMPLED' and 'MAXMESH', has NAs.

The which() function returns the column numbers that have NAs.

```
DATE_SAMPLED                    MAXMESH
           5                         10
```

To eliminate these columns, the variable 'nt_ss_01_filtered' is created and the NA values are removed with the bracketing using the minus '-' sign. The results are presented using the print() function.

```
#08-04/02

# Retrieve the NA columns
print(paste(length(na_cols), "NA columns rejected"))

# Retrieve the columns filtered
print(paste(ncol(nt_ss_01_filtered),"columns present"))
```

The output is as follows:

```
[1] "2 NA columns rejected"

[1] "203 columns present"
```

The new 'nt_ss_01_filtered' variable is equal to the original but with the two columns removed.

8.4 Transforming Data

From the metadata information[2] of the Northern Territory geological dataset, it is possible to verify that when the chemical values are below the limit of detection (LOD) or have not been analysed, the –9999 value was imputed. In the first step, it is better to remove the columns with these values (subset operation), and in the second step, replace –9999 by NAs (transform operation).

[2] https://www.ntlis.nt.gov.au/metadata/export_data?type=html&metadata_id=66C8BD60B438D7F0E050CD9B21441F85

Notice the use of 'nt_ss_02' intermediate variable to perform the subsetting of the rows that have –9999 values. A final 'nt_ss_10' variable is created, from which the –9999 values are replaced by NA.

#08-05

```
# Find the columns with all NA values
nt_ss_02 = nt_ss_01_filtered
na_cols = which(base::colSums(nt_ss_02 == -9999) ==
nrow(nt_ss_02))

# Print the results
print(paste(length(na_cols),"rejected columns with -9999"))

# Subset the data frame to exclude all the NA columns
nt_ss_10 = nt_ss_o2[ ,-na_cols]

# Print the results
print(paste(ncol(nt_ss_10),"columns present"))

# Replace all the -9999 remaining by NA
nt_ss_10[nt_ss_10 == -9999] = NA
```

The results obtained are as follows:

```
[1] "103 -9999 columns rejected"
```

and:

```
[1] "100 columns present"
```

The final dataset 'nt_ss_10' contains 383 observations of 100 variables.

8.5 Plotting the Data

A simple plot allows us to confirm that the data is dispersed across several locations and that it might be from different mineral exploration campaigns.

#08-06

```
# Plot the data
plot(x=nt_ss_10$LONGITUDE, y=nt_ss_10$LATITUDE,
      main = "Rio Tinto Exploration Pty Ltd",
      xlab = "Longitude", ylab = "Latitude")
```

The result of the plot is in Figure 8.1.

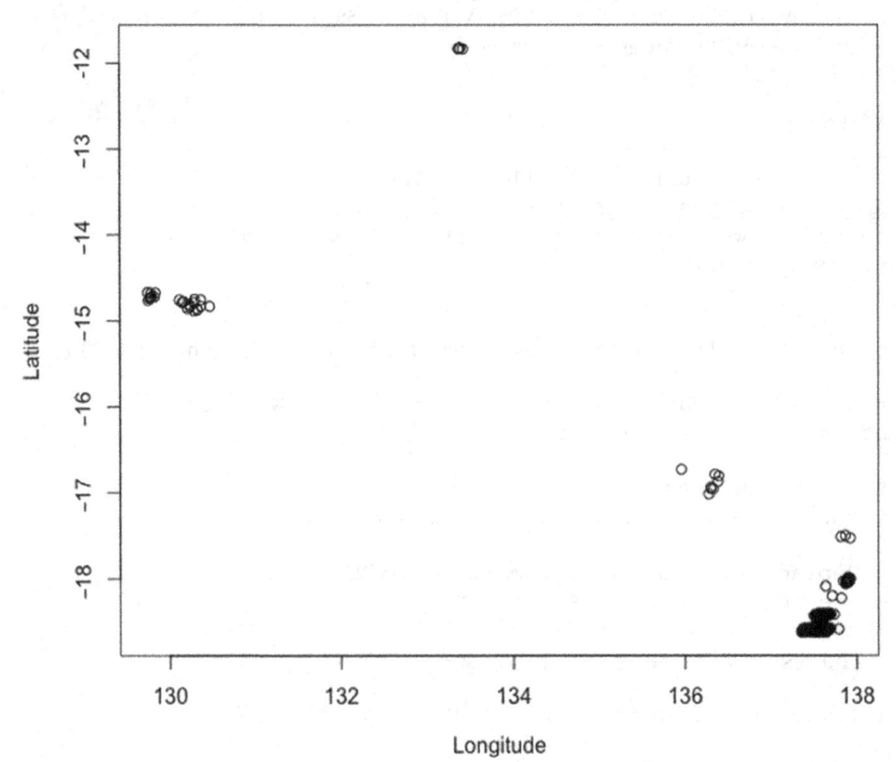

FIGURE 8.1
Plot of the samples from the "Rio Tinto Exploration Pty Ltd" data.

For the next exercises, it is necessary to select some data subsets suitable for spatial analysis. Thus, it is necessary to have a first glimpse of all of the available data. For this task, the previous for() cycle (code snippet #08-02) was adapted to create 2 × 2 plots with the samples collected by each company.

The code is as follows:

#08-07 2 stars

```
# Create a 2x2 layout of plots using par()
par(mfrow = c(2, 2))

# Verifying the companies with more than 1000 samples
for(company_name in unique(nt_ss$COMPANY)) {
  # Select the company by company name
  query = nt_ss$COMPANY == company_name

  if( sum(query)>1000 ){ # Verify if n.  samples > 1000
      # Print the result
```

```
      print( paste(company_name,"with", sum(query), "samples
selected"))

      # Create a data frame with the result
      df = nt_ss[query, ]

      # Plot the data from the data frame
      plot(x=df$LONGITUDE, y=df$LATITUDE, main = company_name,
      xlab = "Longitude", ylab = "Latitude")
   }
}
```

This code creates a large set of graphics with the results from each company in a 2 × 2 layout. Part of the results is shown in Figure 8.2.

It is readily visible that some companies made campaigns in specific streams (e.g. Pancontinental Mining), others covered full areas (e.g. Bureau of Mineral Resources), and others have just sampled individual prospects (e.g. Amoco Minerals).

This first analysis led us to select for our case study example the company 'Bureau of Mineral Resources, Geology and Geophysics' abbreviated to BMRGG. Therefore, a subset with all of the data provided by this company is created. This new dataset has 1534 samples to analyse.

8.6 The Bureau of Mineral Resources Subset

To further analyse the data, the first task is to subset the dataset, and for didactic purposes, we will use the grepl() function to create a logical, i.e. a (TRUE, FALSE) vector, with the company names that the first letters match 'Bureau of Mineral'. The '^' symbol is to indicate to grepl() function that the text must be searched in the beginning of the field.

#08-08 2 stars

```
# Filter the dataset
company_name = grepl("^Bureau of Mineral", nt_ss$COMPANY)
query = nt_ss$COMPANY == company_name

# Create the subset
nt_ss_bmrgg = nt_ss[company_name,]

# Verify if the result is good
title = "Bureau of Mineral Resources, Geology and Geophysics"

# Plot the data
par(mfrow = c(1, 1))
plot(x=nt_ss_bmrgg$LONGITUDE, y=nt_ss_bmrgg$LATITUDE,
      main = title, xlab = "Longitude", ylab = "Latitude")
```

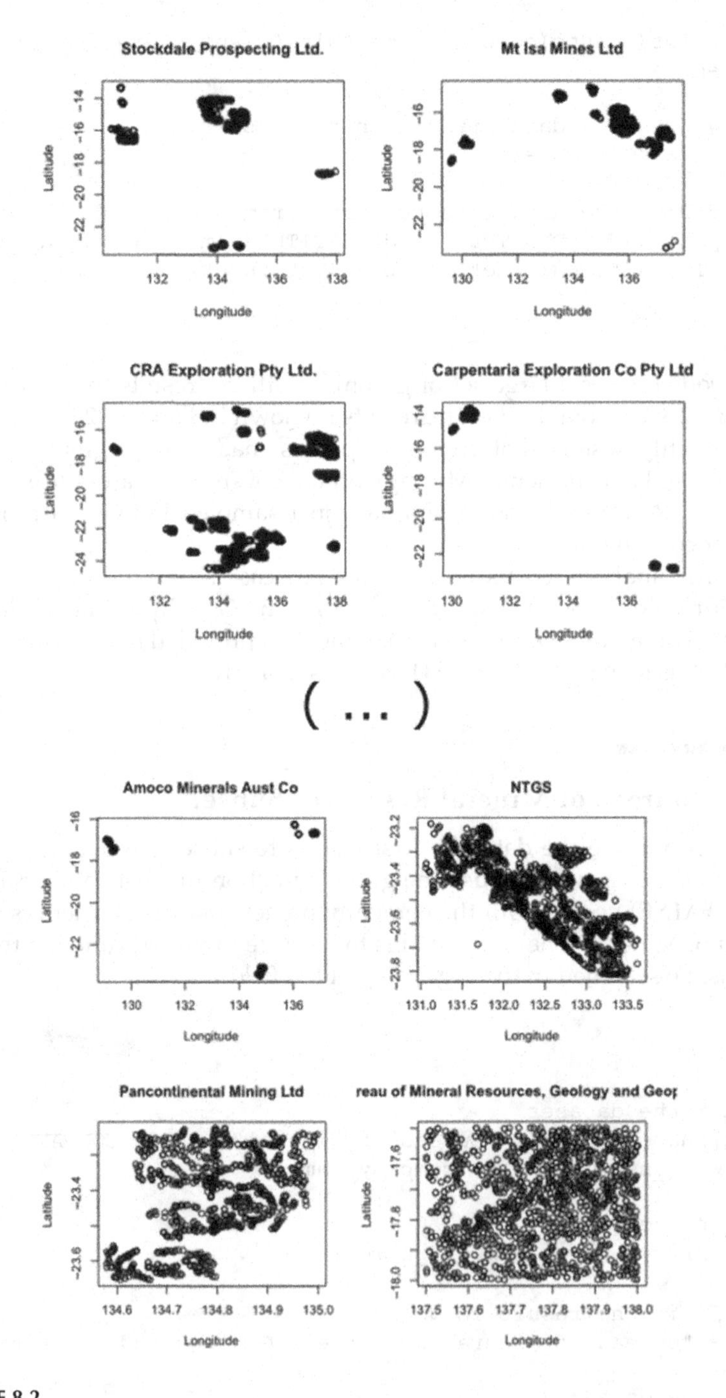

FIGURE 8.2
The 2 × 2 plots produced by the for() cycle for the 20 companies that have more than 1000 samples in the dataset.

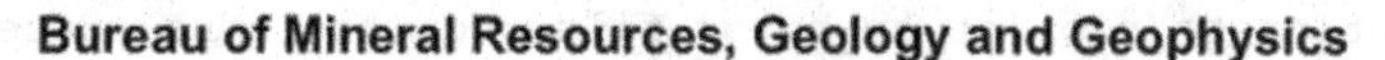

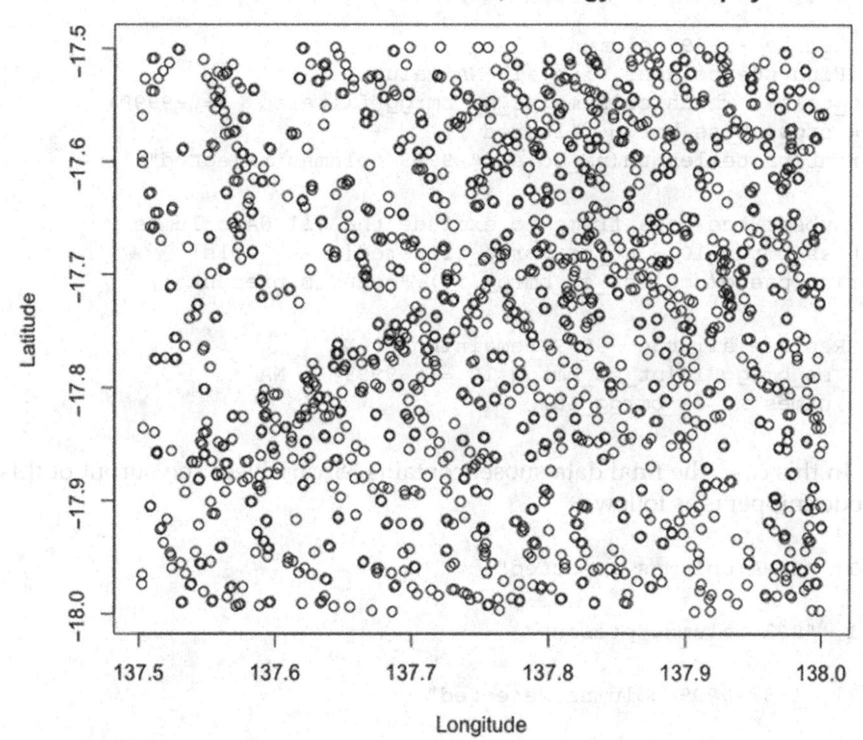

FIGURE 8.3
Plot of the subset of the data from Bureau of Mineral Resources.

The result of the plot is shown in Figure 8.3.

With the colnames() function, it is possible to verify that there are many variables meaningless for a statistical description. Let us just use the 'SAMPLEID' and the chemical analysis that are non-zero. For this, the previous subsetting function is adapted to select the necessary columns. The final result is stored in a working variable called 'nt_ss_bmrgg_10', that contains the NT stream sediments from the BMRGG company.

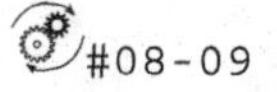#08-09

```
# Remove columns that are NA
na_cols = which(base::colSums(is.na(nt_ss_bmrgg)) ==
nrow(nt_ss_bmrgg))
print(paste(length(na_cols), "NA columns rejected"))

# Subset the data frame to exclude the all-NA columns
nt_ss_bmrgg_filtered = nt_ss_bmrgg[, -na_cols]
```

```
print(paste(ncol(nt_ss_bmrgg_filtered),"columns present"))

# Remove -9999 columns
# Find the columns with all NA values
na_cols = which(colSums(nt_ss_bmrgg_filtered  == -9999)
== nrow(nt_ss_bmrgg_filtered ))
print(paste(length(na_cols),"-9999 columns rejected"))

# Subset the data frame to exclude the all-NA columns
nt_ss_bmrgg_10 = nt_ss_bmrgg_filtered[, -na_cols]
print(paste(ncol(nt_ss_bmrgg_10),"columns present"))

# Replace all the -9999 remaining by NA
nt_ss_bmrgg_10[nt_ss_bmrgg_10 == -9999] = NA
colnames(nt_ss_bmrgg_10)
```

In this case, the final data subset contains 38 variables. The output of this code snippet is as follows:

```
[1] "2 NA columns rejected"

[1] "203 columns present"

[1] "165 -9999 columns rejected"

[1] "38 columns present"

 [1] "UNIQ_ID"         "ID"            "SAMPLEID"
     "SAMPLEREF"       "SAMPLE_TYPE"   "SAMPLE_METHOD"
 [7] "LONGITUDE"       "LATITUDE"      "LITHOLOGY"
     "MINMESH"         "MAXMESH"       "COMPANY"
[13] "ACCURACY"        "TITLE"         "MAP_SHEET_100K"
     "MAP_SHEET_250K"  "REPORT_NO"     "OPEN_FILE"
[19] "JOB_NO"          "COMMENTS"      "AG_PPM"
     "AS_PPM"          "BA_PPM"        "BI_PPM"
[25] "CO_PPM"          "CR_PPM"        "CU_PPM"
     "FE_PCT"          "MN_PPM"        "MO_PPM"
[31] "NI_PPM"          "PB_PPM"        "SN_PPM"
     "TH_PPM"          "TI_PPM"        "U_PPM"
[37] "W_PPM"           "ZN_PPM"
```

From these variables, we are interested in "SAMPLEID" and the chemical analysis, i.e. columns 1 and 21–38. With this information, the Bureau of Mineral Resources dataset is created. Remember to verify if the variables have not changed from this example, as the example database is constantly being updated and might be changed at any moment. The key is to retain the variables with chemical elements and the sample identification.

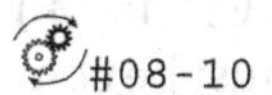#08-10 

```
# Create Bureau Mineral Research dataset
bmrgg = nt_ss_bmrgg_10[,c(3,21:38)]
```

The dataset created is called 'bmrgg', and it contains 1534 observations of 19 variables.

8.7 Descriptive Statistics

The first approach is to make a summary() of the data, which provides a simple and fast overview of its contents.

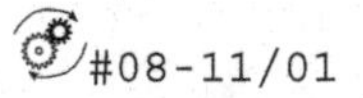 

```
# Summarise the results
summary(bmrgg)
```

The result is as follows:

```
SAMPLEID              AG_PPM              AS_PPM            BA_PPM             BI_PPM
 Length:1534        Min.   :-1.0000   Min.   :-1.00   Min.   :  10.0   Min.   :-2.000
 Class :character   1st Qu.:-1.0000   1st Qu.: 2.00   1st Qu.: 176.2   1st Qu.:-2.000
 Mode  :character   Median :-1.0000   Median : 3.00   Median : 343.0   Median :-2.000
                    Mean   :-0.8872   Mean   : 3.61   Mean   : 410.1   Mean   :-1.098
                    3rd Qu.:-1.0000   3rd Qu.: 4.00   3rd Qu.: 600.8   3rd Qu.:-2.000
                    Max.   : 7.0000   Max.   :22.00   Max.   :1730.0   Max.   :33.000

    CO_PPM          CR_PPM           CU_PPM           FE_PCT           MN_PPM
 Min.   :-1.00   Min.   :  3.00   Min.   :  2.00   Min.   : 0.100   Min.   :   6.0
 1st Qu.: 3.00   1st Qu.: 19.00   1st Qu.:  8.00   1st Qu.: 1.000   1st Qu.:  68.0
 Median : 7.00   Median : 29.00   Median : 13.00   Median : 1.600   Median : 183.0
 Mean   :10.29   Mean   : 47.85   Mean   : 25.43   Mean   : 2.281   Mean   : 299.7
 3rd Qu.:12.00   3rd Qu.: 53.00   3rd Qu.: 26.00   3rd Qu.: 2.600   3rd Qu.: 391.8
 Max.   :88.00   Max.   :479.00   Max.   :325.00   Max.   :14.800   Max.   :6650.0
                 NA's   :1                                           NA's   :6

    MO_PPM        NI_PPM           PB_PPM           SN_PPM           TH_PPM
 Min.   :-3   Min.   : -1.00   Min.   : -2.00   Min.   : -2.00   Min.   : -2.00
 1st Qu.:-3   1st Qu.:  4.00   1st Qu.:  8.00   1st Qu.:  2.00   1st Qu.:  8.00
 Median :-3   Median :  6.00   Median : 11.00   Median :  3.00   Median : 14.00
 Mean   :-3   Mean   : 13.67   Mean   : 13.39   Mean   :  6.48   Mean   : 27.88
 3rd Qu.:-3   3rd Qu.: 12.00   3rd Qu.: 17.00   3rd Qu.:  4.00   3rd Qu.: 33.00
 Max.   :-3   Max.   :128.00   Max.   :178.00   Max.   :449.00   Max.   :666.00

    TI_PPM            U_PPM            W_PPM              ZN_PPM
 Min.   :   400   Min.   :-1.000   Min.   : -3.000   Min.   :  1.00
 1st Qu.:  2600   1st Qu.: 2.000   1st Qu.: -3.000   1st Qu.:  8.00
 Median :  3750   Median : 3.000   Median : -3.000   Median : 16.00
 Mean   :  5095   Mean   : 4.935   Mean   :  1.557   Mean   : 27.14
 3rd Qu.:  5800   3rd Qu.: 6.000   3rd Qu.:  4.000   3rd Qu.: 32.75
 Max.   :127000   Max.   :63.000   Max.   :113.000   Max.   :337.00
```

From this first assessment, all the data is valid despite the Cr ('CR_PPM') and Mn ('MN_PPM') with 1 and 6 NA values, respectively. This must be considered for some types of statistical analysis. Other variables are suspicious, e.g. some min() values are negative, that must be considered as below LOD. In other variables, the median() is equal to the minimum value, e.g. the cases of Ag, Bi, Mo, and W. Moreover, Mo has all values equal to −3.

For a first approach, the suggestion is to create boxplots for all the variables and visually inspect what their values are.

#08-11/02

```
# Create a layout for the plot with 6 column by 3 row
par(mfrow = c(3, 6))

# Loop over each variable and create a boxplot
for (i in 2:19) {
  boxplot(bmrgg[,i], main = colnames(bmrgg)[i])
}
```

The results are displayed in Figure 8.4.

The plots reveal that for the signalled variables, some contain outliers, represented as dots, and all the other values are in a single category that must be below LOD. Consequently, these variables should be considered as mostly below LOD, with some outliers.

For identifying the number of outliers for each variable, it is possible to use the lapply() function to the dataset using the boxplot.stats() function from the 'grDevices' package. This function creates a list with all the variables in the dataset, and for each variable, it calculates some basic statistics, including the 'out' variable that is a vector with all the outliers.

Applying the length() function from the 'base' R package to the 'stats []$out' variable retrieves the number of outliers for each variable. This is done using the simple apply() function, i.e. sapply(), to the 'stats' variable.

#08-12

```
# Retrieve the number of outliers for each variable
# get the summary statistics for each variable
stats = lapply(bmrgg[, 2:19], boxplot.stats)

# Retrieve the number of outliers for each variable
outliers = sapply(stats, function(x) length(x$out))

# Show the results
outliers
```

The combination of lapply() and sapply() functions provides an efficient way to perform and streamline the analysis of outliers across multiple

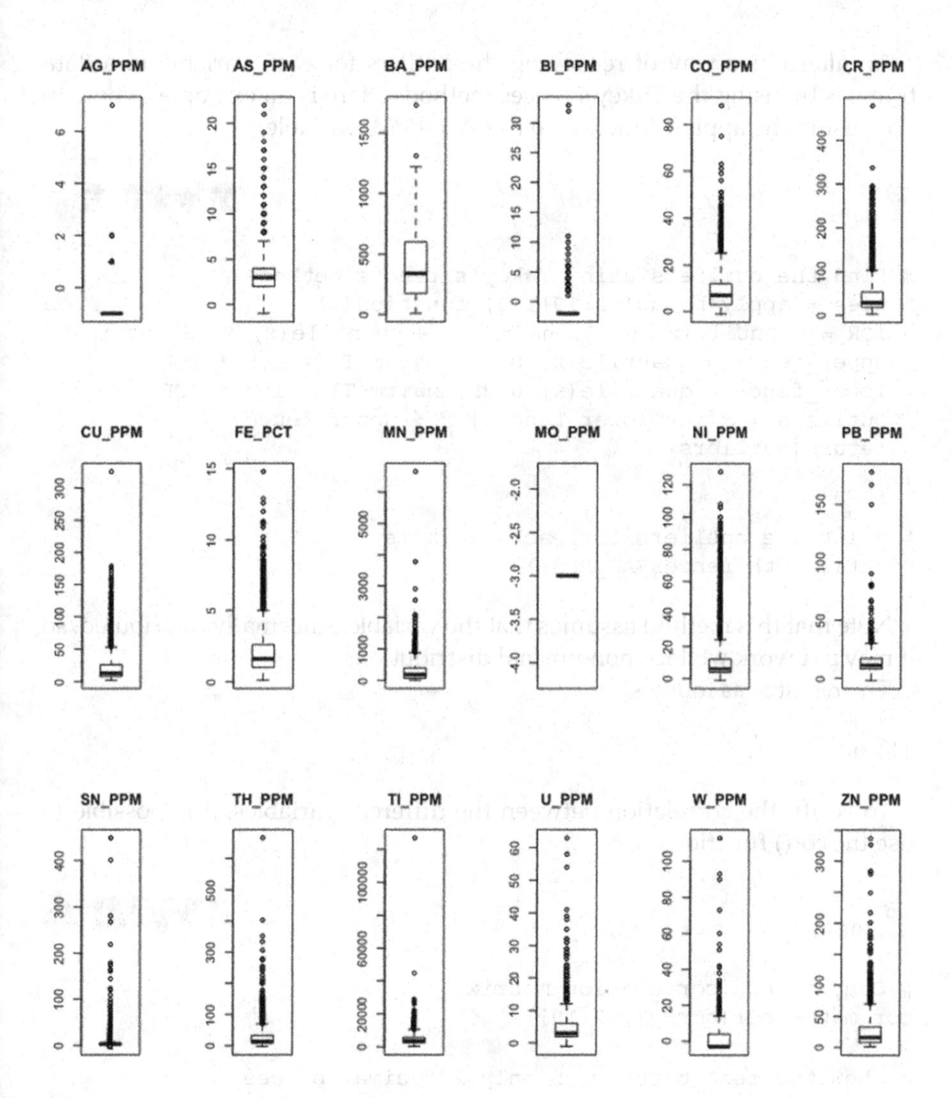

FIGURE 8.4
Boxplots for the variables that are available in the 'bmrgg' dataset.

variables in the dataset; lapply() function ensures that the necessary statistics are calculated for each variable, while the sapply() function simplifies the aggregation and presentation of the specific statistic of interest (the number of outliers), resulting in a concise and readable output.

The obtained result is as follows:

```
AG_PPM AS_PPM BA_PPM BI_PPM CO_PPM CR_PPM CU_PPM FE_PCT MN_PPM
    81     78      2    292    172    162    203    180    101
MO_PPM NI_PPM PB_PPM SN_PPM TH_PPM TI_PPM  U_PPM  W_PPM ZN_PPM
     0    241     65    578    121    123     96     45    132
```

An alternative way of retrieving the outliers for each variable in a data frame is by using the Tukey's fences method.[3] Here is an example of how to do it using the apply() function to the 'AG_PPM' variable.

#08-13

```
# Find the outliers using Tukey's fences method
fences = apply(bmrgg[,2:19], 2, function(x) {
  IQR = quantile(x, 0.75, na.rm=T) - quantile(x, 0.25, na.rm=T)
  upper_fence = quantile(x, 0.75, na.rm=T) + 1.5 * IQR
  lower_fence = quantile(x, 0.25, na.rm=T) - 1.5 * IQR
  outliers = x[x < lower_fence | x > upper_fence]
  return(outliers)
})

# Print the outliers for each variable
print(length(fences$AG_PPM))
```

Note that this method assumes that the variable is normally distributed, so it may not work well for non-normal distributions.

The result is as follows:

```
[1] 81
```

To verify the correlation between the different variables, it is possible to use the cor() function.

#08-14

```
# Compute the correlation matrix
cor_mat = cor(bmrgg[, 2:19])

# Show the result but with only 2 decimal places
round(cor_mat, 2)
```

[3] Tukey's fences method is a statistical technique used to identify outliers in a dataset. It is based on the interquartile range (IQR) of the data, which is the range between the 25th and 75th percentiles. Tukey's fences use this range to define the lower and upper bounds for what is considered to be a 'reasonable' value in the data. The lower bound is defined as the first quartile minus 1.5 times the IQR, and the upper bound is defined as the third quartile plus 1.5 times the IQR. Any value that falls outside these bounds is considered to be an outlier.
Tukey's fences method is useful for identifying outliers because it is not overly sensitive to extreme values and can help to filter out noise in the data. It is commonly used in data cleaning and data pre-processing tasks as well as in exploratory data analysis.

In this example, the round() function from the 'base' R package is used so that the resulting table is more readable. The cor() function retrieves a warning message stating that there are values where the standard deviation is zero. This is due to the problems stated previously in the data.

The resulting matrix is as follows:

```
       AG_PPM AS_PPM BA_PPM BI_PPM CO_PPM CR_PPM CU_PPM FE_PCT MN_PPM
AG_PPM   1.00  -0.01   0.23   0.17   0.24     NA   0.39   0.26     NA
AS_PPM  -0.01   1.00   0.05   0.18  -0.06     NA  -0.04   0.03     NA
BA_PPM   0.23   0.05   1.00   0.28   0.39     NA   0.28   0.26     NA
BI_PPM   0.17   0.18   0.28   1.00  -0.02     NA   0.13  -0.05     NA
CO_PPM   0.24  -0.06   0.39  -0.02   1.00     NA   0.82   0.84     NA
CR_PPM     NA     NA     NA     NA     NA      1     NA     NA     NA
CU_PPM   0.39  -0.04   0.28   0.13   0.82     NA   1.00   0.83     NA
FE_PCT   0.26   0.03   0.26  -0.05   0.84     NA   0.83   1.00     NA
MN_PPM     NA     NA     NA     NA     NA     NA     NA     NA      1
MO_PPM     NA     NA     NA     NA     NA     NA     NA     NA     NA
NI_PPM   0.13  -0.14   0.27  -0.09   0.82     NA   0.75   0.76     NA
PB_PPM   0.13   0.24   0.35   0.26   0.08     NA   0.07   0.02     NA
SN_PPM   0.17   0.07   0.23   0.37  -0.01     NA   0.12  -0.03     NA
TH_PPM  -0.03   0.18   0.28   0.34  -0.07     NA  -0.12  -0.15     NA
TI_PPM   0.22  -0.01   0.19  -0.01   0.64     NA   0.66   0.72     NA
U_PPM    0.05   0.20   0.32   0.50  -0.01     NA   0.01  -0.11     NA
W_PPM    0.00   0.18   0.22   0.33  -0.06     NA  -0.05  -0.10     NA
ZN_PPM   0.31   0.01   0.36   0.01   0.76     NA   0.78   0.82     NA

       MO_PPM NI_PPM PB_PPM SN_PPM TH_PPM TI_PPM U_PPM W_PPM ZN_PPM
AG_PPM     NA   0.13   0.13   0.17  -0.03   0.22  0.05  0.00   0.31
AS_PPM     NA  -0.14   0.24   0.07   0.18  -0.01  0.20  0.18   0.01
BA_PPM     NA   0.27   0.35   0.23   0.28   0.19  0.32  0.22   0.36
BI_PPM     NA  -0.09   0.26   0.37   0.34  -0.01  0.50  0.33   0.01
CO_PPM     NA   0.82   0.08  -0.01  -0.07   0.64 -0.01 -0.06   0.76
CR_PPM     NA     NA     NA     NA     NA     NA    NA    NA     NA
CU_PPM     NA   0.75   0.07   0.12  -0.12   0.66  0.01 -0.05   0.78
FE_PCT     NA   0.76   0.02  -0.03  -0.15   0.72 -0.11 -0.10   0.82
MN_PPM     NA     NA     NA     NA     NA     NA    NA    NA     NA
MO_PPM      1     NA     NA     NA     NA     NA    NA    NA     NA
NI_PPM     NA   1.00  -0.01  -0.04  -0.14   0.59 -0.10 -0.08   0.70
PB_PPM     NA  -0.01   1.00   0.16   0.34   0.03  0.38  0.20   0.24
SN_PPM     NA  -0.04   0.16   1.00   0.14   0.02  0.23  0.20   0.04
TH_PPM     NA  -0.14   0.34   0.14   1.00  -0.08  0.83  0.41  -0.08
TI_PPM     NA   0.59   0.03   0.02  -0.08   1.00 -0.06 -0.04   0.66
U_PPM      NA  -0.10   0.38   0.23   0.83  -0.06  1.00  0.48  -0.03
W_PPM      NA  -0.08   0.20   0.20   0.41  -0.04  0.48  1.00  -0.05
ZN_PPM     NA   0.70   0.24   0.04  -0.08   0.66 -0.03 -0.05   1.00
```

It is possible to verify that the correlation values are spread in the –1 to +1 range. To identify what are the 'strong', 'moderate', and 'low' correlations, we can use some programming. First, it is necessary to decide on cut-off values for correlation coefficients. These cut-off values are subjective and can vary depending on the research question or field of study. For the analysis, let's remove the 'MO_PPM' variable because it is only composed of NAs, which doesn't allow some calculations.

As a general guideline, the following ranges can be used for interpreting the strength of correlation coefficients:

- Strong: 0.7–1.0 (positive or negative).
- Moderate: 0.4–0.69 (positive or negative).
- Low: 0.0–0.39 (positive or negative).

Using these cut-off values, the following code will identify the pairs of variables with strong, moderate, and low correlations:

#08-15 3 stars

```
# Select only the numeric data and excluding the Mo
data = bmrgg[, c(2:10,12:19)]

# Compute the correlation matrix
corr_matrix = cor(data, use="pairwise.complete.obs",
method="pearson")

# Create a matrix to store the correlation categories
corr_categories = matrix(nrow = ncol(data), ncol = ncol(data))

# Loop through each pair of variables in the correlation matrix
for (i in 1:ncol(data)) {
  for (j in 1:ncol(data)) {
      # Check the correlation coefficient for each pair
      corr = corr_matrix[i, j]

      # Categorize based on cutoff
      if (abs(corr) >= 0.7) {
         corr_categories[i, j] = "Strong"
      } else if (abs(corr) >= 0.4) {
         corr_categories[i, j] = "Moderate"
      } else if (abs(corr) >= 0) {
         corr_categories[i, j] = "Low"
      }
  }
}

# Print the correlation categories
rownames(corr_categories) = colnames(data)
colnames(corr_categories) = colnames(data)
print(corr_categories)
```

Breathe, this is a three-star code snippet and take some time to look at this code example. It computes the correlation matrix for a dataset and creates a matrix to store the correlation categories. Next, it loops (for() loop function) through each pair of variables (i, j) in the correlation matrix to categorise the correlation coefficients based on the cut-off values. The resulting matrix will retain the correlation categories for each pair of variables in the dataset.

The result is as follows:

	AG_PPM	AS_PPM	BA_PPM	BI_PPM	CO_PPM	CR_PPM
AG_PPM	"Strong"	"Low"	"Low"	"Low"	"Low"	"Low"
AS_PPM	"Low"	"Strong"	"Low"	"Low"	"Low"	"Low"
BA_PPM	"Low"	"Low"	"Strong"	"Low"	"Low"	"Low"
BI_PPM	"Low"	"Low"	"Low"	"Strong"	"Low"	"Low"
CO_PPM	"Low"	"Low"	"Low"	"Low"	"Strong"	"Moderate"
CR_PPM	"Low"	"Low"	"Low"	"Low"	"Moderate"	"Strong"
CU_PPM	"Low"	"Low"	"Low"	"Low"	"Strong"	"Moderate"
FE_PCT	"Low"	"Low"	"Low"	"Low"	"Strong"	"Moderate"
MN_PPM	"Low"	"Low"	"Low"	"Low"	"Strong"	"Moderate"
NI_PPM	"Low"	"Low"	"Low"	"Low"	"Strong"	"Strong"
PB_PPM	"Low"	"Low"	"Low"	"Low"	"Low"	"Low"
SN_PPM	"Low"	"Low"	"Low"	"Low"	"Low"	"Low"
TH_PPM	"Low"	"Low"	"Low"	"Low"	"Low"	"Low"
TI_PPM	"Low"	"Low"	"Low"	"Low"	"Moderate"	"Moderate"
U_PPM	"Low"	"Low"	"Low"	"Moderate"	"Low"	"Low"
W_PPM	"Low"	"Low"	"Low"	"Low"	"Low"	"Low"
ZN_PPM	"Low"	"Low"	"Low"	"Low"	"Strong"	"Moderate"

	CU_PPM	FE_PCT	MN_PPM	NI_PPM	PB_PPM	SN_PPM
AG_PPM	"Low"	"Low"	"Low"	"Low"	"Low"	"Low"
AS_PPM	"Low"	"Low"	"Low"	"Low"	"Low"	"Low"
BA_PPM	"Low"	"Low"	"Low"	"Low"	"Low"	"Low"
BI_PPM	"Low"	"Low"	"Low"	"Low"	"Low"	"Low"
CO_PPM	"Strong"	"Strong"	"Strong"	"Strong"	"Low"	"Low"
CR_PPM	"Moderate"	"Moderate"	"Moderate"	"Strong"	"Low"	"Low"
CU_PPM	"Strong"	"Strong"	"Moderate"	"Strong"	"Low"	"Low"
FE_PCT	"Strong"	"Strong"	"Moderate"	"Strong"	"Low"	"Low"
MN_PPM	"Moderate"	"Moderate"	"Strong"	"Moderate"	"Low"	"Low"
NI_PPM	"Strong"	"Strong"	"Moderate"	"Strong"	"Low"	"Low"
PB_PPM	"Low"	"Low"	"Low"	"Low"	"Strong"	"Low"
SN_PPM	"Low"	"Low"	"Low"	"Low"	"Low"	"Strong"
TH_PPM	"Low"	"Low"	"Low"	"Low"	"Low"	"Low"
TI_PPM	"Moderate"	"Strong"	"Moderate"	"Moderate"	"Low"	"Low"
U_PPM	"Low"	"Low"	"Low"	"Low"	"Low"	"Low"
W_PPM	"Low"	"Low"	"Low"	"Low"	"Low"	"Low"
ZN_PPM	"Strong"	"Strong"	"Moderate"	"Moderate"	"Low"	"Low"

	TH_PPM	TI_PPM	U_PPM	W_PPM	ZN_PPM
AG_PPM	"Low"	"Low"	"Low"	"Low"	"Low"
AS_PPM	"Low"	"Low"	"Low"	"Low"	"Low"
BA_PPM	"Low"	"Low"	"Low"	"Low"	"Low"
BI_PPM	"Low"	"Low"	"Moderate"	"Low"	"Low"
CO_PPM	"Low"	"Moderate"	"Low"	"Low"	"Strong"
CR_PPM	"Low"	"Moderate"	"Low"	"Low"	"Moderate"
CU_PPM	"Low"	"Moderate"	"Low"	"Low"	"Strong"
FE_PCT	"Low"	"Strong"	"Low"	"Low"	"Strong"
MN_PPM	"Low"	"Moderate"	"Low"	"Low"	"Moderate"

```
NI_PPM "Low"        "Moderate" "Low"        "Low"        "Moderate"
PB_PPM "Low"        "Low"      "Low"        "Low"        "Low"
SN_PPM "Low"        "Low"      "Low"        "Low"        "Low"
TH_PPM "Strong"     "Low"      "Strong"     "Moderate"   "Low"
TI_PPM "Low"        "Strong"   "Low"        "Low"        "Moderate"
U_PPM  "Strong"     "Low"      "Strong"     "Moderate"   "Low"
W_PPM  "Moderate"   "Low"      "Moderate"   "Strong"     "Low"
ZN_PPM "Low"        "Moderate" "Low"        "Low"        "Strong"
```

It is possible to verify that Co, Cr, Cu, Fe, Ni, and Mn are moderate to strongly positively correlated. This is explainable as their geochemical behaviour is common according to the Goldschmidt classification of elements, being mostly siderophile and chalcophile elements.

An attractive way of representing the correlation is using the image() function, from the 'gstats' package, combined with the axis() functions, from the 'graphics' package, for plotting the axis of the image.

#08-16

```
# Create the image for the correlation matrix
par(mfrow = c(1, 1))
image(corr_matrix, main="Correlation Matrix", xaxt='n', yaxt='n')

# Add the axis labels
axis(side = 1, at = seq(0, 1, by=1/(ncol(data) -1)), labels =
colnames(data), las=2, cex.axis=.7)
axis(side = 2, at = seq(0, 1, by=1/(ncol(data) -1)), labels =
colnames(data), las=1, cex.axis=.7)
```

The obtained result is presented in Figure 8.5.

There are other methods, apart from correlation, to investigate dependence between variables. Some of them include the following:

1. Regression analysis: This method involves fitting a linear or non-linear regression model to the data and examining the relationship between the variables.
2. Principal component analysis (PCA): This method involves transforming the data into a new set of uncorrelated variables, known as principal components, and examining the relationships between these components.
3. Factor analysis: This method is similar to PCA but involves finding underlying factors that explain the variability in the data.
4. Canonical correlation analysis (CCA): This method involves finding linear combinations of the variables that are maximally correlated with each other.
5. Cluster analysis: This method involves grouping the data into clusters based on similarities between the variables.

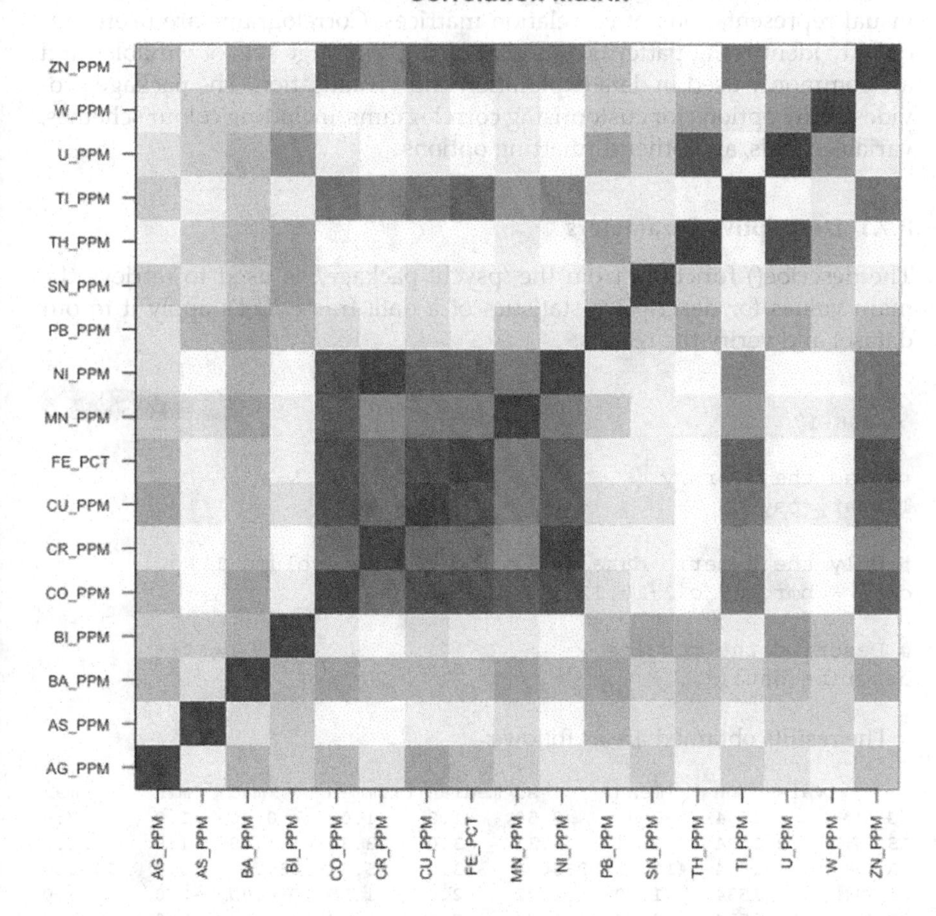

FIGURE 8.5
Image of the correlation matrix. Dark colours represent strong correlations whereas light ones represent low correlations.

For more complex or alternative methods, please consult specialised books, such as Fox & Weisberg (2018).

For the moment, let's turn our attention to some packages that can complement our descriptive statistics and data visualisation and that have not been addressed yet, namely the 'psych' and 'corrgram' packages.

The 'psych' package[4] is a collection of tools for psychological research and related fields, including functions for calculating descriptive statistics, correlation matrices, and factor analysis. It includes many tools for data visualisation and exploration.

[4] http://personality-project.org/r/psych/

The 'corrgram' package[5] is a package for creating correlograms, which are visual representations of correlation matrices. Correlograms are useful for quickly identifying patterns and relationships in large sets of variables and are commonly used in data exploration and visualisation. The package provides many options for customising correlograms, including colour schemes, variable labels, and other formatting options.

8.7.1 Descriptive Parameters

The describe() function, from the 'psych' package,[6] is used to retrieve the main values for descriptive statistics of a data frame. Let's apply it to our dataset and verify the results.

```
#08-17

# Load the library
library(psych)

# Only the numeric data,excluding the Mo (column 11)
data = bmrgg[, c(2:10,12:19)]

# Describe the results
describe(data)
```

The results obtained are as follows:

```
        vars    n    mean      sd median trimmed     mad   min      max
AG_PPM     1 1534   -0.89    0.50   -1.0   -1.00    0.00  -1.0      7.0
AS_PPM     2 1534    3.61    2.33    3.0    3.32    1.48  -1.0     22.0
BA_PPM     3 1534  410.13  266.91  343.0  385.64  284.66  10.0   1730.0
BI_PPM     4 1534   -1.10    2.22   -2.0   -1.55    0.00  -2.0     33.0
CO_PPM     5 1534   10.29   10.73    7.0    8.22    5.93  -1.0     88.0
CR_PPM     6 1533   47.85   51.41   29.0   36.60   19.27   3.0    479.0
CU_PPM     7 1534   25.43   30.78   13.0   18.26    8.90   2.0    325.0
FE_PCT     8 1534    2.28    1.99    1.6    1.91    1.04   0.1     14.8
MN_PPM     9 1528  299.69  379.43  183.0  230.26  197.19   6.0   6650.0
NI_PPM    10 1534   13.67   19.02    6.0    9.18    4.45  -1.0    128.0
PB_PPM    11 1534   13.39   11.09   11.0   11.98    5.93  -2.0    178.0
SN_PPM    12 1534    6.48   24.02    3.0    2.71    1.48  -2.0    449.0
TH_PPM    13 1534   27.88   39.99   14.0   20.08   11.86  -2.0    666.0
TI_PPM    14 1534 5094.52 5000.16 3750.0 4324.43 2001.51 400.0 127000.0
U_PPM     15 1534    4.94    5.28    3.0    3.96    1.48  -1.0     63.0
W_PPM     16 1534    1.56    7.68   -3.0    0.36    0.00  -3.0    113.0
ZN_PPM    17 1534   27.14   33.06   16.0   20.32   14.83   1.0    337.0

          range  skew kurtosis    se
AG_PPM      8.0  5.67    49.20  0.01
```

[5] https://kwstat.github.io/corrgram/
[6] https://personality-project.org/r/psych-manual.pdf

```
AS_PPM      23.0  2.31    10.12   0.06
BA_PPM    1720.0  0.70    -0.21   6.81
BI_PPM      35.0  5.72    69.30   0.06
CO_PPM      89.0  2.04     5.07   0.27
CR_PPM     476.0  2.69     9.10   1.31
CU_PPM     323.0  2.73    10.31   0.79
FE_PCT      14.7  2.00     4.53   0.05
MN_PPM    6644.0  5.15    59.58   9.71
NI_PPM     129.0  2.44     5.98   0.49
PB_PPM     180.0  6.33    74.07   0.28
SN_PPM     451.0 10.80   152.49   0.61
TH_PPM     668.0  5.83    59.24   1.02
TI_PPM  126600.0 10.70   232.98 127.66
U_PPM       64.0  4.14    27.81   0.13
W_PPM      116.0  6.06    63.38   0.20
ZN_PPM     336.0  3.61    20.22   0.84
```

The output consists of 13 columns:

1. var: The name of the variable.
2. n: The number of non-missing values for the variable.
3. mean: The mean of the variable.
4. sd: The standard deviation of the variable.
5. median: The median of the variable.
6. trimmed: The mean of the variable after trimming the top and bottom 20% of the data.
7. mad: The median absolute deviation of the variable.
8. min: The minimum value of the variable.
9. max: The maximum value of the variable.
10. range: The range (difference between max and min) of the variable.
11. skew: The skewness of the variable.
12. kurtosis: The kurtosis of the variable.
13. se: The standard error of the mean of the variable.

Most of these parameters are self-explanatory, with the exception of 'mad', 'skew', and 'kurtosis.

'mad' stands for median absolute deviation, and it is a measure of the variability of the data that is less sensitive to extreme values than the standard deviation. It is calculated as the median of the absolute deviations of the values from their median; mad can be used as a robust alternative to the standard deviation, particularly when the data contains outliers that may heavily influence the standard deviation.

'skew' is an abbreviation from skewness, and it refers to the degree of asymmetry in the distribution of the data (see Figure 8.6). A normal distribution has zero skewness, with the tails of the distribution being equal in both directions. A positive skewness value indicates a distribution with a longer

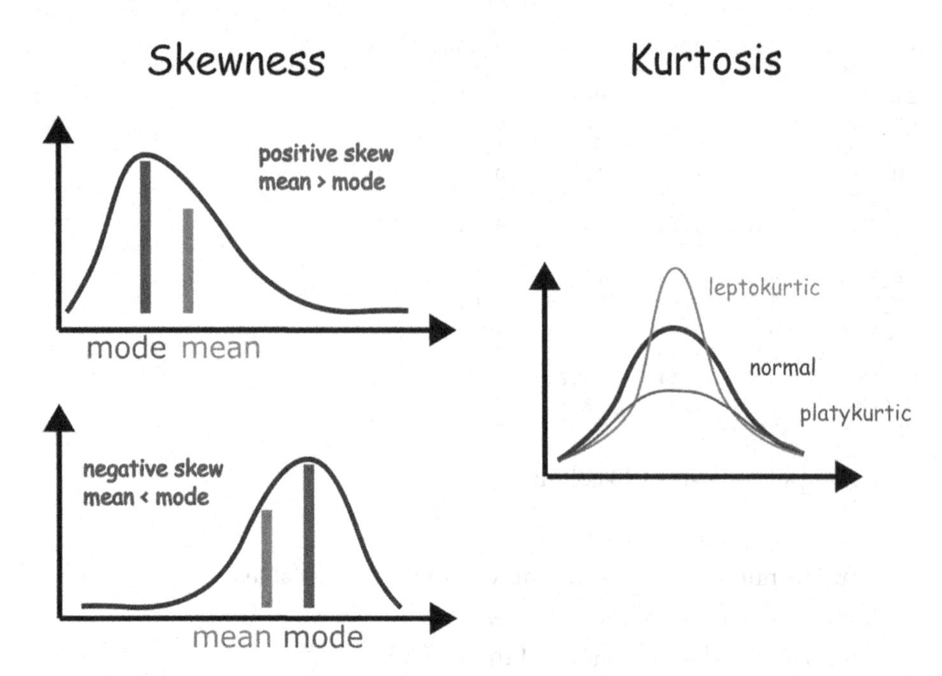

FIGURE 8.6
Illustration of skewness and kurtosis parameters.

right tail, meaning that the majority of values are located on the left side of the distribution. A negative skewness value indicates a distribution with a longer left tail, meaning that the majority of values are located on the right side of the distribution.

'kurtosis' is a measure of the degree of peakedness in the distribution of the data (see Figure 8.6). A normal distribution has a kurtosis value of 3, known as mesokurtic. A distribution with kurtosis greater than 3 is called leptokurtic, meaning that it has a sharper peak and heavier tails than a normal distribution. A distribution with kurtosis less than 3 is called platykurtic, meaning that it has a flatter peak and lighter tails than a normal distribution.

Figure 8.6 presents an illustration of the concepts of skewness and kurtosis.

8.7.2 Correlations Made Beautiful

The corrgram() function is the main function from the 'corrgram' package by Friendly (2002).

The most simple corrgram() creates an image with blue and red squares for positive and negative correlations. The intensity of the colour is proportional to the strength of the correlation.

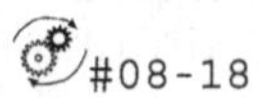
#08-18

```
# Load the library
library(corrgram)
```

```
# Use only the numerical data,excluding the Mo (column 11)
data = bmrgg[, c(2:10,12:19)]

# The corrgram
corrgram(data, main="Correlation for the data from BMRGG")
```

The results are shown in Figure 8.7.

Naturally, the interpretation of the results is similar to the one presented above.

Several charts can be created with the corrgram() function. One that is interesting is the one that shows the plot between variables in the lower left part of the diagram and the correlation values in the upper right part, and the diagonal line represents the density plot function. See the example below.

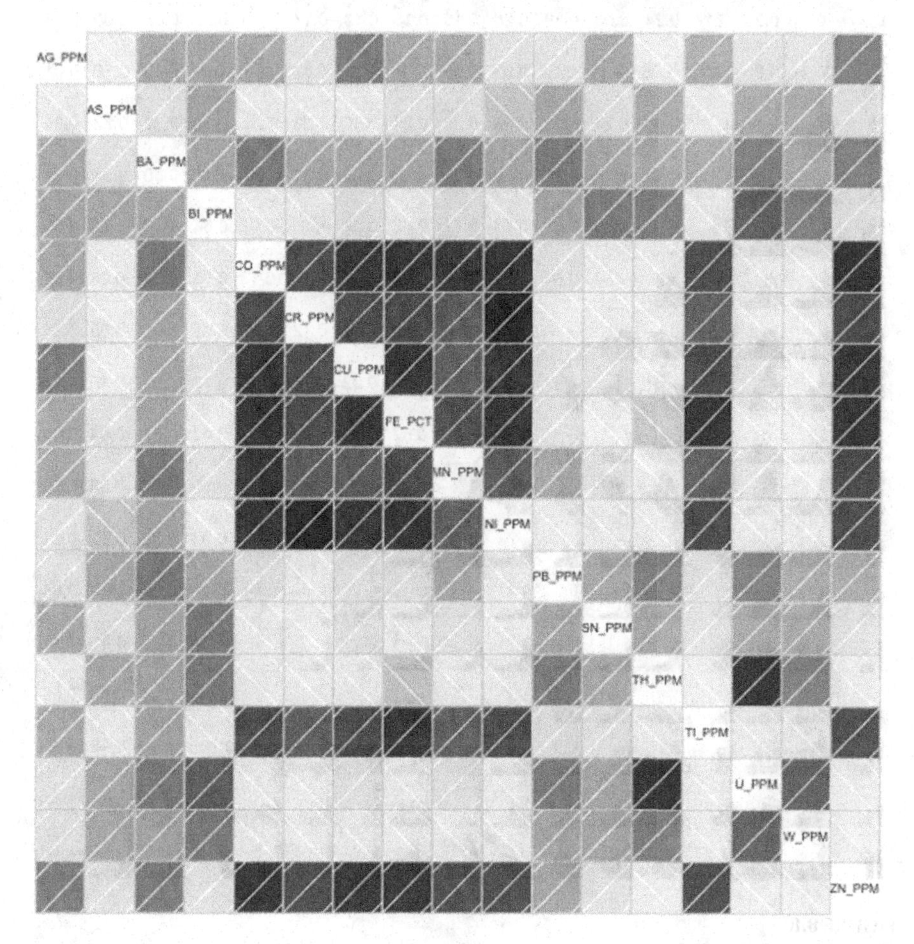

FIGURE 8.7
The corrgram for the 'bmrgg' dataset.

This can be done using the 'lower.panel', 'upper.panel', and 'diag.panel' parameters in the corrgram() function.

#08-19

```
# Demonstrate density panel, correlation confidence panel
corrgram(data,
      main="BMRGG data with example panel functions",
      lower.panel=panel.pts, upper.panel=panel.conf,
      diag.panel=panel.density)
```

The results are shown in Figure 8.8.

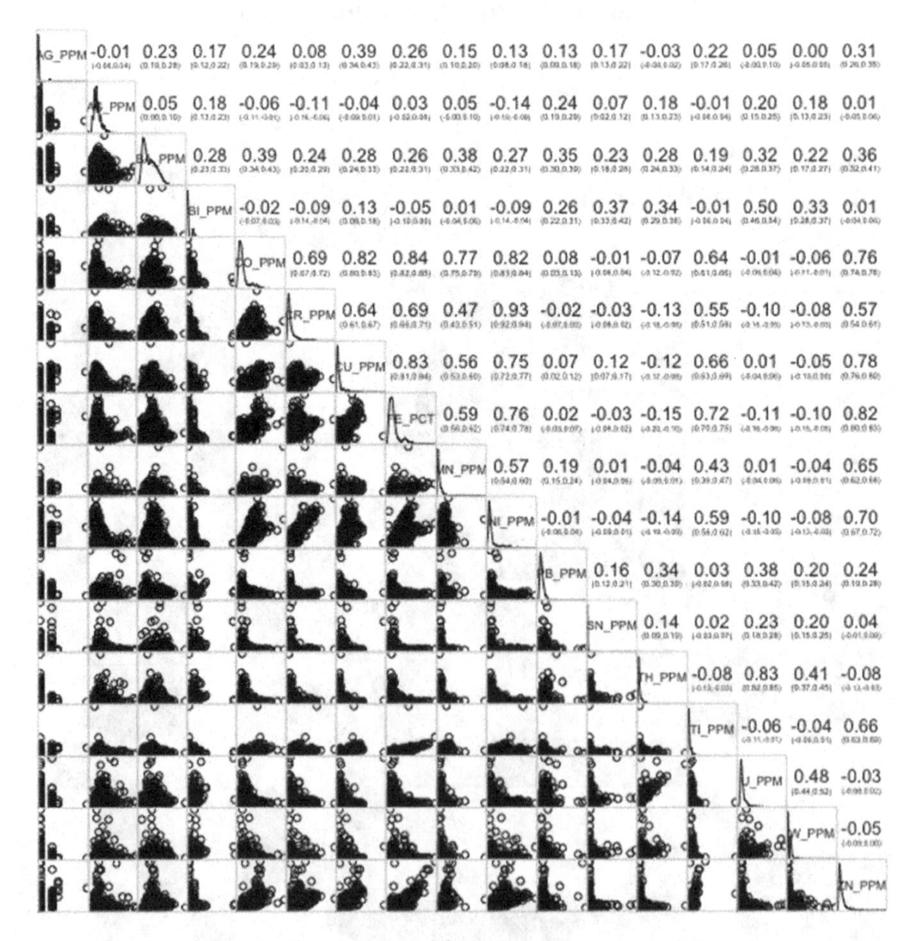

FIGURE 8.8
Corrgram with density plot and confidence values.

8.8 Concluding Remarks

This is just a small part of what can be done for data analysis with R. We have been mainly learning how to deal with some geological data from a statistical viewpoint, but our goal is to go deep in understanding the data because one must acknowledge that part of its information has not yet been dealt with. Its spatial nature can be explored to look for more in-depth explanations of the results. The nature of spatial data and how to deal with it are the purposes of Part II of this book.

References

Fox, J., & Weisberg, S. (2018). An R Companion to Applied Regression. Sage Publications, Thousand Oaks, CA.

Friendly, M. (2002). Corrgrams: Exploratory displays for correlation matrices. The American Statistician, 56, 316–324. https://doi.org/10.1198/000313002533

Part II

Spatial Analysis with R

9

Fundamentals of Spatial Analysis

Spatial analysis is the branch of statistical analysis concerned with the study of data that has a geographic or spatial representation. The goal of spatial analysis is to understand the patterns, relationships, and interactions between entities, such as population, climate, land use, geological unit, or mine locations, all of which have a spatial representation. The data used in spatial analysis can be vector-like data or raster-like data. The vector data might be either point data, such as cities or soil samples; line data, such as rivers or geological faults; or polygon data, such as countries or geological boundaries. The raster data also known as gridded data can be, e.g., an orthophotography, a satellite image, a digital elevation model, or other type of image-like data.

In spatial analysis, to model and analyse the spatial relationships between objects, one can use several technologies and different approaches, depending on the objectives defined, might them be to model the spatial autocorrelation of geological faults, acknowledge the existence of spatial clustering in mineral deposits, or understand the spatial dependence of zinc in soil samples to a certain elevation in the terrain. Therefore, spatial analysis is used in many different fields, as for example geography, geology, or urban planning. Examples of its use, in geography, include the understanding of the patterns of land use, population distribution, and urban growth. In geology, one can mention the knowledge of the distribution of natural resources, might it be mineral deposits or water resources or its relations with other geological features such as faults or rock formations. In urban planning, it can be used for specific tasks such as defining the routing for police brigades or for broad tasks such as the evaluation of the impact of urban development on the environment and to plan future growth.

There is a large number of tools available for conducting spatial analysis, including Geographic Information Systems (GIS) software such as QGIS, ArcGIS, or SAGA GIS; nonetheless, R provides an approach that enables the user to fine-tune and better control all the aspects of a spatial data analysis project. The wide range of data retrieving and cleaning, visualisation, and the analytical capabilities provides the researchers with tools to a deeper understanding of the relationships between the spatial entities.

DOI: 10.1201/9781032651880-11

9.1 Key Concepts and Techniques

A wide range of tools and techniques are commonly used to analyse and understand the relationships, patterns, and interactions between spatial data. Here is a list of definitions of important concepts and techniques that provide an overview of these analytical instruments:

GIS serves as a foundational discipline, enabling the capture, storage, analysis, and visualisation of geographic data. It constitutes a common interface for creating maps and conducting spatial investigations.

Terrain analysis offers insights into earth's surface, utilising elevation data to generate models such as slope, aspect, and hillshade. Its applications span environmental sciences, geology, and geography, shedding light on landform characteristics, hydrology, erosion, and land use.

Remote sensing employs modern tools and methods to collect data from elevated viewpoints (satelites, planes, helicopters or drones), contributing to the creation of detailed cartographic visualisations that enhance our understanding of physical and human features.

Point pattern analysis is employed to study the distribution of points in a given space, revealing patterns and relationships in sample distribution or environmental attributes.

Spatial statistics encompasses techniques like spatial autocorrelation, or spatial regression, unveiling concealed patterns, trends and relatioships within spatial data.

Spatial interpolation predicts values at unsampled locations using available data, aiding environmental and geospatial analyses.

Geostatistics employs spatial autocorrelation and regression to uncover patterns within geographic data.

Multivariate spatial analysis examines multiple variables concurrently within a spatial framework, searching for relationships and patterns related with a geological setting or geographic landscape.

In this book, we will address several of these spatial analysis techniques within a geological context. Nonetheless, for certain cases, we employ illustrative or simpler examples for ease of comprehension of the coding options.

9.2 Applications in Geology and Earth Sciences

As it can be foreseen, the applications of spatial analysis in geology and earth sciences are numerous and far-reaching, including every subject that is imaginable for a geoscientist, from geomorphological analysis, structural geology, natural hazard assessment, mineral exploration, or groundwater management. There is a plethora of examples of spatial analysis applications. Let's cover some illustrative examples from academic studies to more technical approaches.

Spatial analysis in geomorphological studies (e.g. Slaymaker et al., 2009; Goudie & Viles, 2010) allow the exploration of spatial relationships and patterns in terrain attributes. Using spatial statistics, geostatistics, and spatial modelling, geomorphologists can analyse the distribution of various landforms and their relations with the underlying geologic structures. Another example is the identification and characterisation of processes, such as erosion, deposition, or mass movements, which can inform our understanding of landscape evolution. Moreover, spatial analysis techniques can be used to assess the impacts of natural and anthropogenic processes on landscapes and to develop predictive models for future changes.

In the case of structural geology (e.g. Ragan, 2009; Rowland et al., 2021), the spatial analysis techniques allow for the identification and quantification of geological features such as faults, folds, and fractures as well as the measurement and modelling of orientation and spatial relationships between geological structures. Spatial analysis can also be used to investigate the distribution and spatial patterns of geological properties such as chemical rock properties and composition, providing insights into the evolution and deformation of geological systems over time.

Spatial analysis provides powerful tools for evaluating the risks associated with natural hazards (e.g. Bryant, 2005; Pine, 2008) such as earthquakes, volcanic eruptions, and tsunamis. By using geospatial data to model and analyse the likelihood of these events, geologists can develop strategies for reducing the risk of damage and loss of life.

Another important application of spatial analysis in geology is mineral exploration (e.g. Wellmer, 1998; Pohl, 2011). The discovery of minerals such as gold, silver, and copper is crucial to the development of many economies. Spatial analysis provides geologists with the tools they need to identify the best areas to explore and develop new mineral deposits. This can involve analysing large amounts of geospatial data, including remote sensing images and geologic maps, to identify areas with the highest potential for mineral deposits.

The management of groundwater resources (e.g. Hiscock & Bense, 2014; Brassington, 2017) is another example where spatial analysis plays a critical role. Groundwater is an essential resource for many communities, providing drinking water, irrigation water, and other essential services. Spatial analysis provides hydrogeologists with the tools they need to understand the distribution of groundwater resources and to develop strategies for their management. This can include developing models of groundwater flow, evaluating the impacts of human activities on groundwater resources, and monitoring changes in groundwater levels over time.

These are just a few examples of the possible applications of spatial analysis in geology. It is our intention to provide the tools to conduct your own investigations and develop new approaches to spatial data analysis.

References

Brassington, R. (2017). Field Hydrogeology, fourth edition. Wiley. https://www.perlego.com/book/991077/field-hydrogeology-pdf

Bryant, E. (2005). Natural Hazards. Cambridge University Press. Cambridge, UK.

Goudie, A., & Viles, H. (2010). Landscapes and Geomorphology: A Very Short Introduction. OUP, Oxford.

Hiscock, K., & Bense, V. (2014). Hydrogeology, second edition. Wiley. https://www.perlego.com/book/995190/hydrogeology-principles-and-practice-pdf

Pine, J. (2008). Natural Hazards Analysis: Reducing the Impact of Disasters. CRC Press, Boca Raton, FL.

Pohl, W. (2011). Economic Geology: Principles and Practice. John Wiley & Sons. West Sussex, UK.

Ragan, D. (2009). Structural Geology. An Introduction to Geometrical Techniques. Cambridge University Press, Cambridge, UK, p. 602.

Rowland, S., Duebendorfer, E., & Gates, A. (2021). Structural Analysis and Synthesis: A Laboratory Course in Structural Geology. John Wiley & Sons, West Sussex, UK.

Slaymaker, O., Spencer, T., & Embleton-Hamann, C. (2009). Geomorphology and global environmental change. Cambridge University Press. Cambridge, UK.

Wellmer, F. (1998). Statistical evaluations in exploration for mineral deposits. Springer Science & Business Media. Springer Berlin, Heidelberg.

10

Spatial Objects

Spatial objects are representations of geographic features in a digital format. These objects are typically used to store, manipulate, and analyse spatial data. There are several types of spatial objects, each representing different features, such as points, lines, polylines, polygons, and grids. Figure 10.1 is an illustration of the most used types of vectorial objects and its connection with the corresponding attributes.

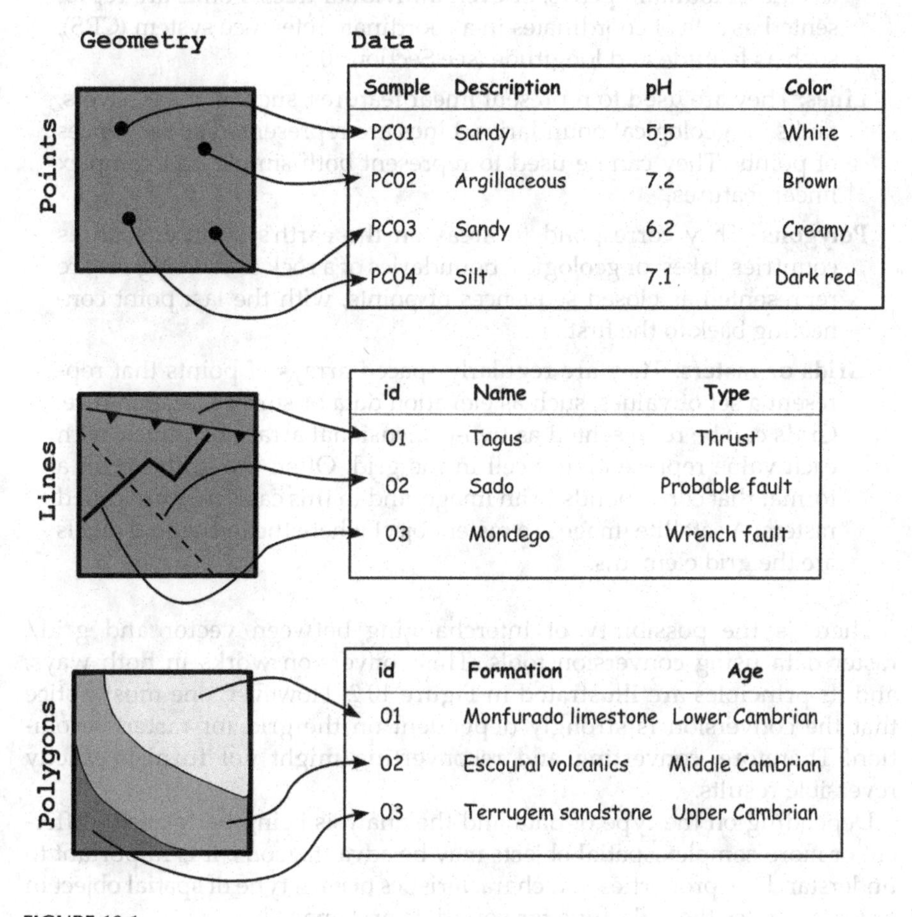

FIGURE 10.1
The different types of vector-like spatial objects.

DOI: 10.1201/9781032651880-12

The 'sf' package (described in detail in Section 10.4.2) is the kernel of many spatial analysis workflows. Apart from the basic vector types, i.e. points, lines (named linestrings in 'sf' package) and polygons, the 'sf' package also defines the objects multipoints, multilinestrings, multipolygons, and a mixed type named geometrycollection. These are what Pebesma & Bivand (2023) call the big seven of spatial data types.

A thorough comprehension of these entities is essential for effectively handling spatial data. In addition to the fundamental vector types that provide localised information, one must consider another type, named the gridded type (or raster), which, in clear contrast to the vector type, contains information from a surface area. The definitions of these object types could be as follows:

Points: They represent individual locations on the earth's surface, such as cities, mountain peaks, or even individual trees. Points are represented as sets of coordinates in a coordinate reference system (CRS), such as latitude and longitude (see Section 10.2).

Lines: They are used to represent linear features, such as roads, rivers, faults, or geological boundaries. Lines are represented as sequences of points. They can be used to represent both simple and complex linear features.

Polygons: They correspond to areas on the earth's surface, such as countries, lakes, or geological boundaries of a rock unit. Polygons are represented as closed sequences of points, with the last point connecting back to the first.

Grids or rasters: They are regularly spaced arrays of points that represent a set of values, such as elevation data or surface temperature. Grids can be represented as two-dimensional arrays of values, with each value representing a cell in the grid. Often the grids are in a format that corresponds to an image, and in this case, they are called rasters. A satellite image is a raster object where the individual pixels are the grid elements.

There is the possibility of interchanging between vector and grid/raster data using conversion tools. This conversion works in both ways, and its principles are illustrated in Figure 10.2. However, one must notice that the conversion is strongly dependent on the grid (or raster) resolution. Therefore, converting and reconverting might not furnish exactly reversible results.

Depending on the type of data and the analysis being performed, different or more complex spatial objects may be advantageous. It is important to understand the properties and characteristics of each type of spatial object in order to choose the right type for your data and analysis.

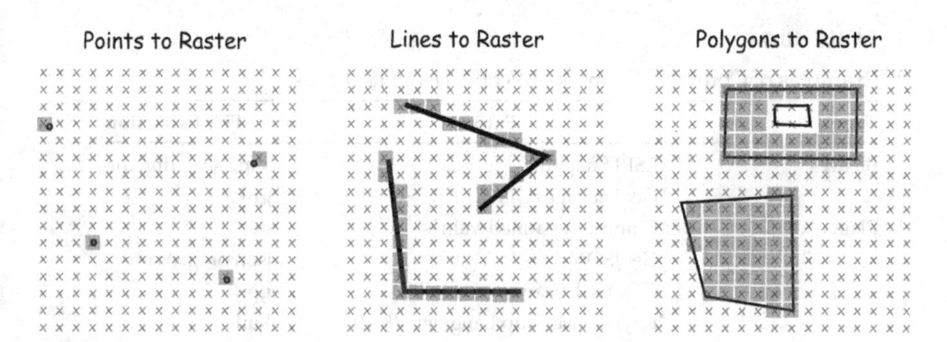

FIGURE 10.2
Illustration of the conversion between vector and grid/raster data. The vectors are in bold and black colour whereas the grid elements (or pixels) are in grey. The crosses represent spatial positions.

The 'sp' package, introduced in 2005, was the first in R to establish a uniform vector layer classification system (see Section 10.4.1). Along with 'rgdal' and 'rgeos', the 'sp' package played a dominant role in the field of spatial analysis in R for a number of years. Nevertheless, new packages are developed every year, and it is advisable that one maintains its knowledge base up to date in relation to the innovations in spatial analysis.

For example, the 'rgdal' and 'rgeos' packages were discontinued in 2023, and therefore newer packages, like 'sf', 'terra', and 'stars', are already available to replace most of its functions.

Whenever considered convenient, in the text of this book examples for the use of both the 'sp' and 'sf' packages will be provided and signalled.

10.1 Vectorial Objects

The spatial data must, at some point, be stored in a device, either on a local computer or a remote server. This electronic representation of the spatial data are the files, which are accessed and utilised through various software applications or computer programs. These files can have different formats, extensions, and appearance depending on the data format that is being used. The most common spatial data types and corresponding file types are listed in Table 10.1.

For storing the spatial objects in a disk or in the cloud, there are several common file formats used for vector layers, including binary formats like the ESRI Shapefile and plain text formats like GeoJSON. Vector layers can also be stored in spatial databases, such as PostgreSQL/PostGIS. Table 10.1 summarises the more common types of file formats.

TABLE 10.1

Examples of Common Formats of Vector Data File Types

Type	Format	File Extension
Binary	ESRI Shapefile	.shp, .shx, .dbf, .prj, …
	GeoPackage (GPKG)	.gpkg
Plain text	Comma Separated Values	.csv
	GeoJSON	.json or .geojson
	GPS Exchange Format (GPX)	.gpx
	Keyhole Markup Language (KML)	.kml
Spatial databases	PostGIS/PostgreSQL	

The 'sp' package (Bivand et al., 2013) that is used mainly to deal with vector objects defines six primary classes for vector layers as listed in Table 10.2. These classes include the following:

- Three classes only with geometry (points, lines, and polygons), without attributes.
- Three classes with geometry and with attributes (data frames).

The 'sf' package is as well mainly concerned with vector objects (Pebesma & Bivand, 2023) and is the latest alternative to the 'sp' package for working with spatial data (see Section 10.4.2). It introduces a modern and more efficient approach to handling spatial vector data. Unlike the 'sp' package, the 'sf' package primarily revolves around a single class called 'sf' that combines both geometry and attributes in a single object.

The 'sf' class in the 'sf' package encompasses both the geometric information (points, lines, polygons, etc.) and the attribute data associated with each spatial feature. It leverages the data frame structure to store attribute data, allowing for seamless integration with other R functions and packages.

With the 'sf' package, we don't need separate classes for geometry-only and attribute-only data. Instead, we work with a unified 'sf' object that holds both

TABLE 10.2

Classes of Vector Layers Defined by the 'sp' Package

Class	Geometry type	Attributes
SpatialPoints	Points	—
SpatialPointsDataFrame		data.frame
SpatialLines	Lines	—
SpatialLinesDataFrame		data.frame
SpatialPolygons	Polygons	—
SpatialPolygonsDataFrame		data.frame

the geometric and data components. This integration simplifies the workflow and enables efficient and streamlined spatial data analysis.

10.1.1 Shapefiles

A shapefile[1] is the most common format for storing vectorial spatial data, developed by ESRI, a company that is a provider of geographic information system (GIS) software. A shapefile consists of several files with different extensions that store complementary parts of the spatial data. The three main files in a shapefile are as follows:

1. File with '.shp' extension: This is the main file that stores the spatial data in the form of vector geometry, such as points, lines, and polygons. It contains the coordinates of the vertices that define the shape as well as the metadata that describes the attributes of the shape.
2. File with '.shx' extension: This is the index file that stores the offsets of the records in the shapefile. It allows for faster access to the records in the. shp file.
3. File with '.dbf' extension: This is the attribute table file that stores the non-spatial data associated with the shapes, such as the names of cities, the chemical values of soil samples, or any other information that is relevant to the data being represented. It is a dBASE format file, which contains records that correspond to the individual shapes in the '.shp' file.

There are also other optional files that can be included in a shapefile, depending on the data being represented. For example:

4. File with '.prj' extension: This is the projection file that contains the CRS of the spatial data in the shapefile. It is not always included, but it is necessary if you want to use the shapefile in a GIS or mapping software that requires the CRS information.
5. Files with '.sbn' and '.sbx' extensions: These are spatial index files that can be used to speed up queries on large shapefiles.
6. File with '.shp.xml' extension: This is an XML metadata file that can store additional metadata about the shapefile, such as its author, date of creation, and keywords.

It is the combination of these files that make up what hereafter is called a shapefile.

---

[1] www.esri.com/content/dam/esrisites/sitecore-archive/Files/Pdfs/library/whitepapers/pdfs/shapefile.pdf

10.1.2 GeoPackages

GeoPackage is an open, non-proprietary, platform-independent format for storing geospatial data, developed by the Open Geospatial Consortium (OGC).[2] A GeoPackage file is a single file that can contain both the vector and raster data, along with its attributes and metadata. The file extension for a GeoPackage is '.gpkg'.

A GPKG file can contain multiple tables, each with its own schema and geometry. These tables are accessed using SQL queries or can be read and written using geospatial software that supports the GeoPackage standard.

A GeoPackage file consists of the following main components:

Metadata: It contains information about the GeoPackage file, such as its version, the projection used, and the author.

Spatial tables: Tables that store the geometric data for the features in the GeoPackage file.

Attributes tables: Tables to store the non-spatial attributes for the features in the GeoPackage file.

Extensions: They are used to store additional data, such as vector, raster, time series, or three-dimensional data.

The metadata section is always present in a GeoPackage file. The spatial and attribute tables are optional, but they are usually present. The extensions section is also optional, but it can be used to store additional data that is not supported by the core GeoPackage format.

Some advantages of using GeoPackage files over other geospatial file formats include the following:

Cross-platform compatibility: Files can be used on a variety of platforms, including desktop, mobile, and web applications.

Easy to share: Files are self-contained, meaning that all the data and metadata are stored in a single file, making it easy to share with others.

Scalability: Files can store very large datasets, making them ideal for managing big geospatial data.

Standardisation: It is an open standard format, meaning that it is not owned by a specific software vendor, and is widely adopted by many GIS software vendors.

Although GeoPackage files are not the main industry and academy standards, they provide an efficient and standardised way to store, manage, and share geospatial data.

---

[2] https://www.ogc.org/

10.1.3 GeoJSON

GeoJSON is a text-based format for encoding geospatial data, developed by the Internet Engineering Task Force.[3] GeoJSON is based on the JSON (JavaScript Object Notation) data format and is designed to be lightweight, easy to read and write, and compatible with web technologies.

A GeoJSON file can contain different types of vectorial data, such as points, lines, polygons, and multi-geometries, along with their associated attributes. The file extension for a GeoJSON file is '.geojson' or '.json'.

A GeoJSON file consists of a single object, which contains two main components:

Geometry Object: It describes the spatial location and shape of the geospatial data. The geometry object can be a single geometry (e.g., point, line, and polygon) or a collection of geometries (e.g., MultiPoint, MultiLineString, and MultiPolygon).

Properties Object: It describes the non-spatial attributes of the geospatial data. These can include any information that is relevant to the data being represented, such as the name of a city, the population of a country, or any other information that is associated with the data.

In addition to these components, a GeoJSON file may also include a 'crs' object that specifies the CRS of the data and a 'bbox' object that provides the bounding box for the geospatial data.

Some advantages of using GeoJSON over other geospatial file formats include the following:

Easy to read and write: The JSON format is human-readable and easy to write and parse.

Interoperability: It is widely supported by many web-based mapping libraries, making it easy to share geospatial data across different web-based platforms.

Lightweight: It can be easily transmitted over the web, making it ideal for web-based mapping applications.

GeoJSON is particularly used in web-based mapping applications, and it provides a simple and efficient way to encode and transmit geospatial data.

10.1.4 GPX Files

GPX (GPS Exchange Format) is an XML-based file format for storing GPS data, such as waypoints, tracks, and routes. GPX files are used primarily in

[3] https://datatracker.ietf.org/wg/json/about/

outdoor activities like hiking, biking, and boating, where location data is important. The file extension for a GPX file is '.gpx'.

A GPX file consists of a series of elements, including:

Metadata: It provides information about the file, such as the name of the file, the author, the creation date, and a description of the data.

Waypoints: It represents a specific point on the earth's surface, such as a landmark, a campsite, or a trailhead. Each waypoint includes a latitude and longitude, along with any associated attributes, such as the name, elevation, and description of the point.

Tracks: It represents a recorded path or route. Each track consists of a series of trackpoints that record the location and time of the recorded path.

Routes: It represents a set of waypoints that define a path or a route. Each route consists of a series of route points that are connected by straight lines.

In addition to these elements, a GPX file may also include extensions that provide additional information, such as heart rate, cadence, and temperature.

Some advantages of using GPX files over other geospatial file formats include the following:

Compatibility: GPX files are compatible with a wide range of GPS devices and mapping software, making it easy to share and use data across different platforms.

Simplicity: GPX files are easy to create and edit using a variety of tools and software, making it ideal for outdoor enthusiasts who want to create and share their own GPS data.

Standardisation: GPX is an open standard format that is widely adopted by many GPS device manufacturers and mapping software vendors.

GPX provides a simple and efficient way to store and share GPS data, particularly for outdoor enthusiasts who want to record and share their GPS tracks and waypoints.

10.1.5 KML Files

KML (Keyhole Markup Language) is also an XML-based file format for storing geospatial data, developed by Keyhole, Inc., which was later acquired by Google. KML files are commonly used in Google Earth and other mapping applications and can be used to store a wide range of geospatial data, including points, lines, polygons, and 3D models. The file extension for a KML file is '.kml'.

KMZ is a compressed version of KML that stores all of the same data but in a smaller file size due to its zipped format. The file extension for a KMZ file is '.kmz', and yes the z stands for zip.

A KML file consists of a series of elements, including:

Document: It is the top-level element that contains all of the other elements in the file.

Placemarks: It represents a specific location on the earth's surface, such as a landmark, a point of interest, or a route. Each placemark includes a latitude and longitude, along with any associated attributes, such as the name, description, and style of the point.

Paths and Polygons: They are used to represent lines and areas, respectively. They consist of a series of vertices that define the shape of the line or polygon.

3D Models: KML and KMZ files can also include three-dimensional models, which can be used to represent buildings, terrain, or other structures.

In addition to these elements, KML files can also include styles, which can be used to customise the appearance of the data, as well as ground overlays, which can be used to display images or maps as an overlay on the earth's surface.

Some advantages of using KML files over other geospatial file formats include the following:

Compatibility: It can be viewed and edited using a variety of mapping software, including Google Earth, ArcGIS, and QGIS.

3D Support: It supports three-dimensional models, making it possible to create and share complex three-dimensional geospatial data.

Visualisation: It can be used to create visually compelling maps and visualisations, making them ideal for presentations and public outreach.

KML and KMZ provide a flexible and powerful way to store and share geospatial data, particularly for applications that require 3D visualisation and custom styling.

10.2 Gridded and Raster Objects

In spatial analysis, gridded and raster objects are common data types for representing spatially continuous data across a regular grid or pixel-based structure. Gridded data is commonly used to model and analyse geographic, environmental, climatic, or geological phenomena. Raster data is a type of gridded data that uses a regular grid of cells to discretise continuous spatial information. Each cell in the raster grid represents a geographic region and contains a single value, such as elevation, temperature, or land cover classification. This gridded representation allows for efficient storage, visualisation, and analysis of spatial data, making raster files the prevalent format in many spatial analysis workflows.

In particular, raster files are used in a wide range of applications, such as remote sensing, GIS, terrain analysis, and climate modelling. They are particularly well-suited for capturing continuous variations over large areas and enable the assessment of spatial patterns, interpolation, and computation of derived quantities like slope, aspect, and distance. The raster datasets offer versatility in representing complex geographical phenomena, enabling the study, modelling, and predicting of various environmental and geographical processes with precision and efficiency.

When working with gridded or raster data, there are several common file types (Table 10.3) used for storing such data.

While the primary focus of the 'sp' package is on handling spatial vector data (Section 10.4.1), it does have some basic capabilities to read and work with raster data through the 'raster' and 'grid' classes. However, it's worth noting that these capabilities are limited compared to dedicated raster packages like

TABLE 10.3

Examples of Common Formats of Gridded and Raster File Types.

Type	Description	File Extension
GeoTIFF	GeoTIFF is the most popular file format that allows for the storage of both raster data and georeferencing information, such as coordinate system, spatial extent, and pixel resolution. It is widely supported by GIS software and is commonly used for various types of gridded data.	.tif
NetCDF	NetCDF (Network Common Data Form) is a data format commonly used for storing large scientific datasets, including gridded data. It allows for efficient storage and retrieval of multi-dimensional arrays and supports metadata to describe the data.	.nc
HDF5	HDF5 (Hierarchical Data Format version 5) is a versatile data format that can store large and complex datasets, including raster data. It provides a hierarchical structure to organise data and is widely used in scientific and engineering applications.	.h5
ENVI	ENVI (Environment for Visualizing Images) is a file format commonly used in remote sensing and image processing applications. It stores raster data along with metadata and can handle various data types and multi-band images.	.envi
ASCII Grid	ASCII Grid is a simple and widely supported format for storing raster data. It uses plain text to represent gridded data, where each cell value is stored as a text value in the file.	.asc
Esri GRID	Esri GRID is a proprietary raster data format used by Esri's ArcGIS software. It can store both single-band and multi-band raster data along with metadata and attribute tables.	.grd or .asc

'raster' and 'terra.' The 'raster' class in the 'sp' package allows to store raster data in a 'SpatialGridDataFrame' or a 'SpatialPixelsDataFrame' object, which includes both the raster values and the spatial information. However, the 'sp' package does not provide advanced raster data processing functions or extensive raster-specific functionalities.

The 'raster' package is the most used tool for working with raster data in spatial analysis and remote sensing applications (Section 10.4.3). It provides a comprehensive set of functions and classes for reading, writing, manipulating, and analysing raster datasets. The package allows users to handle large, multi-layered raster files efficiently, perform various raster operations, and extract valuable information from spatially continuous data. With its extensive capabilities, the 'raster' package facilitates a wide range of tasks, including terrain analysis, land cover classification, interpolation, and creating raster-based visualisations.

The 'terra' package is the newer package to handle gridded and raster data and is designed to provide high-performance capabilities, making it well-suited for analysing big data and complex geospatial processes (Section 10.4.4). This package introduces its own classes, such as 'SpatRaster', 'SpatVector', and 'SpatExtent', to represent raster and vector data along with relevant spatial metadata. It supports various raster operations, including mathematical calculations, focal functions, reclassification, and resampling.

Additionally, 'terra' seamlessly integrates with other popular packages, allowing users to leverage its functionalities in conjunction with a wide range of statistical and geospatial tools. With its focus on memory efficiency and speed, the 'terra' package is the essential tool for professionals and researchers dealing with gridded, raster, and vector spatial data, providing a solid foundation for efficient spatial data analysis and modelling.

10.3 Coordinate Reference Systems (CRS)

A coordinate system is a reference framework used to identify the location of objects in a two- or three-dimensional space. In geospatial analysis, coordinate systems are used to represent the location of geographic features and to perform spatial operations, such as distance calculations, overlay analysis, or map projections.

A coordinate system is defined by a set of parameters, including a datum, a projection, and units of measurement (Thomas, 2017).[4] The datum defines the reference surface for the coordinate system, while the projection defines how the curved surface of the earth is represented on a two-dimensional map or computer screen. Units of measurement, such as metres or degrees, are used to express the distance or size of geographic features.

[4] Available in https://csegrecorder.com/articles/view/datums-projections-and-coordinate-systems

In spatial analysis, coordinate systems are essential for accurate and meaningful analysis. Geographic features are represented by sets of coordinates, and the accuracy of these coordinates depends on the choice of coordinate system. When performing spatial analysis, the coordinate system used for a dataset must be consistent with the coordinate system used for other datasets or analysis tools.

The EPSG (European Petroleum Survey Group) system is a widely used coordinate system database that provides a standardised way to identify and reference coordinate systems. Each coordinate system in the EPSG system is identified by a unique number, known as an EPSG code.[5] For example, the EPSG code for the commonly used WGS84 geographic coordinate system (GCS) is 4326.

The EPSG code is important because it provides a standardised way to identify and reference coordinate systems, which is essential for ensuring interoperability and consistency in geospatial analysis. It also provides a way to easily convert between different coordinate systems, making it possible to combine and analyse data from different sources that may use different coordinate systems.

There are several types of coordinate systems used in geospatial analysis. Some of the most commonly used types are as follows:

GCS: A GCS is a coordinate system that uses latitude and longitude to define locations on the earth's surface. Latitude measures distance north or south of the equator while longitude measures distance east or west of the prime meridian. GCS is useful for representing global data, such as satellite imagery or weather data.

PCS: A PCS is a coordinate system that uses x and y coordinates to represent locations on a two-dimensional surface. PCS are commonly used for mapping and analysis, as they allow for accurate distance and area calculations. PCS can be based on different projections, such as Mercator, Lambert Conformal Conic, or Albers Equal Area.

Local coordinate system (LCS): An LCS is a coordinate system that is used to represent locations in a small, localised area, such as a building or a construction site. LCS are often used in mineral exploration, surveying, and engineering applications.

Geocentric coordinate system (GEOCS): A GEOCS is a coordinate system that uses x, y, and z coordinates to represent locations in three-dimensional space. GEOCS are used for global positioning and satellite imagery.

Datum-based coordinate system: A datum-based coordinate system is a coordinate system that is based on a specific geodetic datum, which is a mathematical model of the earth's shape and orientation. Examples of datums include NAD27, NAD83, and WGS84.

---

[5] To know more about EPSG systems refer to https://epsg.io/

Cartesian coordinate system: A cartesian coordinate system is a two-dimensional coordinate system that uses x and y coordinates to represent locations on a plane. Cartesian coordinates are often used in mathematics and computer science.

Although there are many coordinate systems, the more common ones are i) WGS84 (World Geodetic System 1984) and ii) UTM (Universal Transverse Mercator), where:

WGS84 is a GCS that uses latitude and longitude to define locations on the earth's surface. WGS84 is widely used as a standard for GPS and other global navigation systems. It is based on a reference ellipsoid that approximates the shape of the earth and is defined by a set of parameters that describe the position and orientation of the ellipsoid relative to the earth. The WGS84 datum is commonly used to represent global data, such as satellite imagery or weather data.

UTM is a PCS that divides the earth into 60 zones, each 6 degrees of longitude wide. UTM uses a Transverse Mercator projection to represent locations in each zone. UTM coordinates are expressed in metres and are based on a specific zone's central meridian and a false easting and northing value. UTM is commonly used for mapping and analysis at a regional scale, such as for urban planning, transportation, and environmental analysis. UTM is particularly useful for making accurate distance and area measurements and for minimising distortion in the area of interest.

UTM and WGS84 are often used together in spatial analysis. For example, GPS devices typically use the WGS84 datum to record coordinates while maps and GIS software may use UTM coordinates to represent locations on a map. In such cases, it is important to correctly transform (see Section 12.1 in Chapter 12) the coordinates between the two coordinate systems to ensure accurate analysis and visualisation.

10.4 Spatial Analysis Packages

There are several packages that can be used to manipulate spatial objects. Among these, the most common include 'sp', 'sf', 'raster', or 'terra'. These packages are the basis for expanding the R capacities to manipulate and analyse spatial data. Complemented with more specialised and content related packages, such as the 'spatstat' or 'gstat' packages, they constitute the core of any spatial analysis project in R.

10.4.1 The 'sp' Package

The 'sp' package provides several functions and data structures for working with spatial data (Figure 10.3).

The 'SpatialPoints' and 'SpatialPointsDataFrame' classes represent individual points in a two- or three-dimensional space. It contains x, y, and optional z coordinates, enabling the storage and manipulation of point data in a spatial context. 'SpatialPoints' objects are crucial for capturing discrete point locations, such as cities, sampling sites, or sensor measurements, and

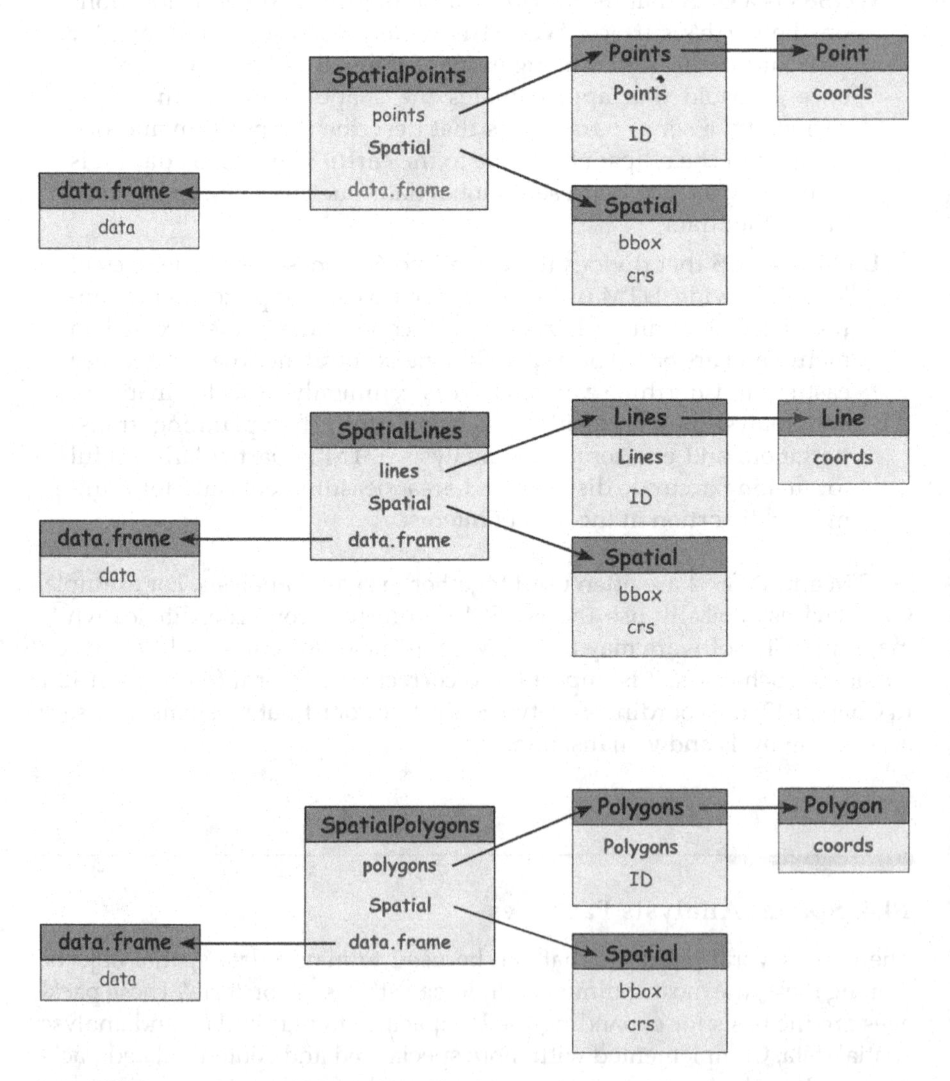

FIGURE 10.3
Representation of the basic vectorial data structures and classes in the 'sp' package.

provide a foundation for spatial analysis tasks like clustering, interpolation, and nearest neighbour calculations. The 'SpatialPointsDataFrame' class extends 'SpatialPoints' by combining the point data structure with a data frame that contains the data attributes. This enables the seamless integration of non-spatial data, such as temperature or population values, with spatial points. 'SpatialPointsDataFrame' objects are invaluable for handling datasets with point locations and associated attribute information, allowing for comprehensive spatial data exploration and analysis.

'SpatialLines' and 'SpatialLinesDataFrame' are designed for representing linear features, such as geological faults, stream flows, roads or rivers, in a two- or three-dimensional space. They store a series of x, y, and optional z coordinates that define the vertices of each line segment. 'SpatialLines' objects are used for managing linear data, enabling tasks such as network analysis, route planning, and length measurements along line features. The 'SpatialLinesDataFrame' class extends 'SpatialLines' by combining the line data structure with its attributes. This allows for the integration of non-spatial data, such as road names or river widths, with spatial lines. 'SpatialLinesDataFrame' objects are particularly valuable for handling datasets with linear features and their associated attribute information, enabling comprehensive and meaningful geospatial analysis.

'SpatialPolygons' and 'SpatialPolygonsDataFrame' are used for representing closed shapes, such as geological units, administrative boundaries, or country borders, in a two- or three-dimensional space. Similar to 'SpatialLines', 'SpatialPolygons' are composed by a series of x, y, and optional z coordinates that define the vertices of each polygon's boundary. 'SpatialPolygons' objects handle area-based spatial phenomena, such as land use patterns, administrative regions, or ecological zones. The 'SpatialPolygonsDataFrame' class extends 'SpatialPolygons' by combining the polygon data structure with the data attributes. This integration allows for the incorporation of non-spatial data, such as geological unit age, population, or land cover classifications, with spatial polygons. 'SpatialPolygonsDataFrame' objects are indispensable for managing datasets with polygon features and their associated attribute information, facilitating insightful spatial data exploration and analysis.

Complementary to the vectorial data structures, the 'sp' package also provides gridded and raster data structures.

The 'SpatialGrids' and 'SpatialGridDataFrame' represent regular or irregular grids of cells in a two- or three-dimensional space. Specifically designed for handling raster data, where each cell in the grid contains a single value, such as temperature, elevation, or land cover class. 'SpatialGrids' objects are suitable for efficiently storing and analysing gridded data, enabling operations like resampling, reprojection, and extraction of data at specific locations. They are widely used in various fields, including remote sensing, climate modelling, and terrain analysis, where continuous surface data is discretised into raster grids for efficient processing and analysis. The 'SpatialGridDataFrame' class extends 'SpatialGrids' by combining the spatial grid data structure with a data frame of attributes, similar to 'SpatialPolygonsDataFrame' but for raster data. 'SpatialGridDataFrame' objects provide a seamless integration of spatial and attribute data, streamlining raster data analysis and facilitating information-rich visualisation.

'SpatialPixels' and 'SpatialPixelsDataFrame' are similar to 'SpatialGrids', representing pixel-based spatial data, but it covers a different use case. The key difference lies in how they handle spatial indexing and memory usage, making spatial pixels more flexible for certain types of spatial analyses where the grid may be conceptualized as a set of points (pixels) rather than a continuous surface. Each pixel is defined by its x, y, and optional z coordinates, and the pixel values are stored in a corresponding raster layer. As in the other cases, 'SpatialPixelsDataFrame' combines the spatial pixel data structure with its attributes, similar to 'SpatialPolygonsDataFrame' and 'SpatialGridDataFrame', but for pixel-based data. These data structures are useful for handling point-based raster data, where the location of each pixel is not part of a predefined grid but is instead based on specific measurement or observation locations. 'SpatialPixels' and 'SpatialPixelsDataFrame' are versatile tools for managing and analysing datasets with pixel-based spatial data and associated attribute information.

These data structures provide the foundation for working with spatial data, and the 'sp' package provides a variety of functions for working with and analysing these data structures.

The list of attributes of an 'sp' data structure include the following:

coords: It is a matrix or data.frame that contains the coordinates of the spatial object.

bbox: It is a matrix containing the minimum and maximum values of the coordinates for each dimension.

proj4string: A character string describing the CRS projection of the data in Proj4 format.

These attributes are the kernel for working with spatial data in R.

Furthermore, the 'sp' package provides several functions that are used to handle spatial data, including:

bbox(): It returns the bounding box, or the minimum and maximum coordinates, of a spatial object.

coordinates(): It returns the matrix of coordinates of a spatial object.

gridded(): It returns a logical value indicating whether a spatial object is gridded.

is.projected(): It returns a logical value indicating whether a spatial object is projected.

proj4string(): It returns or sets the CRS based on the PROJ.4 string that defines the CRS of a 'Spatial*' object.

spTransform(): It is used for map projection and datum transformation.

spDists(): It computes a vector of distances between a set of two-dimensional points, represented by a matrix with longitude values in the first column and latitude values in the second column, and a single two-dimensional point. The distances are calculated using either the Euclidean method or the great circle distance method, which considers the WGS84 ellipsoid to provide more accurate distance measurements on the earth's curved surface.

over(): It returns a logical matrix indicating whether the coordinates of a set of points fall within the polygons of a SpatialPolygons object.

spplot(): It extends the plot() function to plot a spatial object.

These are some of the main methods provided by the 'sp' package. To learn more about its functions and applications, it is recommended to consult the 'sp' package reference in Bivand et al. (2013).

10.4.2 The 'sf' Package

The 'sf' package, where 'sf' stands for 'simple features', is the newest package for handling vector data in R (Pebesma & Bivand, 2023). It provides a modern and convenient interface for working with spatial data, making it a popular choice for spatial data analysis. It uses the Simple Features standard,[6] which is the commonly accepted standard for geographic data representation.

One of the key data structures in the 'sf' package is the 'sf' object (Figure 10.4). An 'sf' object contains both attribute data (similar to a regular data frame) and geometric information which define the spatial shape and location of each observation. The geographic component is stored as a set of 'simple features', which are simple geometric shapes such as points, lines, or polygons.

The 'sf' object provides a convenient way to store and manipulate geographic data and is compatible with many other packages for geospatial data analysis.

Another important data structure in the 'sf' package is the 'st_geometry' column. This is a special column in the 'sf' object that holds the geographic component of the data. It can be a single geometry, such as a point, line, or polygon, or it can be a collection of multiple geometries. The 'st_geometry' column is implemented as an S3[7] class in R, which means that it can be treated as a separate object with its own properties and methods. This allows for convenient manipulation and analysis of the geographic component of the data.

---

[6] https://www.ogc.org/standards/

[7] S3 (Simple S Object System) is a basic and informal class system used for object-oriented programming. It is one of the early systems for object-oriented programming in R, predating S4, which is a more formal and rigorous system.

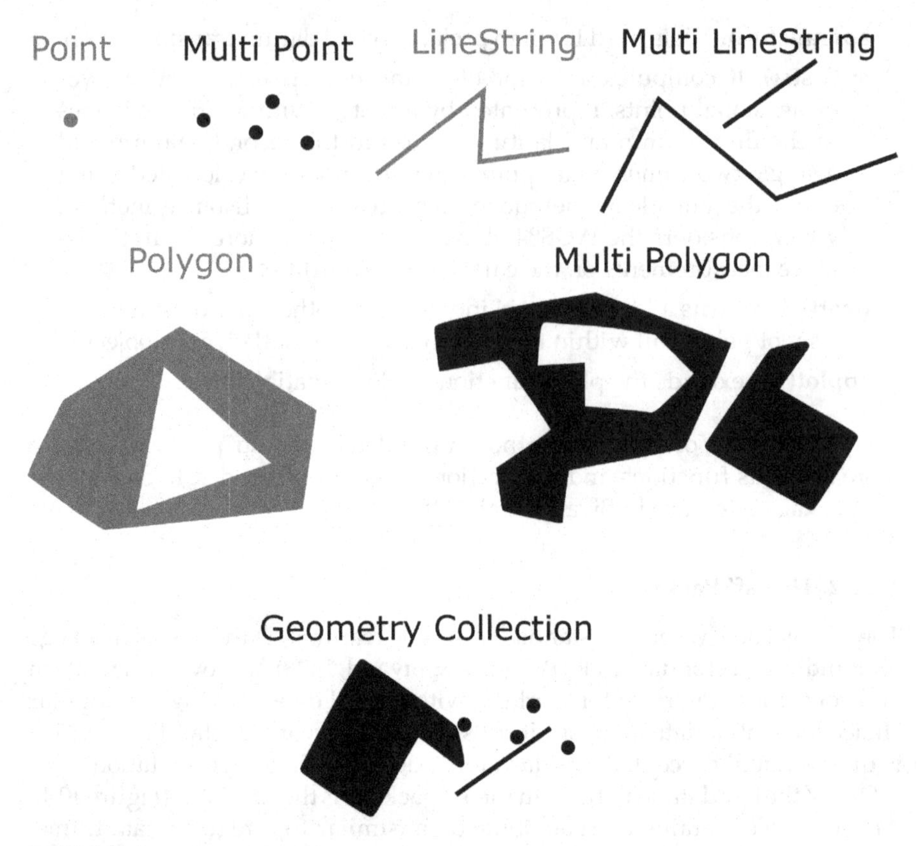

FIGURE 10.4
The Simple Features geometries for objects.

In addition to the 'sf' object and 'st_geometry' column, the 'sf' package also provides several other data structures and classes for working with geospatial data. These include the following:

st_point: It is a data structure for representing a single point.

st_linestring: It is a data structure for representing a sequence of points as a line.

st_polygon: It is a data structure for representing a closed shape defined by a set of points.

st_multipoint: It is a data structure for representing multiple points.

st_multilinestring: It is a data structure for representing multiple lines.

st_multipolygon: It is a data structure for representing multiple polygons.

st_geometrycollection: It is a data structure for representing a collection of different geometries objects. A geometry collection is a type of geometry that can contain multiple types of spatial objects like points, lines, and polygons, all within a single object.

The 'sf' package has several functions for reading, writing and handling geospatial data from and to various file formats. Among the most used functions, we can list the following:

st_read(): This function reads geospatial data from a variety of file formats, including Shapefiles, GeoJSON, and KML.

st_write(): This function writes geospatial data to a variety of file formats, including Shapefiles, GeoJSON, and KML.

st_transform(): This function transforms geospatial data from one CRS to another.

st_buffer(): This function creates a buffer around a spatial object.

st_intersects(): This function tests whether two spatial objects intersect.

st_join(): This function performs a spatial join between two spatial objects.

The constellation of 'sf' package functions makes it easy to read, write, and manipulate geospatial data in R. They are also easy to use, and they provide a wide range of functionality. When working with geospatial data, one must be familiar with the functions principles and working details.

Notice that the 'sp' and 'sf' packages are both used for handling spatial data in R; however, they have some differences in their approach and functionality:

Data structure: The 'sp' package uses a list-based structure for spatial objects called Spatial* classes, while the 'sf' package uses a data frame-based structure for spatial objects, using simple features.

Memory usage: The 'sp' package is memory-efficient for large datasets but can be slower for some operations compared to the 'sf' package. On the other hand, the 'sf' package can use more memory but is faster for some operations.

Compatibility with other packages: The 'sp' package has been around for a longer time and is more widely used, so it is compatible with many other packages. The 'sf' package is newer and still gaining popularity but has the advantage of being more closely tied to the 'tidyverse' and 'ggplot2' packages.

Functionality: The 'sp' package has more functions for working with spatial data, such as for spatial overlay operations and spatial analysis. The 'sf' package, on the other hand, has a simpler interface and is easier to use for basic operations but may have less functionality for advanced spatial analysis.

10.4.3 The 'raster' Package

The 'raster' package is an all-encompassing library for managing raster data (Hijmans, 2023). It provides functions for reading and writing data from a wide range of formats, including GeoTIFF, ASCII Grid, and netCDF; it also provides functions for data manipulation and processing, including cropping, resampling, projection, and masking.

The core data structure in the 'raster' package is the 'RasterLayer' class. It stores gridded data and its associated metadata, including the spatial extent, resolution, and CRS. The data stored in a 'RasterLayer' is organised in a two-dimensional matrix, where each cell represents a unique geographic location and the associated value for a given variable.

Other important data structures in the 'raster' package are the 'RasterStack' and 'RasterBrick' classes that store multiple 'RasterLayer' as a single object, allowing to handle multi-band data, such as multispectral satellite imagery.

The 'raster' package also provides functions for spatial analysis, including interpolation, overlay operations, and distance-based analysis. Additionally, its capabilities include a set of tools for visualising raster data, such as plotting, contouring, and colour mapping.

Resuming, the data structures for spatial data representation, using the 'raster' package, are as follows:

RasterLayer: It is a two-dimensional array that stores raster cell values and associated metadata (e.g. spatial extent, resolution, and projection).

RasterBrick: It stores multiple 'RasterLayer' objects with the same spatial extent and resolution, typically used to manage multi-layer datasets such as multi-band satellite images or time series from the same sensor. Each layer within a 'RasterBrick' must align perfectly, making it an optimized, efficient structure for handling closely related raster datasets.

RasterStack: It is a container object that groups multiple 'RasterLayer' objects into a single, layered data structure. Each layer in a 'RasterStack' can be considered independently and can have different resolutions or extents, offering more flexibility than a 'RasterBrick'.

Each of these data structures has a set of attributes that describe the spatial characteristics of the data. These structures allow the manipulation of raster objects, and its comprehensive list includes the following:

Extent: It is the spatial extent of the raster data, which defines the boundaries and coverage area of the raster in the CRS units.

Resolution: It is the cell size or spatial resolution of the raster, indicating the width and height of each raster cell in the CRS units.

Projection: It is the CRS used to represent the raster data, specifying how the spatial coordinates are mapped to the Earth's surface.

Number of Rows and Columns: It is the dimensions of the raster, specifying the number of rows and columns in the two-dimensional RasterLayer or in each layer of the three-dimensional RasterBrick.

Number of Layers: For RasterBrick and RasterStack, the number of layers indicates the total number of raster layers contained within the data structure.

Data Type: It is the type of data stored in the raster cells, such as integer or floating-point values, representing different information like elevation, temperature, or land cover.

NA (Not Available) Value: It is the value that represents missing or non-valid data in the raster.

Categorical Values: For categorical rasters, these are the unique values that represent different classes or categories.

Temporal Information: For RasterBrick, additional temporal attributes, such as timestamps or dates, may be associated with each layer to represent different time steps.

These attributes can be accessed using the '@' operator, e.g. 'raster@extent'. However, it's generally recommended to use the specific accessor functions provided by the raster package for this purpose, such as extent(raster) to get the raster's extent or crs(raster) for its CRS, as this approach is safer and less prone to errors. Additionally, the dim() and extent() functions can be used to retrieve information on the dimensions of the raster data.

Among the functions included in the 'raster' package, we highlight the following:

aggregate(): It is used to aggregate the values of cells in a raster layer to a coarser resolution.

crop(): It allows for cropping a raster layer to a smaller extent, defined by a bounding box or another spatial object.

extract(): It is used to extract cell values from a raster layer for a set of points or polygon objects.

mask(): It is used to apply a binary mask to a raster layer, effectively setting values outside of the mask to NA.

overlay(): It is used for the application of a function over the values of multiple layers where they overlap. It's essential for combining rasters based on spatial operations.

resample(): It is used to resample a raster layer to a different resolution.

rasterize(): It is used to convert vector data, such as points or polygons, into a raster layer.

merge(): It is used to merge two or more raster layers into a single layer, combining the values of overlapping cells.

The combination of 'sp' and 'raster' packages have been the basis tools for spatial data analysis and manipulation. Figure 10.5 provides a synthesis of these objects and referring packages.

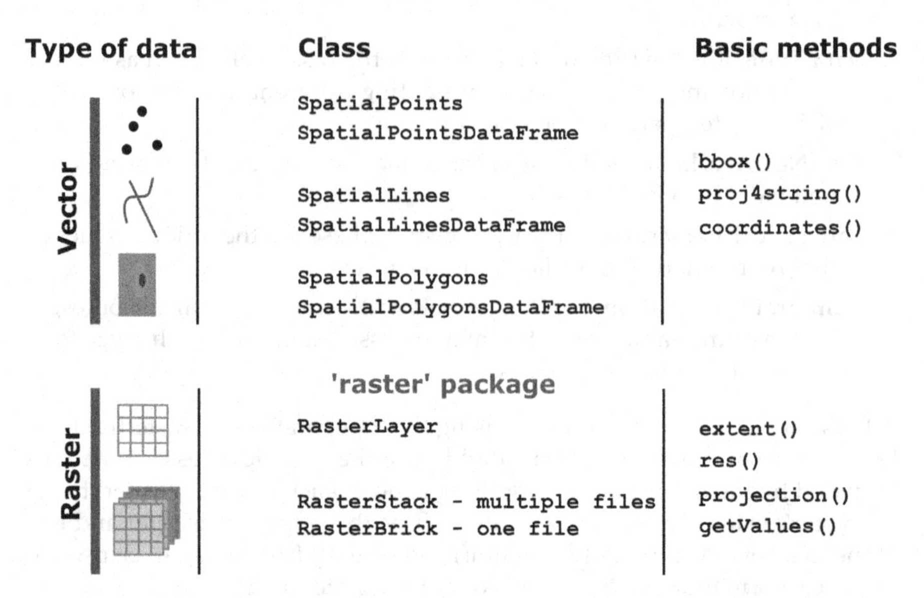

FIGURE 10.5
Visual synthesis of 'sp' and 'raster' packages objects and sample methods.

10.4.4 The 'terra' Package

The 'terra' package is a recent geospatial package that provides a flexible, high-performance environment for working with large-scale spatial data. Its data types, data structures, and methods or functions include the following:

Data types: Data types include raster, vector, and point clouds, where 'raster' data can be represented as 'SpatRaster' objects while vector data can be represented as 'SpatVector' objects. The package also provides specialised data types for working with specific types of geospatial data, such as 'SpatRasterNC' for working with 'netCDF' data.

Data structures: The key data structure used is the 'SpatRaster' object, which is a container for raster data that includes information about the spatial extent, resolution, and projection of the data. 'SpatRaster' objects can include multiple layers (or bands) of data, such as multispectral images or time series data. The package also provides specialised data structures for working with vector and point cloud data.

Methods or functions: Methods for manipulating and analysing geospatial data are available including functions for reprojecting, resampling, and aggregating raster data, as well as those for clustering, interpolating, and modelling spatial data. The package also provides tools for working with time series data and supports a wide range of statistical and machine learning models for spatial data analysis.

Examples of commonly used methods or functions are as follows:

Reading and writing data:

rast(): Create a raster object.

terraOptions(): Set and query options for 'terra'.

writeRaster(): Write raster data to various formats like GeoTIFF.

Basic raster manipulation:

subset() or **clip()**: Crop or subset a raster based on spatial extent.

mask(): Mask a raster with another raster or vector mask.

Reprojection and resampling:

project(): Reproject a raster to a different CRS.

resample(): Change the resolution or grid of a raster.

Zonal statistics:

extract(): Extract values from raster based on zones (e.g., polygons).

zonal(): Calculate zonal statistics such as mean and sum.

Overlay analysis:

rasterize(): Convert vector data to raster format.

intersect(): Perform intersection of raster layers.

union() and other set operations.

Distance and buffering:

distance(): Calculate distance to features in a raster.

buffer(): Create a buffer around features in a raster.

Most of these functions will be introduced along with the examples in the next chapters.

10.4.5 See Also

On the internet, one can find lots of web pages where the authors list their favourite R packages. These are just some examples of links to web pages with these types of collections. Any search on the internet can provide you with new ones.

- https://roelverbelen.netlify.app/resources/r/packages/
- https://support.posit.co/hc/en-us/articles/201057987-Quick-list-of-useful-R-packages

- https://github.com/qinwf/awesome-R
- https://couttsgeodata.netlify.app/post/2021-02-28-r_geoscience/
- https://homepage.univie.ac.at/dieter.mader/geofloss3.md.html

10.5 Concluding Remarks

The understanding of the data structures that can be evoked in spatial analysis and its features, components, and functions is a key aspect for performing any spatial analysis. Vectorial and gridded data present themselves in many "flavours" and "colours" depending on the package that is being used; therefore, it is good advice to fully understand the potential of each of the packages and its corresponding data types.

Another key issue is the understanding and knowledge of what file types exist, its applications, and how to handle them. Different applications, instruments, and software require different file types. R is empowered to handle the most common and some uncommon data types.

Finally, please pay attention to the coordinate systems of every spatial object in any analysis project. My experience is built in many hours spent on errors generated by trying to intersect features defined in different coordinate systems or project coordinates that are decimal degrees, or even degrees, minutes and seconds that should be decimal degrees.

References

Bivand, R., Pebesma, E., & Gomez-Rubio, V. (2013). Applied Spatial Data Analysis with R, Second Edition. Springer, New York.

Hijmans, R. (2023). raster: Geographic Data Analysis and Modeling. R package version 3.6-20, https://CRAN.R-project.org/package=raster

Pebesma, E., & Bivand, R. (2023). Spatial data science: With applications in R. CRC Press, Boca Raton, FL, p. 300.

Thomas, A. (2017). Datums, Projections and Coordinate Systems. Recorder, V42,2. Canadian Society for exploration geophysicists. https://csegrecorder.com/articles/view/datums-projections-and-coordinate-systems

11

Going Vectorial

As previously stated, vector data refers to spatial data that is stored as a series of points, lines, and polygon features in a digital format. This data structure is in contrast to raster data, which is stored as a grid of cells, each with a unique value. Vector data is mostly used to represent features such as roads, rivers, geochemical soil samples, or geological boundaries.

In R, vector data can be primarily managed using the 'sp' or 'sf' packages, which provide a range of functions and classes to store, manipulate, and analyse vector data. The main classes for handling vector data are 'Spatial*' and 'sf', respectively, which are used to represent terrain features and the associated attributes.

Once vector data has been loaded into the R environment, it can be visualised, transformed, and analysed in a number of ways. For example, the 'leaflet' package can be used to create interactive maps, whereas the 'ggplot2' package can be used to create static maps; other packages can be used for other operations in spatial analysis and modelling.

In geological projects, vector data can be used to represent various features such as faults, mineral deposit's locations, or geological units, encompassing all potential elements for geological applications. This data can be analysed to identify relationships between geological features, predict the presence of mineral deposits, and model geological processes such as tectonic deformation or erosion.

The following two chapters (Chapters 11 and 12) will deal mainly with vector data, followed by a chapter (Chapter 13) for wrapping all the knowledge with a holistic example of handling vectorial spatial data in a geological case.

11.1 Sources of Vectorial Spatial Data

Vectorial spatial data is present in many forms in the earth sciences and geological applications. It is collected on a daily basis in any geological campaign, e.g. a stream sediments or soil sampling location, a geological and structural mapping, a magnetic survey, a marine bathymetric campaign, or a drone photogrammetry campaign.

Commonly, for most of these data acquisition campaigns, there are protocols or procedures for data preparation and treatment, after which the

DOI: 10.1201/9781032651880-13

data is stored in files in a computer where it can be managed using a spatial analysis approach.

Depending on the source of the data, the steps for its analysis might be different or as in many of the cases the data is mingled in a unique spatial data analysis project (see example in Chapter 13). Herein, we describe some everyday examples of the vector data acquisition and what are its constraints and applications.

GPS data: GPS (Global Positioning System) data is a type of geospatial data that uses satellite signals to determine the location of a device on the earth's surface. GPS data is often collected by mobile devices, such as smartphones or GPS receivers, and can be used to track the location of samples, geological features or terrain features, and infrastructures. GPS data is often stored in a vector format, with each point representing a unique location and time.

Geochemical survey: Thanks to the miniaturisation of many detectors, it is now feasible to directly measure the chemical properties of soils and rocks during field campaigns. Equipment such as portable X-ray fluorescence devices and LIBS (Laser-Induced Breakdown Spectroscopy) often come equipped with a GPS receiver, enabling the collection of georeferenced geochemical data. This data can be immediately utilised in spatial analysis projects, with or without undergoing pre-processing steps.

Geophysical survey: Depending on the type of geophysical campaign, often the equipment (e.g. field gravimeter or magnetometer) collects data from not only the physical properties of the terrain but also its coordinates by locating each data point with a GPS. This data often needs to be pre-processed before it is used in geological interpretations.

LiDAR data: LiDAR (Light Detection and Ranging) data is a type of geospatial data that uses laser technology to measure the distance between the sensor and the earth's surface or any light-reflecting object. LiDAR data is collected from aerial or terrestrial platforms and can be used to create high-resolution three-dimensional models of the earth's surface. LiDAR data is often stored in point cloud format, with each point representing a unique location and elevation. LiDAR data is useful for terrain analysis, flood modelling, or mass movement characterisation.

Cadastral data: Cadastral data is a type of geospatial data that includes information about land ownership and boundaries. Cadastral data is often stored in a vector format, with each polygon representing a unique parcel of land and its ownership information. Cadastral data is useful for land use and land cover projects or framing a region or a study area.

11.2 Read and Write Spatial Data

Reading and writing spatial data can be done using various of the previously cited packages such as 'raster' or 'sf'. These packages provide functions to import and export a wide range of spatial data formats including shapefiles, KML, and GeoJSON.

11.2.1 Reading Spatial Data Files

11.2.1.1 Shapefiles

Shapefiles (see Section 10.1.1 in Chapter 10) store vector data format in GIS and are composed of multiple files with different extensions, including '.shp', '.shx', and '.dbf'. The 'sf' package provides a function, the st_read() function that is used for reading shapefiles, and all its metadata available in the related files. It receives as arguments the 'path' and 'filename' and the 'driver' to state the file type.

One can list the available drivers using the st_driver() function without arguments. The list of the most commonly used drivers include the following:

- **Shapefile**: "ESRI Shapefile"
- **GeoJSON**: "GeoJSON"
- **KML (Keyhole Markup Language)**: "KML"
- **GPKG (GeoPackage)**: "GPKG"
- **GPX (GPS Exchange Format)**: "GPX"
- **CSV (Comma Separated Values)**: "CSV"
- **WKT (Well-Known Text)**: "WKT"
- **SQLite (Structured Query Language format)**: "SQLite"
- **NetCDF (Network Common Data Form)**: "NetCDF"

Here's an example of its use:

```
# Load the library
library(sf)

# Read shapefile
my_shp = st_read("path/to/my/shapefile.shp", driver = "ESRI
Shapefile")
```

The 'raster' package also possesses the shapefile() function that similarly reads a file in the ESRI shapefile format. It reads the shapefile as a 'Spatial*' object from the 'sp' package. Its use is straightforward:

Here's the example:

```
# Load the library
library(raster)
```

```
# Read shapefile
my_shp = shapefile("path/to/my/shapefile.shp")
```

11.2.1.2 KML files

KML is a file format used to display geographical data in Google Earth, Google Maps, and other GIS applications. In R, KML files can be read using the 'sf' package's st_read() function. For example:

```
# Read a KML file
kml_file = st_read("path/to/kml_file", driver = "KML")
```

Reading most of the common formats of vector files is possible using these functions. Please use the RTD rule (i.e. Read The Docs) to verify its application to a specific type of vector file. Anyway, in most of the cases, it is quite simple to find out how to read the intended file type.

11.2.2 Writing Spatial Data Files

11.2.2.1 Shapefiles

As above, writing spatial data in R can be done using the st_write() function from the 'sf' package. For example, writing a shapefile:

```
# Write the sf data to a shapefile
st_write(sf_data, dsn = "path/to/myfile.shp", driver = "ESRI
Shapefile")
```

As for reading, the shapefile() function from the 'raster' package can write a shapefile file type. Here is the example:
Here's an example of its use:

```
# Write shapefile
shapefile(my_shp, "path/to/my/shapefile.shp")
```

11.2.2.2 KML Files

To write a KML file with the sf package, you can use the st_write() function with the 'driver = "kml"' parameter to specify the KML format. Here's an example:

```
# Write KML file
st_write(sf_data, dsn = "path/to/myfile.kml", driver = "kml")
```

This will create a KML file named 'myfile.kml' in the specified directory. The resulting KML file will contain placemarks for the elements in the input data.

Yes, you guessed it! This syntax is common for most of the other file types available, making it straightforward to write the code to save the intended file type.

11.2.3 Converting between Spatial File Types

Often, it is necessary to convert between data types and formats to perform some specific spatial analysis. There are several available packages and functions that allow these conversions, and there are several reasons why data conversion between spatial types is important.

Firstly, data conversion allows us to work with different data sources and integrate them into a single project. For example, we may want to combine data from a CSV file with point data from a GPX file. By converting the GPX data to a data frame or an 'sf' object, we can easily merge the two datasets and perform spatial analysis that includes both datasets.

Secondly, different types of spatial data may require different data types and formats for efficient processing and visualisation. For example, 'sf' objects are efficient for working with vector data while GPX and KML files are more suitable for working with GPS data and web mapping. By converting between these data types, we can optimise workflows and improve performance.

Thirdly, data conversion enables us to share our results with others who may be using different software or tools. For example, we may want to share our results with colleagues who are using GIS software that requires a specific data format. By converting our data to the required format, we can ensure that our colleagues can use and reproduce our results.

Finally, data conversion can help us to standardise our data and ensure that it is compatible with other software and tools. By converting our data to a common format such as GeoJSON or KML, we can ensure that it can be used by other software and tools that support these formats.

Let's assume that we have the following data for a set of geographic coordinates:

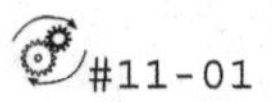

```
# Load the library
library(sp)
```

```
# Create a data frame
data = data.frame(name = c("New York City", "Los Angeles",
"Chicago", "Houston"),
                   lat = c(40.7128, 34.0522, 41.8781, 29.7604),
                   lng = c(-74.0060, -118.2437, -87.6298,
-95.3698))
```

11.2.3.1 Convert Data Frame to Spatial Object*

This data represents the name, latitude, and longitude of four cities in the United States. We can convert this data between different formats. First let's transform the data frame to the 'Spatial*' format from the 'sp' package. For this, the coordinates() function defines the variables that contain the coordinates, hence converting the data frame to a 'SpatialPointsDataFrame' object. Additionally, don't forget to identify the coordinate reference system (CRS) for this data. In this case, the CRS() method is used to define the WGS84 CRS.

#11-02 1 star

```
# Create a sp object
coordinates(data) = c("lng", "lat")

# Define the crs
proj4string(data) = CRS("+proj=longlat +datum=WGS84 +no_defs")

# Verify the data type
str(data)
```

The result is a 'SpatialPointsDataFrame' that contains all the data.

```
Formal class 'SpatialPointsDataFrame' [package "sp"] with 5 slots
 ..@ data       :'data.frame':    4 obs. of  1 variable:
 .. ..$ name: chr [1:4] "New York City" "Los Angeles"
"Chicago" "Houston"
 ..@ coords.nrs : int [1:2] 3 2
 ..@ coords     : num [1:4, 1:2] -74 -118.2 -87.6 -95.4 40.7 ...
  .. ..- attr(*, "dimnames")=List of 2
  .. .. ..$ : NULL
  .. .. ..$ : chr [1:2] "lng" "lat"
  ..@ bbox        : num [1:2, 1:2] -118.2 29.8 -74 41.9
  .. ..- attr(*, "dimnames")=List of 2
  .. .. ..$ : chr [1:2] "lng" "lat"
  .. .. ..$ : chr [1:2] "min" "max"
  ..@ proj4string:Formal class 'CRS' [package "sp"] with 1 slot
  .. .. ..@ projargs: chr "+proj=longlat +datum=WGS84 +no_defs"
```

```
.. .. ..$ comment: chr "GEOGCRS[\"unknown\",\n DATUM[\"World
Geodetic System 1984\",\n     ELLIPSOID[\"WGS
84\",6378137,298.25722"| __truncated__
```

A 'SpatialPointsDataFrame' is a class that represents a set of geographic points, each of which has associated attribute data (see Section 10.3 in Chapter 10). A 'SpatialPointsDataFrame' object has several slots (identified by the '@' symbol) that store different types of data.

11.2.3.2 Converting Between 'sp' and 'sf'

The 'sp' and 'sf' are the two most common packages used in spatial analysis in R. To convert the previous 'SpatialPointsDataFrame' named 'data' to an object in the 'sf' format, we can use the st_as_sf() function from the 'sf' package. The following example shows the code:

#11-03 

```
# Load the Library
library(sf)

# Convert to sf
data_sf = st_as_sf(data, coords = c("lng", "lat"), crs = 4326)

# Verify the results
str(data_sf)
```

The resulting object has the structure as follows:

```
Classes 'sf' and 'data.frame':    4 obs. of  2 variables:
 $ name     : Factor w/ 4 levels "Chicago","Houston",..: 4 3 1 2
 $ geometry:sfc_POINT of length 4; first list element:  'XY'
num  -74 40.7
 - attr(*, "sf_column")= chr "geometry"
 - attr(*, "agr")= Factor w/ 3 levels "constant","aggregate",..: NA
  ..- attr(*, "names")= chr "name"
```

The 'sp' package, one of the earliest R packages for spatial data, structures its data using S4 objects,[1] where data components are organised into slots. In contrast, the 'sf' package, a more recent addition designed to work with the 'Simple Features' standard, utilises a different approach, organising data in a way that is consistent with modern R practices, particularly by integrating closely with 'data.frame' objects and using the $symbol for data access, similar to how columns in a data frame are accessed.

[1] S4 objects provide a structured and formal way to organize data and functions in R, making it easier to manage complex programs and ensure that data is handled correctly.

11.2.3.3 Converting Between 'sf' and JSON

To convert the 'data' to JSON format, we can use the 'jsonlite' package[2]:

```
#11-04/01

# Load the library
library(jsonlite)

# Convert to JSON
json_data = toJSON(data_sf)

# Verify the structure
str(json_data)
```

The results obtained are as follows:

```
'json' chr "[{\"name\":\"New York City\", \"geometry\":
{\"type\":\"Point\", \"coordinates\":[-74.006,40.7128]}},
{\"name\":\"L"| __truncated__
```

The data can be visualised in the new format as follows:

```
#11-04/02

# View the data
json_data
```

and the data is shown as follows:

```
[{"name":"New York City","geometry":{"type":"Point","coordinates":
[-74.006,40.7128]}},{"name":"Los Angeles","geometry":
{"type":"Point","coordinates":[-118.2437,34.0522]}},
{"name":"Chicago","geometry":{"type":"Point","coordinates":
[-87.6298,41.8781]}},{"name":"Houston","geometry":
{"type":"Point","coordinates":[-95.3698,29.7604]}}]
```

11.2.3.4 Converting Between 'sf' and GPX

To convert the 'data_sf' object to GPX format, we can use the 'sf' package and write it to the disk. There is no specific object to store GPX files in the R environment; therefore, it only can be saved to disk. For converting an 'sf' object to a GPX file, just write it on disk with the st_write() function:

```
#11-05

# Save the data
st_write(data_sf, "gpx_data_file.gpx", driver = "GPX")
```

[2] Remember to install the package with install.packages('jsonlite').

11.2.3.5 Converting Between 'sf' and KML

As for the GPX example, the KML format is not a specific type of object in the R environment, but as above, an 'sf' object can be saved to disk using the st_write() function with the required format using the "driver =" option.

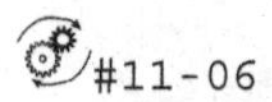#11-06 

```
# Save the data
st_write(data_sf, "kml_data_file.kml", driver = "KML")
```

11.3 OpenStreetMap: A Source of Geographic Data

OpenStreetMap (OSM) is a collaborative project that aims to deliver a free, editable map of the world. One of the great features of OSM is that it is open source and free to use, making it a valuable resource for geographic data with a variety of applications. In combination with R, it is possible to extract and use OSM data for spatial analyses, creation of maps, and other applications.

One of the packages available for extracting OSM data is 'osmdata', which provides a simple interface to download and extract OSM data directly from the OSM API (Application Programming Interface). This package allows to extract the data in different formats, including 'sf' objects, which can be easily integrated with other spatial data.

11.3.1 The 'osmdata' Package

The 'osmdata' package[3] provides an interface to the Overpass API, which allows users to retrieve OSM data in a flexible and efficient way.

In previous versions,[4] it had to be installed from a repository different from the Comprehensive R Archive Network (CRAN). In this case, we needed to have the 'remotes' package also installed. This package allows the installation of packages that are not necessarily in the CRAN repository.

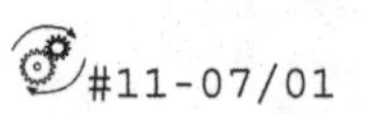 

```
# Install remotes package
install.packages("remotes")

# Install the 'osmdata' package
remotes::install_github ("ropensci/osmdata")
```

[3] https://github.com/ropensci/osmdata

[4] At the moment, while making the final revision of the book (February 2024) I verified that 'osmdata' is already available in Comprehensive R Archive Network (CRAN). I decided to maintain the original text to explain the possibility of using the 'remotes' package for packages outside CRAN. This is useful for the R packages that are not in CRAN, especially many experimental ones that are available via GitHub.

In our case, we get a message to upgrade the installed packages and we choose the '1: All' option.

If everything goes right, we have a fresh installation of 'osmdata' package. The message in the console asks to update packages. Do so by selecting the "1: All" option.

```
Downloading GitHub repo ropensci/osmdata@HEAD
These packages have more recent versions available.
It is recommended to update all of them.
Which would you like to update?

1: All
2: CRAN packages only
3: None
4: cpp11 (0.4.5 -> 0.4.6) [CRAN]
```

After successful installation verify if everything is okay by loading the 'osmdata' library.

To retrieve OSM data for a specific location, it is necessary to define a bounding box that covers the area of interest. The bounding box can be specified as a set of coordinates or as a place name that is recognised by the Nominatim Geocoding Service.[5] This service provides a huge dataset of names and its locations on the planet.

Here's an example of how to retrieve the 'peak' and 'volcano' features in the 'Ilha do Pico Island' in Azores, Portugal.

In this code, getbb("Ilha do Pico, Portugal") retrieves the bounding box for Ilha do Pico using the Nominatim service. The opq() function creates an Overpass API query object with the specified bounding box. The add_osm_feature() function specifies that we want to retrieve OSM objects with the key "natural" and the values "volcano" and "peak".

The osmdata_sf() function is used to retrieve the data from the OSM servers and store it in the 'pico_volcano_peak' variable.

#11-07/02

```
# Load the library
library(osmdata)
library(sf)

# Create the query
osm_query = opq(bbox = getbb("Ilha do Pico, Portugal"))

# Define the features in the query
osm_query_point = add_osm_feature(osm_query, key="natural",
value = c("volcano","peak"))

# Retrieve the results
pico_volcano_peak = osmdata_sf(osm_query_point)
```

[5] https://nominatim.org/

```
# View the results
pico_volcano_peak
```

The resulting output is as follows:

```
Object of class 'osmdata' with:
             $bbox : 38.3820698,-28.5418993,38.5614291,-28.0281952
    $overpass_call : The call submitted to the overpass API
             $meta : metadata including timestamp and version
numbers
       $osm_points : 'sf' Simple Features Collection with 25
points
        $osm_lines : NULL
     $osm_polygons : 'sf' Simple Features Collection with 0
polygons
   $osm_multilines : NULL
$osm_multipolygons : NULL
```

The output provides a detailed overview of the data retrieved from OSM, including the geographical area covered, the specifics of the query, relevant metadata, and the types of geographical features obtained. Specifically, it highlights that the data encompasses a specified bounding box area, details the exact query made to the Overpass API, includes metadata like the query's timestamp and version numbers of the data for context, and outlines the retrieved geographical features, which in this case are 25 point locations. However, it also notes that no line or polygon features were found within the specified area or according to the query parameters. This information is essential for conducting spatial analysis and understanding what data is available for use within the defined geographical scope.

It is also possible to retrieve general data without a specific OSM tag. For example, to retrieve all the keys that are "natural", we can modify the code as follows:

#11-07/03

```
# Create the query
osm_query = opq(bbox = getbb("Ilha do Pico, Portugal"))

# Define the features in the query
osm_query_all =  add_osm_feature(osm_query, key="natural")
```

After creating the query to retrieve the information to an R object, the osmdata_sf() function is used. The following code does this.

#11-07/04

```
# Retrieve the results
pico_island_all = osmdata_sf(osm_query_all)
```

```
# Show the results
pico_island_all
```

If the server is too busy, there might be some friendly warning message like the one below.

```
Request failed [504]. Retrying in 1.9 seconds...
```

If the attempt is successful, the 'osm' object might look like the one below. However, notice that the retrieved information might slightly change, as the OSM database is constantly being updated.

The console output is as follows:

```
Object of class 'osmdata' with:
          $bbox : 38.3820698,-28.5418993,38.5614291,-28.0281952
 $overpass_call : The call submitted to the overpass API
          $meta : metadata including timestamp and version
numbers
    $osm_points : 'sf' Simple Features Collection with 21866
points
     $osm_lines : 'sf' Simple Features Collection with 184
linestrings
  $osm_polygons : 'sf' Simple Features Collection with 285
polygons
$osm_multilines : NULL
$osm_multipolygons : 'sf' Simple Features Collection with 6
multipolygons
```

This states that the retrieved data includes 21866 points, 184 linestrings, 285 polygons, and 6 multipolygons.

The components of an 'osm' object include the following:

bbox: It is a bounding box that defines the geographic extent of the OSM data. In this example, the bounding box is defined by four coordinates in decimal degrees (longitude and latitude): 38.3820698, –28.5418993, 38.5614291, –28.0281952.

overpass_call: It is the text of the Overpass API call that was used to retrieve the OSM data. This is useful for troubleshooting and understanding the query that was used to retrieve the data.

meta: It is the metadata about the OSM data, such as the timestamp of the query and the version of the data.

osm_points: It is an 'sf' (simple features) collection of point features, represented as an R data frame with a geometry column and attributes in the other columns.

osm_lines: It is an 'sf' collection of line features, represented as an R data frame with a geometry column and attributes in the other columns.

osm_polygons: It is an 'sf' collection of polygon features, represented as an R data frame with a geometry column and attributes in the other columns.

osm_multilines: It is an 'sf' collection of multi-line features, represented as an R data frame with a geometry column and attributes in the other columns.

osm_multipolygons: It is an 'sf' collection of multipolygon features, represented as an R data frame with a geometry column and attributes in the other columns.

The retrieved 'sf' object needs to be split into several components to become useful. To use the points, we create a new variable, 'pico_volcano'. This variable will be an 'sf' object, but the data was retrieved without a CRS; hence, it is necessary to attribute it with the function st_crs().

Although the data has no CRS defined, its coordinates are in the CRS system 'EPSG = 4326' (WGS84 remember); therefore, the code is straightforward.

The example is to extract the simple points retrieved from the 'peak' and 'volcano' query (named 'pico_volcano_peak', see code snippet #11-07/02) that has 25 points.

#11-07/05

```
# Create a points data
pico_volcano = pico_volcano_peak$osm_points

# Set the crs
st_crs(pico_volcano) = 4326
```

To conclude this example, let's examine everything combined. We'll create an appealing plot that includes the peaks and volcanoes of Pico Island. To enhance the visualisation, the code also retrieves the coastal boundary of Pico Island, represented as a 'multipolygon', and incorporates it into the plot. This approach allows to visualise both the island's peak features and its coastal outline in a single, comprehensive map.

#11-07/06

```
# Use opq to define the query
osm_query = opq(bbox = getbb("Ilha do Pico, Portugal"))

# Define the features to extract
pico_land = add_osm_feature(osm_query, key="name", value =
"Ilha do Pico")
```

```
# Retrieve the results
pico = osmdata_sf(pico_land)

# Create a polygons data
pico_out = pico$osm_multipolygons$geometry

# Set the crs
st_crs(pico_out) = 4326

# Plot the results
plot(pico_out, col="green")
points(as_Spatial(pico_volcano), pch=2, col="red")
```

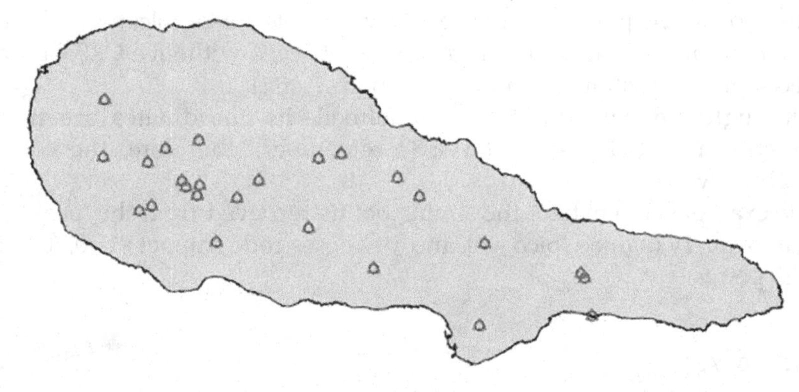

FIGURE 11.1
Contours of Pico Island with its peaks and volcanoes represented as triangles.

The results are displayed in Figure 11.1.

The 'osmdata' package provides many other options for retrieving OSM data, such as filtering by 'user' or 'timestamp' or retrieving data in a specific format. For more information on the package, you can consult the package documentation[6] and the Overpass API documentation.[7]

11.4 Visualising Spatial Data

The previous example is a nice plot of the island boundaries and of some of its structures; however, it hardly could be used in a report or geological publication. Let us explore a little more about visualising spatial information, thus creating publishable plots and maps.

Visualising and exploring spatial data are usually done by creating a map or a plot to relate certain features. Some of the most commonly used packages

[6] https://cran.r-project.org/web/packages/osmdata/osmdata.pdf
[7] https://osmlab.github.io/learnoverpass//en/

described in this book include 'ggplot2' and 'leaflet'. Each of these packages provide different functionalities, and use different data types, so choosing the right one for a particular project will depend on its specific needs.

As always in R, there are many more packages for spatial data visualisation. It is recommended that, when necessary, read the literature and find what are the new and more fit for purpose packages.

11.4.1 A First on 'ggplot2'[8]

The 'gg' acronym from the package 'ggplot2' stands for "Grammar of Graphics", which is a conceptual framework for creating visualisations developed by Leland Wilkinson (Wilkinson et al., 2005). The 'ggplot2' package is mainly used for data visualisation and is particularly well-suited for exploring spatial data. The package allows users to create a variety of different plots, including the most common scatterplots, line plots, bar plots, and histograms.

To use 'ggplot2' for spatial analysis, the data must be in a format that the package can understand, such as the 'SpatialPoints', 'SpatialPolygons', 'data.frame', or 'sf' objects. Once the data is in the correct format, the user can create the plot using the ggplot() function, followed by a series of geom_*() functions to specify the type of plot to be created.

The core idea behind 'ggplot2' is that a plot can be regarded as a set of abstraction layers that, when added or removed, create a final visualisation. These layers can include data, scales, and geometric objects, each with their own set of aesthetics and properties. Moreover, 'ggplot2' provides a set of functions and operators that allow users to define these abstractions in a consistent and flexible way.

Please notice that the kernel function from the 'ggplot2' package is the ggplot() function, not ggplot2(). The basic syntax of a 'ggplot2' call is as follows:

```
ggplot(data, aes(x = …, y = ...)) +
  geom_...() +
  scale_...() +
  theme_...()
```

Here, 'data' is the data frame or data source to be plotted, aes() specifies the aesthetics (e.g. x and y axis) of the plot, and geom_*() specifies the geometric object (e.g. point, line, and bar) to be plotted. scale_*() and theme_*() functions are used to customise the scales and appearance of the plot, respectively.

Notice that each function is cumulatively called using the '+' sign.

[8] For more information, consult https://ggplot2.tidyverse.org/

One of the key strengths of 'ggplot2' is its ability to create complex visualisations by layering different geometric objects and aesthetic definitions on top of each other using the + sign. For instance, to create a scatter plot of two variables with a line of best fit and error bars, we start with the ggplot() function, specifying our dataset and the aes() function to map our 'x' and 'y' variables (variable1 and variable2). Next, we add a geom_point() layer for the scatter plot, followed by geom_smooth(method = "lm") to add a trend line based on a linear model ("lm"). Lastly, we include geom_errorbar() with aes(ymin = variable2 – sd, ymax = variable2 + sd) to add error bars that represent the variability around variable2, defined by a standard deviation (sd) value, and set a specific width for these bars.

```
ggplot(data, aes(x = variable1, y = variable2)) +
  geom_point() +
  geom_smooth(method = "lm") +
  geom_errorbar(aes(ymin = variable2 - sd, ymax = variable2 +
sd), width = 0.2)
```

Note that in this illustrative example, 'sd' represents the standard deviation and must be calculated based on your data.

Another strength of 'ggplot2' is the ability to customise the appearance of the plot using themes and scales. For example, we can change the colour scheme, axis labels, and font size of the plot as follows:

```
ggplot(data, aes(x = variable1, y = variable2)) +
  geom_point() +
  geom_smooth(method = "lm") +
  geom_errorbar(aes(ymin = variable2 - sd, ymax = variable2 +
sd), width = 0.2) +
  scale_color_brewer(palette = "Blues") +
  labs(x = "X Axis Label", y = "Y Axis Label", title = "Plot
Title") +
  theme_bw() +
  theme(text = element_text(size = 12))
```

In this case, we use the scale_color_brewer() function to change the colour scheme of the plot and the labs() function to add axis labels and a title. Also, the theme_bw() function is called to change the background of the plot to white and the theme() function to set the font size to 12 points.

After creating a plot, 'ggplot2' provides the specific function – ggsave() – to write the result on disk. The code below is an example of saving a plot using 'ggplot2'.

```
# Create a scatterplot of random data using ggplot2
my_plot = ggplot(data.frame(x = rnorm(50), y = rnorm(50)),
aes(x, y)) +
  geom_point() + labs(title = "My Scatterplot", x = "X", y = "Y")

# Save the plot to a PDF file
ggsave("the_plot.pdf", my_plot, width = 8, height = 6, units =
"in", dpi = 300)
```

In this example, we then create a scatter plot using 'ggplot2' by specifying the data frame, aesthetics, and geometric elements using the ggplot(), aes(), and geom_point() functions. We also add a title and axis labels using the labs() function. In this case, the plot is not presented on the screen but stored in a variable. The plot is stored in the 'my_plot' variable and can be visualised or further twitched.

Finally, we use the ggsave() function to save the plot to a PDF file named 'the_plot' with a width of 8 inches, a height of 6 inches, and a resolution of 300 DPI. We specify the output filename as the first argument and the plot object as the second argument. Additional arguments can be used to control the output format, size, and resolution of the plot.

Broadly, the grammar of graphics allows users to layer different geometric objects, customise scales and themes, and create complex visualisations with a few lines of code.

Figure 11.2 represents the concept behind creating a plot with 'ggplot2'.

11.4.2 A First on 'leaflet'

The 'leaflet' package is another way for visualising spatial data, this one specific for creating interactive maps. The package provides a simple and intuitive interface for creating maps and allows access to a variety of different base maps, markers, and popups that can be customised to meet the needs of the project.

As for the 'ggplot2' package, the data must be in a format that the package can understand, such as a 'SpatialPoints' or 'SpatialPolygons'. Once the data is in the correct format, the user can create a map using the leaflet() function, followed by a series of add*() functions to specify the type of map elements to be added. In the current versions of 'leaflet' to R, it is preferable to use data in the 'sp' package format, i.e. as Spatial* objects.

The basic syntax of a 'leaflet' call is as follows:

```
leaflet() %>%
  addTiles() %>%
  addMarkers(lng = ..., lat = ..., popup = ...)
```

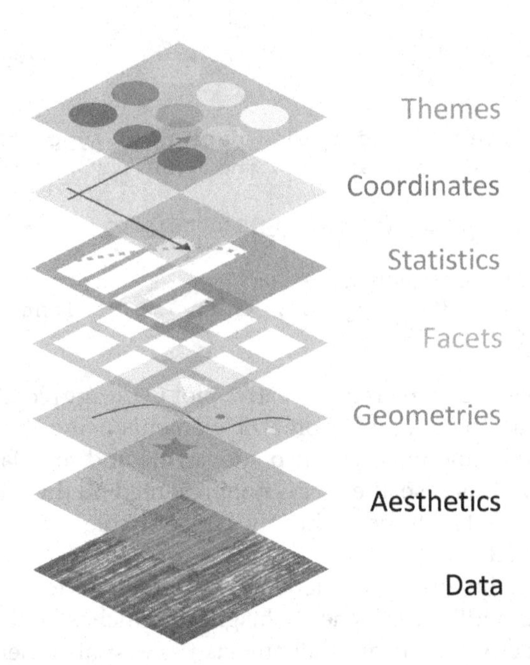

FIGURE 11.2
The concept of grammar of graphics underlying the 'ggplot2'.

In this non-functional example, leaflet() creates a new map object, addTiles() function adds the base map (i.e. the background), and addMarkers() function adds markers to the map at the specified longitude and latitude coordinates ('lng' and 'lat' parameters). The popup argument specifies the content of the popup window that appears when the user clicks on the marker.

Notice that unlike in the 'ggplot2' case where the '+' sign is used to superimpose layers, in the 'leaflet' case, the pipe ('%>%') operation is used.

One of the key strengths of 'leaflet' is the ability to customise the appearance of the map using a wide variety of options. For example, we can change the tile layer, add custom markers, and control the zoom level and centre of the map as follows:

```
leaflet() %>%
  addProviderTiles(providers$OpenStreetMap.Mapnik) %>%
  addMarkers(lng = ..., lat = ..., icon = customIcon, popup =
...) %>%
  setView(lng = ..., lat = ..., zoom = ...)
```

In this example, we use the addProviderTiles() function to change the tile layer to OSM and the addMarkers() function to add custom markers with a specified icon. We use the setView() function to set the centre and zoom level of the map.

Another strength of 'leaflet' is the ability to add layers and controls to the map. For example, we can add a layer control to allow users to toggle different layers on and off, and we can add a search bar to allow users to search for specific locations:

```
leaflet() %>%
  addProviderTiles(providers$OpenStreetMap.Mapnik) %>%
  addMarkers(lng = ..., lat = ..., popup = ...) %>%
  addLayersControl(overlayGroups = c("Group 1", "Group 2")) %>%
  addSearchOSM(options = providerOSM())
```

In this case, we use the addLayersControl() function to add a layer control with two overlay groups, and we use the addSearchOSM() function to add a search bar that uses the OSM Nominatim Geocoding Service.

'leaflet' is an excellent choice for creating web-based maps and for integrating maps into other R-based data analysis workflows.

11.4.3 Plotting Spatial Points

The following code uses the ggplot() function to exemplify how to display the points from the 'meuse' dataset. When the data from 'meuse' is loaded, it corresponds to a 'data.frame' object. In this case, it is not necessary to convert it to a spatial object. By defining the aesthetics with the aes() function, the data is correctly plotted.

```
#11-08

# Load the packages
library(sp)
library(ggplot2)

# Load the meuse dataset
data(meuse)

# Plot the SpatialPointsDataFrame using ggplot2
ggplot(data = meuse, aes(x = x, y = y)) +
  geom_point() +
  coord_equal() +
  ggtitle("Meuse Dataset")
```

The resulting plot is shown in Figure 11.3.

With 'leaflet', the results are more impressive as the data is combined with a background map. The code starts by transforming the 'meuse' data frame

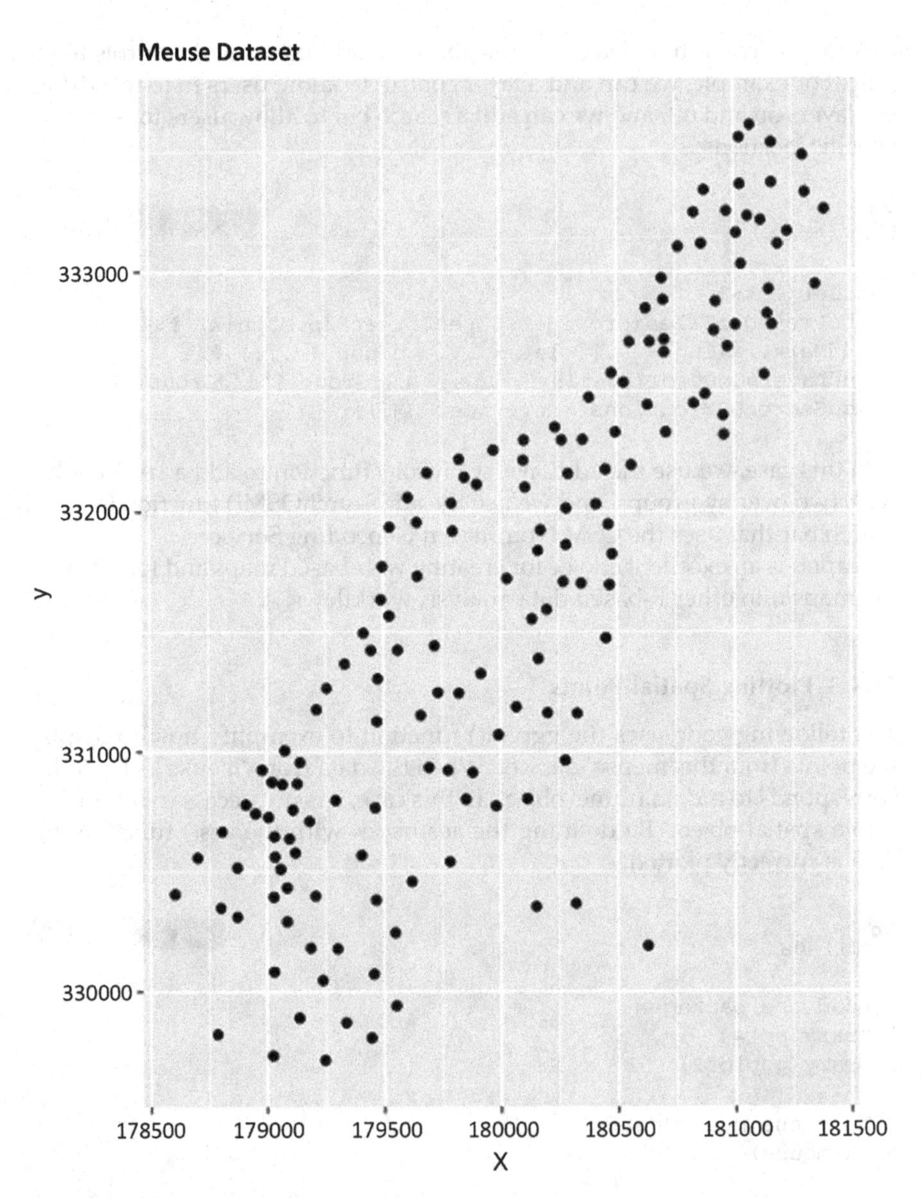

FIGURE 11.3
The plot of the 'meuse' dataset.

to an 'SpatialPointsDataFrame' that is an 'sp' object, by calling the coordinates() function. The CRS is set using the CRS() function.

For plotting with leaflet, it is necessary to transform the CRS of the 'meuse' data to the WGS84 system (i.e. EPSG: 4326); this is done by calling the spTransform() function. After this, the reprojected data is copied to the 'meuse_map' variable. Finally, the leaflet() function is used to draw the map.

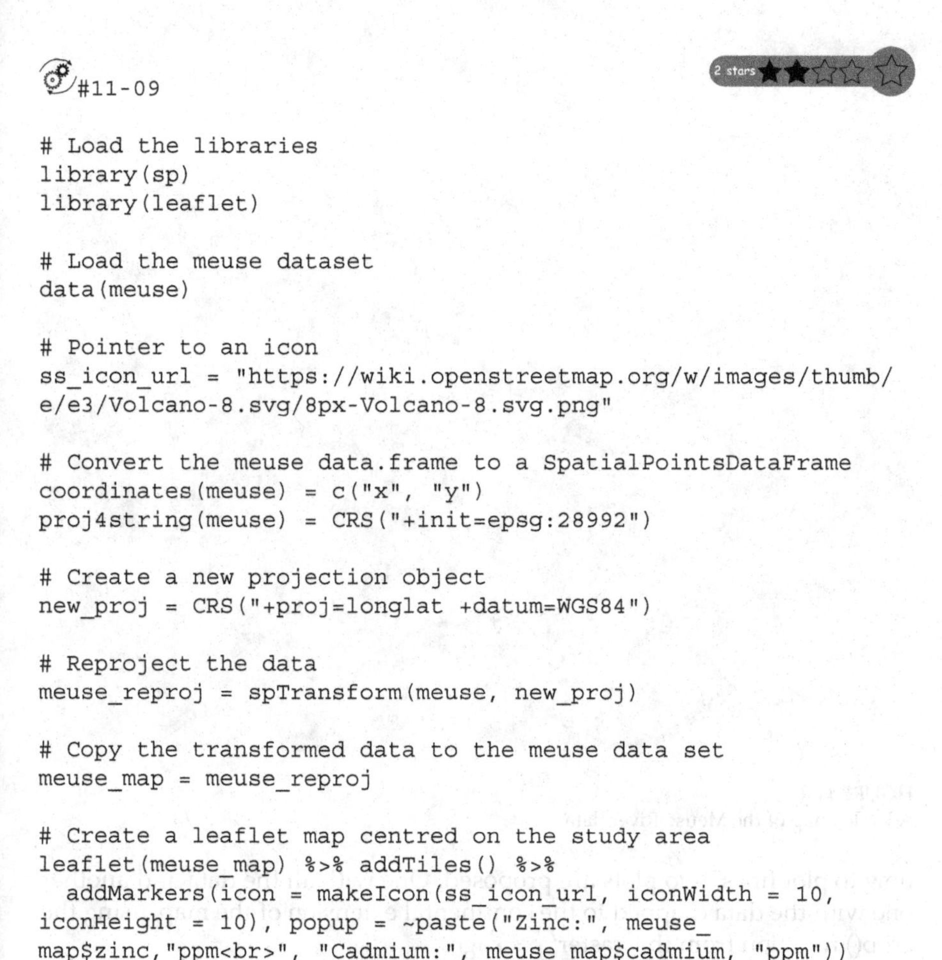


```
#11-09

# Load the libraries
library(sp)
library(leaflet)

# Load the meuse dataset
data(meuse)

# Pointer to an icon
ss_icon_url = "https://wiki.openstreetmap.org/w/images/thumb/
e/e3/Volcano-8.svg/8px-Volcano-8.svg.png"

# Convert the meuse data.frame to a SpatialPointsDataFrame
coordinates(meuse) = c("x", "y")
proj4string(meuse) = CRS("+init=epsg:28992")

# Create a new projection object
new_proj = CRS("+proj=longlat +datum=WGS84")

# Reproject the data
meuse_reproj = spTransform(meuse, new_proj)

# Copy the transformed data to the meuse data set
meuse_map = meuse_reproj

# Create a leaflet map centred on the study area
leaflet(meuse_map) %>% addTiles() %>%
  addMarkers(icon = makeIcon(ss_icon_url, iconWidth = 10,
iconHeight = 10), popup = ~paste("Zinc:", meuse_
map$zinc,"ppm<br>", "Cadmium:", meuse_map$cadmium, "ppm"))
```

The visualisation, showcased in Figure 11.4, incorporates interactive markers for each data point on the map. When a user clicks on any of these markers, a popup label appears, providing detailed information about the specific location. Specifically, the popup displays the concentrations of 'zinc' and 'cadmium' at that point, presented in parts per million (ppm). These markers are visually represented by custom icons, which are generated through the addMarkers() function using the makeIcon() function. The icon itself is sourced from an external URL pointing to a 'png' image hosted on the OSM wiki, depicted as a triangle. This approach not only enhances the map's interactivity but also allows for the inclusion of detailed environmental data in a user-friendly manner.

11.4.4 Plotting Spatial Lines

The following example is an adaptation of a previous example (see Section 3.2 in Chapter 3) where we downloaded the Portuguese roads. For exemplifying

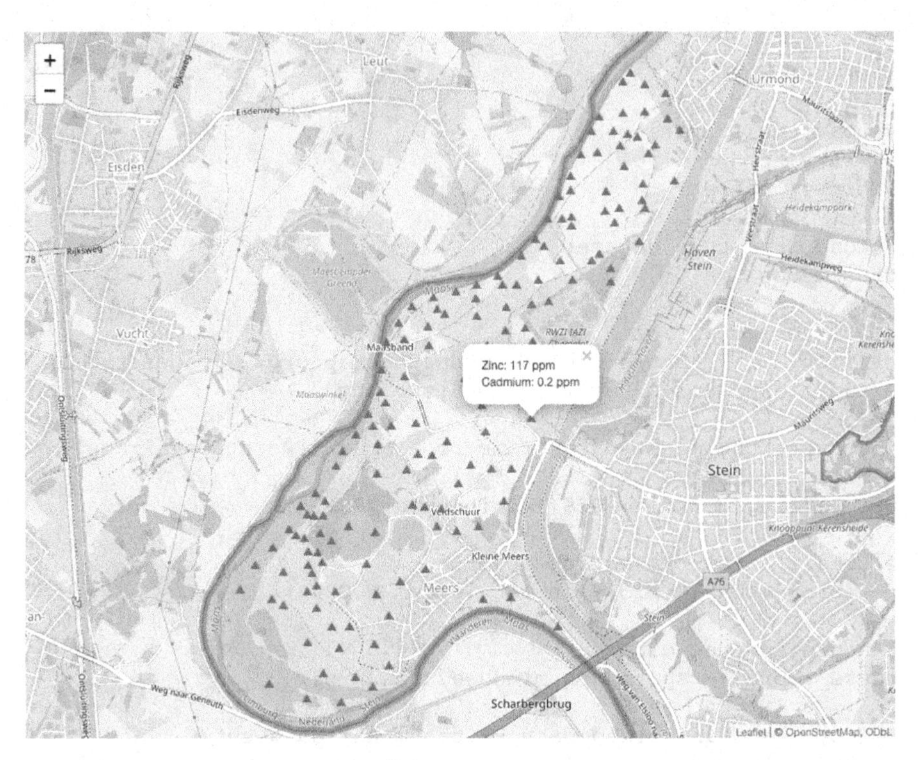

FIGURE 11.4
A leaflet map of the Meuse River data.

how to plot lines, two plots are proposed. One with all the data and another one with the data cropped to the continental extension of the map, using the crop() function from the 'raster' package.

In this example, the shapefile() function from the 'raster' package is used to read the data as a 'SpatialLinesDataFrame' object.

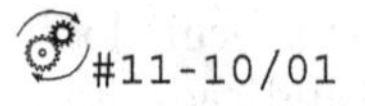

```
#11-10/01

# Load the libraries
library(raster)
library(sp)

# Define the working folder
setwd("~/Spatial/Data/")

# Define the link to the data
url ="http://github.com/pnogas67/spatialData/blob/main/Roads_Pt.
zip?raw=TRUE"

# Download the data
download.file(url, destfile = "Roads_Pt.zip")
```

```
# Unzip the file
unzip(zipfile = "Roads_Pt.zip", exdir = 'Roads Pt')

# Read the shapefile
roads_pt = shapefile("Roads Pt/PRT_roads.shp")
```

After downloading and importing the data, the next step is to plot the data with the ggplot() function accompanied by the geom_line() function.

#11-10/02 2 stars

```
# Improved plot with ggplot
ggplot(data = roads_pt, aes(x = long, y = lat, group=group)) +
  geom_line() +
  coord_fixed() +
  labs(title = "Portugal Roads",
       subtitle = "Continent and Islands")
```

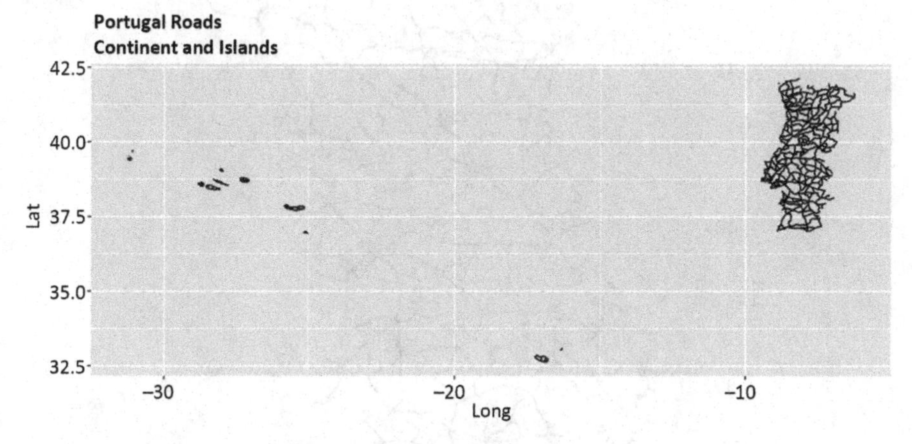

FIGURE 11.5
Plot of the Roads of Portugal, including Atlantic Islands.

Figure 11.5 shows the obtained image.

The use of the crop() function from the 'raster' package (see Section 12.3.1 in Chapter 12) allows the extraction of part of the data based on a rectangular area defined by the extent() function, also from the 'raster' package.

#11-11

```
# Crop the data
roads_cont = crop(roads_pt, extent(-10,-6,36.25,42.5))

# Improved plot with ggplot
```

```
ggplot(data = roads_cont, aes(x = long, y = lat, group=group)) +
  geom_line() + coord_fixed() +
  labs(title = "Portugal Mainland Roads", subtitle = "Continent
only")
```

Figure 11.6 displays the results obtained.

Adapting this code snippet for creating an interactive map with 'leaflet' is straightforward.

The code is as follows:

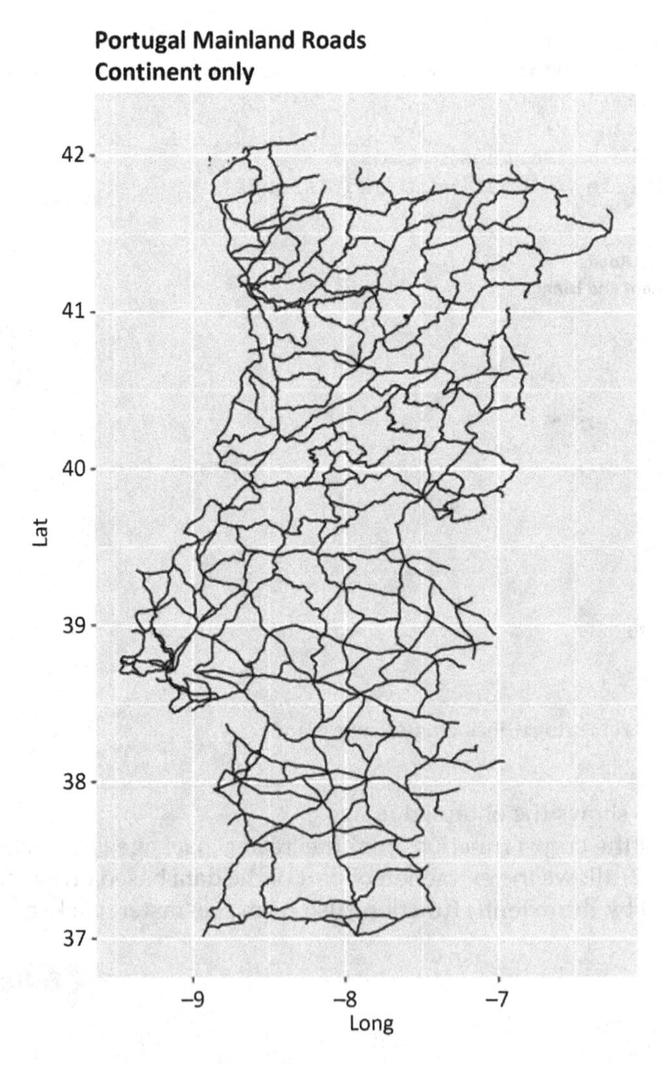

FIGURE 11.6
Plot of the roads of mainland Portugal.

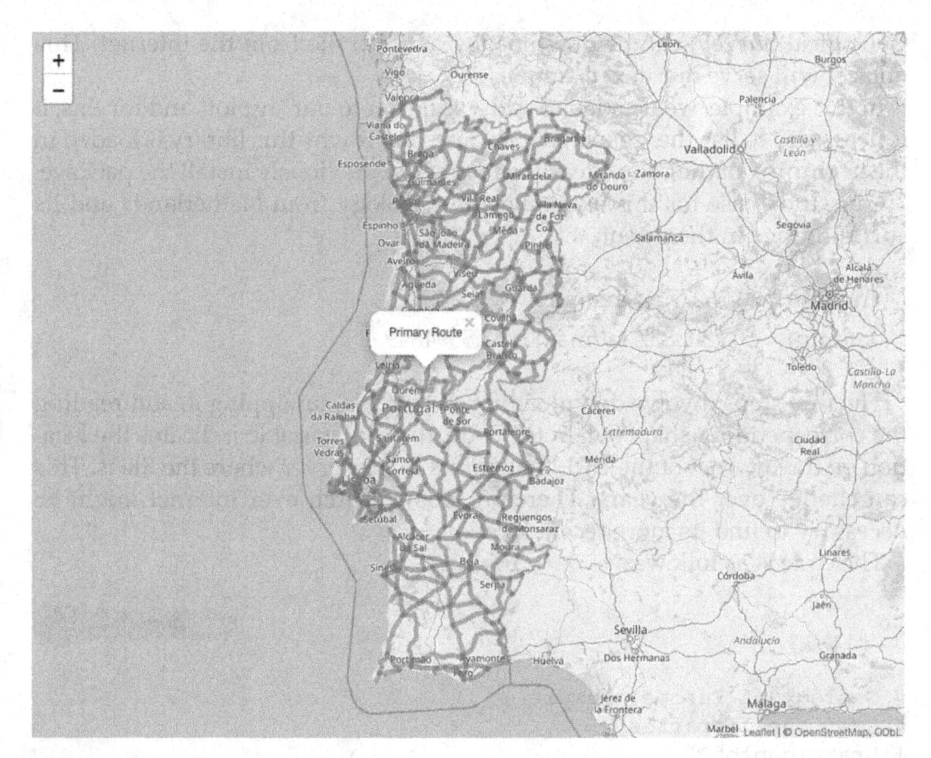

FIGURE 11.7
Map of Portuguese mainland roads using the 'leaflet' package.

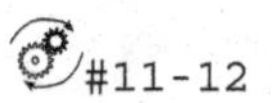#11-12

```
# Create the roads leaflet map
leaflet(roads_cont) %>%
  addTiles() %>%
  addPolylines(color = "red", popup = ~paste(roads_cont$RTT_
DESCRI))
```

The map created is shown in Figure 11.7.

11.4.5 Plotting Spatial Polygons

Polygons as lines and points can also be represented using the 'ggplot2' package.

For this example, we will use another dataset that is freely available online. The geological map of Europe that is published by the USGS[9] (United States

[9] https://certmapper.cr.usgs.gov/data/apps/world-maps/

Geological Survey) can be downloaded as a zip file from the internet. This dataset will serve our next example.

In this example, we decided to add a scale bar to our 'ggplot', and for this, it is necessary to use the 'ggspatial' package; that is why this library is loaded in the beginning of the code. Please remember to previously install the package.

Our study case for this example is the geology from Netherlands and its surroundings in the extent:

lng(min) = 3.3627721, lng(max) = 7.2272718
lat(min) = 50.7503837, lat(max) = 53.5594849

The first steps involve downloading the zip file, unzipping it, and reading the corresponding shapefile. In the code, the 'url' variable indicates the location on the internet of the URL that points to the server where the file is. This can change over the years. Therefore, some search over internet might be necessary to find its more recent location.

The code is as follows:

#11-13/01 2 stars

```
# Geology of Europe example
# Read the libraries
library(ggplot2)
library(sp)
library(sf)
library(ggspatial)
library(raster)

# Define the url
url = "https://www.sciencebase.gov/catalog/file/get/60abc880d3
4ea221ce51e621?facet=geo4_21"

# Download the file
europe_geo = download.file(url, destfile = "Europe geology.zip" )

# Unzip the file
unzip(zipfile = "Europe geology.zip", exdir = 'Europe Geology')

# Read the Shapefile
geo = shapefile("Europe Geology/geo4_21.shp")
```

The plot of all the data (more than 58 Mbytes) is shown in Figure 11.8. Notice that it is not recommended trying to plot it if your computer has limited memory and disk resources.

However, this particular plot doesn't serve our needs effectively. Instead, let's construct a more detailed map focused on the geology of the Netherlands region for our example.

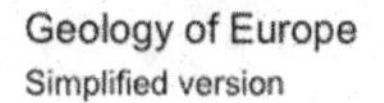

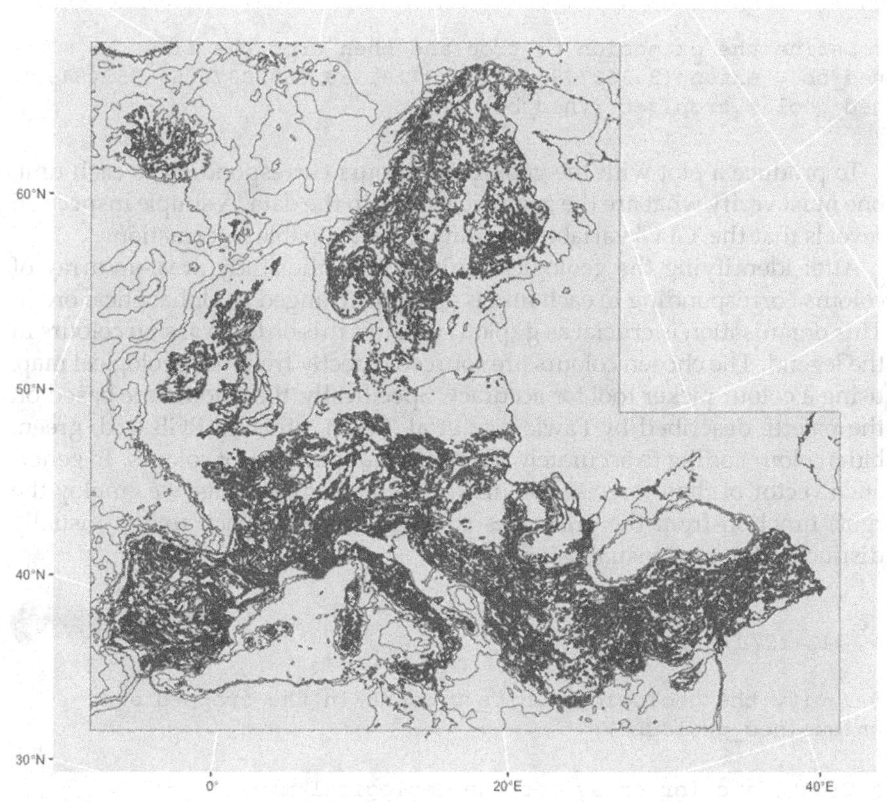

FIGURE 11.8
Plot of the shapefile with the geology of Europe.

To focus our analysis on the geology of the study region, we must first ensure that the 'geo' variable is in the WGS84 CRS, also known as 'EPSG:4326'. It's important to understand that spatial operations, such as cropping, require all involved spatial objects to share the same CRS (see Section 12.1.1 in Chapter 12). To achieve this, we create a text variable specifying the desired projection system and then use the spTransform() function to convert the 'geo' variable to this CRS. Following the transformation, we employ the crop() function (see Section 12.3.1 in Chapter 12) in conjunction with the previously introduced extent() function to precisely define and crop the area of interest.

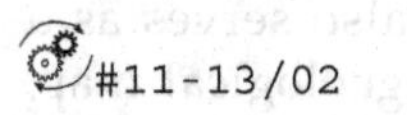

```
# Create a new projection object
wgs84 = CRS("+proj=longlat +datum=WGS84")
```

```
# Reproject the data
geo_r = spTransform(geo, wgs84)

# Define the extension to crop and then crop the file
ned_bb = extent(3.3627721, 7.2272718, 50.7503837, 53.5594849)
ned_geol = crop(geo_r,ned_bb)
```

To produce a plot with the geological colours corresponding to each unit, one must verify what are the geological units in the data. A simple inspection reveals that the 'GLG' variable contains this invaluable information.

After identifying the geological units within the study area, an array of colours corresponding to each unit is created, arranged in alphabetical order. This organisation is crucial as ggplot() will use this order to assign colours in the legend. The chosen colours are sourced directly from the geological map, using a colour picker tool for accuracy. Specifically, the colours are based on the palette described by Pawlewicz et al. (2002), utilising RGB (red, green, blue) colour coding to accurately represent the spectrum of colours. To generate a vector of these precise colours for the geological units, we employ the rgb() function from the grDevices package, ensuring each unit is visually distinct in the map visualisation.

#11-13/03

```
# Verify the geological units present in the cropped area
unique(ned_geol$GLG)

# Create a color array for the Geological map
color_array = c(
            rgb(165,172,195, maxColorValue = 255),
            rgb(251,248,167, maxColorValue = 255),
            rgb(140,138,183, maxColorValue = 255),
            rgb(241,246,250, maxColorValue = 255),
            rgb(210,223,177, maxColorValue = 255),
            rgb(246,237,159, maxColorValue = 255),
            rgb(236,226,199, maxColorValue = 255),
            rgb(188,173,193, maxColorValue = 255),
            rgb(252,250,213, maxColorValue = 255),
            rgb(235,242,249, maxColorValue = 255),
            rgb(164,199,212, maxColorValue = 255),
            rgb(227,201,73, maxColorValue = 255))
```

The result from the crop produced 12 geological units listed in the results. This code snippet not only assigns colours but also serves as a guide for replicating the visual style of the referenced geological map, enhancing the interpretability and aesthetic appeal of the resulting map visualisation.

```
[1] Seal Q  H2O  K     Tr   C      Pg     D      N     Cm    Pzm  Tv
55 Levels: C CD Cm Cz CzMzi CzMzv Czv D DS H2O i ice J JTr K
KJ m Mz Mzi Mzm MzPz MzPzm Mzv N O OCm oth P ... unk
```

The 'ned_geol' is a 'SpatialPolygonsDataFrame' object, and for plotting, the cropped data needs to be converted to 'sf' object with the function as() from the 'sf' package.

The function to create the scale bar is annotation_scale() from the 'ggspatial' package.

#11-13/04

```
# Convert sp to sf
ned_geol_sf = as(ned_geol,"sf")

# The plot is made with the geom_sf() function.
ggplot(data = ned_geol_sf) +
 geom_sf(aes(fill = ned_geol_sf$GLG)) +
 scale_fill_manual(values=color_array, name="Geological units") +
 annotation_scale(location = "bl", width_hint = 0.25, text_cex
= 0.75) +
 labs(title = "Geology of the Netherlands")
```

The resulting map is shown in Figure 11.9.

This is a more usable plot of a geological map. And yes, the map has a scale bar at the bottom left.

The same example can as well be plotted using leaflet() function. In this case, the functions addPolygons(), addLegend(), and addScaleBar() are used.

#11-13/05

```
# Create a leaflet map
leaflet() %>%
  addTiles() %>%
  addPolygons(data = ned_geol_sf,
        fillColor = ~color_array[as.integer(factor(ned_geol_
sf$GLG))],
        color = "black", weight = 1, opacity = 1, fillOpacity
= .7,
        group = "Geological Units") %>%
  # Add a legend to the map
  addLegend(position = "bottomright", title = "Geological Units",
        colors = color_array, labels = levels(factor(ned_geol_
sf$GLG)), opacity = 0.7) %>%
  # Add a scale bar to the map
  addScaleBar(position = "bottomleft",
         options = scaleBarOptions(imperial = FALSE))
```

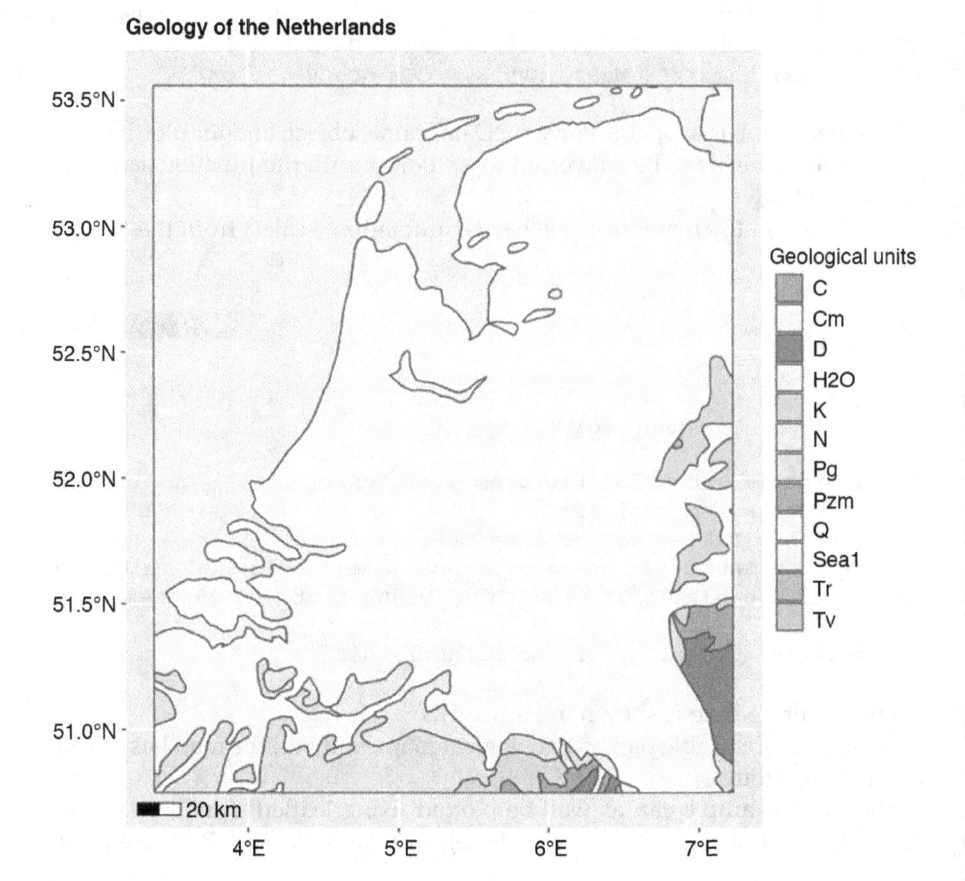

FIGURE 11.9
The simplified geology of the Netherlands and surroundings.

The resulting plot is shown in Figure 11.10.

11.5 Concluding Remarks

Yes, after ten chapters, we created the first maps in a visually appealing format!

Understanding and using vectorial data is a pivotal aspect of spatial data analysis. Foremost, as for any type of analysis, our most wrecking nerves and time-consuming tasks involve data pre-processing, including cleaning, transforming, and subsetting (see Chapter 12). In this chapter, we skim over these subjects. In the next ones, we will delve deeper into the topics.

The project OSM is the Wikipedia for mappers as it provides free, fast, and curated cartographic information from all over the world, often with

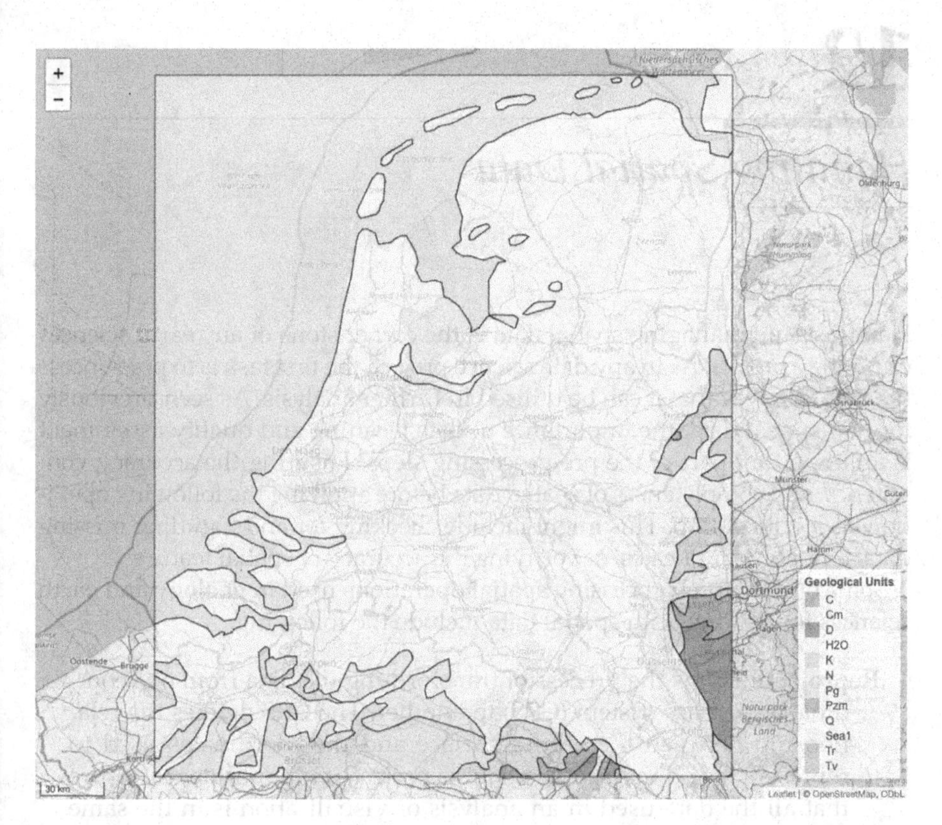

FIGURE 11.10
The simplified geology of the Netherlands and surroundings plotted using the 'leaflet' package.

precious detail and invaluable metadata. Using its information, one can retrieve the opening hours of the local coffee shop or identify the existing pharmacies in Manhattan.

As an appetiser for visualising spatial data, the 'ggplot' and 'leaflet' packages are explained remembering that its uses are in essence distinct, one creating static maps and the other interactive, internet-ready maps.

References

Pawlewicz, M. J., Steinshouer, D. W., & Gautier, D. L. (2002). Map showing geology, oil and gas fields, and geologic provinces of Europe including Turkey: U.S. Geological Survey Open-File Report 97-470-I, p. 14. https://doi.org/10.3133/ofr97470I

Wilkinson, L., Wills, D., Rope, D., Norton, A., & Dubbs, R. (2005). The Grammar of Graphics (2nd ed.). Springer, Chicago, IL.

12

Handling Spatial Data

Handling and managing spatial data is the corner stone of any earth sciences or geology project. As in any data science project, the first task is to pre-process the data in a way that it can be utilised in further analysis. As seen previously for non-spatial data, the importance of data cleaning and quality assessment is an important part of the pre-processing steps. Ensuring the accuracy, consistency, and completeness of spatial data before applying the following operations is a critical step. This might include checking for and handling missing data, removing duplicates, or verifying the accuracy of spatial features.

Some of these pre-processing spatial operations used in geology and earth sciences for dealing with spatial data include the following:

Reprojection: It is the process of transforming the data from one coordinate reference system (CRS) to another. The CRS defines the relationship between the earth's surface and the coordinates used to represent it, and in a spatial analysis project, it is important to ensure that all the data used in an analysis or visualisation is in the same CRS. The 'sp' and 'sf' packages have the core functions for reprojection such as spTransform() and st_transform(), respectively.

Subsetting: Cropping and clipping are the processes of removing features or parts of features that fall outside a defined area. This is useful when one wants to focus the work on a portion of the data. For example, one may want to select only the stream sediment samples that are inside a defined geological unit. Among the functions for achieving this type of tasks, there is the st_intersect() function, from the 'sf' package, that can be used for clipping spatial data based on another geometric object.

Buffering: It is the process of creating a feature at a constant distance around a feature or set of features. For example, a buffer can be created around a soil sample point to identify the area of influence within a certain distance of that point. For the "sf' package, the buffering operation is obtained using the st_buffer() function.

Querying: Spatial queries are operations that enable analysis and extraction of information from spatial data. Its use includes functions for topological relationships, spatial joins, spatial overlays, point-in-polygon tests, or merging. One example is to identify, in a geological map, what geological units are crosscut by a fault.

 DOI: 10.1201/9781032651880-14

The choice of the necessary operation for spatial analysis will depend on the data and the desired outcome.

Mathematically, the functions used to manipulate spatial objects are referred to as topological functions, and there is a plethora of such functions available in the 'sf' and 'sp' packages (Pebesma & Bivand, 2023).

Table 12.1 provides a list of these topological functions along with a simplified description.

TABLE 12.1

Some of the Most Important Topological Functions in the 'sf' Package

Function	Description
Logical	
st_contains	Tests if one geometry contains another.
st_touches	Tests if two geometries touch.
st_crosses	Tests if two geometries cross.
st_overlaps	Tests if two geometries overlap.
st_within	Tests if one geometry is within another.
st_intersects	Tests if two geometries intersect.
Numeric	
st_centroid	Computes the centroid of a geometry.
st_area	Computes the area of a geometry.
st_length	Computes the length of a geometry.
st_distance	Computes the distance between two geometries.
Geometric	
st_sym_difference	Computes the symmetric difference of two geometries.
st_difference	Computes the difference of two geometries.
st_intersection	Computes the intersection of two geometries.
st_union	Computes the union of a set of geometries.

Another spatial data operation that is important is the conversion between R specific spatial data types. For this, it is possible to use the as() function to convert between 'sp' and 'sf' objects. Here are some generic examples that illustrate how to use the as() function for these operations:

To convert an 'sp' package object to an 'sf' object, one can use the following code.

```
#12-01/01

# Load the libraries
library(sp)
library(sf)

# Create an 'sp' package object
coords = matrix(c(1, 2, 3, 4), ncol = 2)
sp_points = SpatialPoints(coords)
```

```
# Convert to 'sf' object
sf_points = as(sp_points, "sf")
```

In this example, we first create an 'sp' package object called 'sp_points' that consists of a matrix of coordinates. We then use the as() function to convert 'sp_points' to an 'sf' object called 'sf_points'.

For the inverse operation, to convert an 'sf' object to an 'sp' package object, the as() function can as well be used.

In this example, we first create an 'sf' object called 'sf_points' that consists of a data frame with two columns representing the x and y coordinates. We then use the as() function to convert 'sf_points' to an 'sp' object called 'sp_points'.

#12-01/02

```
# Create an 'sf' object
coords = matrix(c(1, 2, 3, 4), ncol = 2)
sf_points = st_as_sf(data.frame(x = coords[, 1], y = coords[,
2]), coords = c("x", "y"))

# Convert to 'sp' package object
sp_points = as(sf_points, "Spatial")
```

To verify the 'sp_points' and 'sf_points variables structure, we use the following code:

#12-01/03

```
# Verify the sp_points
str(sp_points)

# Verify the sf_points
str(sf_points)
```

The output is as follows:

```
Formal class 'SpatialPointsDataFrame' [package "sp"] with 5 slots
  ..@ data        :'data.frame':    2 obs. of  0 variables
  ..@ coords.nrs  : num(0)
  ..@ coords      : num [1:2, 1:2] 1 2 3 4
  .. ..- attr(*, "dimnames")=List of 2
  .. .. ..$ : NULL
  .. .. ..$ : chr [1:2] "coords.x1" "coords.x2"
  ..@ bbox        : num [1:2, 1:2] 1 3 2 4
  .. ..- attr(*, "dimnames")=List of 2
  .. .. ..$ : chr [1:2] "coords.x1" "coords.x2"
```

```
.. .. ..$ : chr [1:2] "min" "max"
..@ proj4string:Formal class 'CRS' [package "sp"] with 1 slot
.. .. ..@ projargs: chr NA

Classes 'sf' and 'data.frame':    2 obs. of  1 variable:
 $ geometry:sfc_POINT of length 2; first list element:  'XY'
num  1 3
 - attr(*, "sf_column")= chr "geometry"
 - attr(*, "agr")= Factor w/ 3 levels "constant","aggregate",..:
  ..- attr(*, "names")= chr(0)
```

The as_Spatial() function is an alias for the more generic as(object, "Spatial") function. Note that in the as() function for 'sf' to 'sp' conversion, we need to specify the class of the desired output object as "Spatial". Similarly, in the as() function for 'sp' to 'sf' conversion, we need to specify the class of the desired output object as "sf".

By this moment, it is clear that both 'sp' and 'sf' packages share many functionalities. In many cases, it is indifferent which package one is using, but remember that the 'sp' package is ageing, and at some point, it might be discontinued. For a matter of precaution, it is advisable to use, as much as possible, the functions and objects from the 'sf' package.[1]

12.1 Geometries and Projection

The possibility to switch between different coordinate systems of a dataset has already been explained previously (Chapter 11) as a necessity to illustrate an example. It is common in geological projects to work with data from different origins, hence with different coordinate systems; hereafter, these transformations are explained with more detail as handling these transformations is of paramount importance.

Before performing a reprojection, one must be assured that the objects are in the right type, i.e. 'Spatial*' or 'sf' formats, that they have a CRS and that they have a geometry.

12.1.1 Defining Coordinate Reference Systems

The proj4string() is a function from the 'sp' package that sets or retrieves the CRS of a 'sp' object. For example, proj4string(meuse_sp) would return the CRS of the 'meuse_sp' object. For the 'sf' package, the st_crs() function is

[1] Disclaimer: I love the 'sp' package and use it as often as I can. But I am also ageing.

used in the same way but returns the crs in WKT-2 format[2] that is a standard format for representing CRSs.

The CRS() function from the 'sp' package is used to create a CRS object based on a Proj.4 string or an European Petroleum Survey Group (EPSG) code. An example of a Proj.4 string for the WGS84 system is "+proj=longlat +datum=WGS84 +no_defs".

The st_crs() function from the 'sf' package is used to define and transform spatial coordinates in a standardised way.

These formats are used in geographic information system (GIS) and other spatial analysis software and applications.

It's worth mentioning that the proj4string() function, while still functional, is being gradually superseded by newer standards and functions that align with the PROJ library (version 6 and above) and its handling of CRS definitions, particularly in the 'sf 'package. Remember to stay updated with the latest practices and documentation of these packages.

12.1.2 Defining the Geometry

The CRS in the 'sp' and 'sf' packages has different ways of defining the geometry, i.e. the coordinates of the 'Spatial*' and 'sf' objects.

The coordinates() function from the 'sp' package is used to retrieve and assign the coordinates to an 'sp' object. In this way, a data frame with two columns that correspond to coordinates (e.g. x and y) can be transformed into a 'SpatialPointsDataFrame' object.

Let's look at the following example. Create from the 'meuse' dataset a data frame named 'meuse_df' that contains the 'meuse' data. If we inspect the data type and structure of 'meuse_df', we verify that originally is a data frame and that there are two variables x and y that should contain the coordinates.

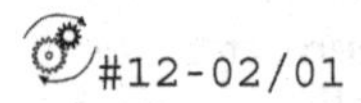

```
# Use meuse dataset
data(meuse)
```

[2] WKT-2 stands for Well-Known Text, version 2, and is a text-based format that is easy to read and write. It is also machine-readable, which makes it easy for software to parse and understand the information. A WKT-2 string is made up of a number of elements, including the following:

Type of CRS: It is a string that identifies the type of CRS. For example, the type of the EPSG:28992 CRS is "PROJCS".

Parameters of the CRS: These are a set of numbers that specify the parameters of the CRS. They are equivalent to the PROJ.4 string. For example, the parameters of the EPSG:28992 CRS are "+proj=utm +zone=28 +datum=WGS84 +units=m +no_defs".

Datum of the CRS: It is a reference point for the CRS. For example, the datum of the EPSG:28992 CRS is WGS84.

Units of the CRS: The units of measurement used by the CRS. For example, the units of the EPSG:28992 CRS are metres.

```
# Create data frame
meuse_df = meuse

# Verify structure of the dataset
str(meuse_df)
```

The output is as follows:

```
'data.frame':    155 obs. of  14 variables:
 $ x        : num  181072 181025 181165 181298 181307 ...
 $ y        : num  333611 333558 333537 333484 333330 ...
 $ cadmium: num  11.7 8.6 6.5 2.6 2.8 3 3.2 2.8 2.4 1.6 ...
 $ copper : num  85 81 68 81 48 61 31 29 37 24 ...
 $ lead   : num  299 277 199 116 117 137 132 150 133 80 ...
 $ zinc   : num  1022 1141 640 257 269 ...
 $ elev   : num  7.91 6.98 7.8 7.66 7.48 ...
 $ dist   : num  0.00136 0.01222 0.10303 0.19009 0.27709 ...
 $ om       : num  13.6 14 13 8 8.7 7.8 9.2 9.5 10.6 6.3 ...
 $ ffreq  : Factor w/ 3 levels "1","2","3": 1 1 1 1 1 1 1 1 1
1 ...
 $ soil   : Factor w/ 3 levels "1","2","3": 1 1 1 2 2 2 2 1 1
2 ...
 $ lime   : Factor w/ 2 levels "0","1": 2 2 2 1 1 1 1 1 1 1 ...
 $ landuse: Factor w/ 15 levels "Aa","Ab","Ag",..: 4 4 4 11 4
11 4 2 2 15 ...
 $ dist.m : num  50 30 150 270 380 470 240 120 240 420 ...
```

To transform the data frame to a 'SpatialPointsDataFrame', we simply define the coordinates to the 'meuse_df' object.

#12-02/02 

```
# Create a new sp variable
meuse_sp = meuse_df

# Assign the coordinates
coordinates(meuse_sp) = c("x","y")

# Verify the new data type
str(meuse_sp)
```

The output is as follows:

```
Formal class 'SpatialPointsDataFrame' [package "sp"] with 5 slots
  ..@ data     :'data.frame':     155 obs. of 12 variables:
  .. ..$ cadmium: num [1:155] 11.7 8.6 6.5 2.6 2.8 3 3.2 2.8
2.4 1.6 ...
```

```
  .. ..$ copper : num [1:155] 85 81 68 81 48 61 31 29 37 24 ...
  .. ..$ lead   : num [1:155] 299 277 199 116 117 137 132 150
133 80 ...
  .. ..$ zinc   : num [1:155] 1022 1141 640 257 269 ...
  .. ..$ elev   : num [1:155] 7.91 6.98 7.8 7.66 7.48 ...
  .. ..$ dist   : num [1:155] 0.00136 0.01222 0.10303 0.19009
0.27709 ...
  .. ..$ om    : num [1:155] 13.6 14 13 8 8.7 7.8 9.2 9.5 10.6
6.3 ...
  .. ..$ ffreq  : Factor w/ 3 levels "1","2","3": 1 1 1 1 1 1
1 1 1 1 ...
  .. ..$ soil   : Factor w/ 3 levels "1","2","3": 1 1 1 2 2 2
2 1 1 2 ...
  .. ..$ lime   : Factor w/ 2 levels "0","1": 2 2 2 1 1 1 1 1
1 1 ...
  .. ..$ landuse: Factor w/ 15 levels "Aa","Ab","Ag",..: 4 4 4
11 4 11 4 2 2 15 ...
  .. ..$ dist.m : num [1:155] 50 30 150 270 380 470 240 120
240 420 ...
  ..@ coords.nrs : int [1:2] 1 2
  ..@ coords   : num [1:155, 1:2] 181072 181025 181165 181298
181307 ...
  .. ..- attr(*, "dimnames")=List of 2
  .. .. ..$ : chr [1:155] "1" "2" "3" "4" ...
  .. .. ..$ : chr [1:2] "x" "y"
  ..@ bbox     : num [1:2, 1:2] 178605 329714 181390 333611
  .. ..- attr(*, "dimnames")=List of 2
  .. .. ..$ : chr [1:2] "x" "y"
  .. .. ..$ : chr [1:2] "min" "max"
  ..@ proj4string:Formal class 'CRS' [package "sp"] with 1 slot
  .. .. ..@ projargs: chr NA
```

Note that the 'proj4string' slot (signalled with a '@' at the bottom of the list) is empty, i.e. NA. For changing this, one could define it using the proj4string() function and passing to it the corresponding CRS().

To acknowledge the proj.4 string to use with a certain CRS, one can use the EPSG site. For example, to know the string for the EPSG = 28992, use this link.[3]

#12-02/03

```
# Assign a crs to meuse_sp
proj4_meuse = "+proj=sterea +lat_0=52.1561605555556 +lon_0=
5.38763888888889 +k=0.9999079 +x_0=155000 +y_0=463000
+ellps=bessel +towgs84=565.4171, 50.3319, 465.5524, 1.9342,
-1.6677, 9.1019, 4.0725 +units=m +no_defs +type=crs"

proj4string(meuse_sp) = CRS(proj4_meuse)
```

[3] https://epsg.io/28992

Now the 'meuse_sp' variable possesses a CRS that is the 'EPSG:28992'. We know this is the correct one by consulting the metadata for the 'meuse' dataset[4] or Section 6.1.1 in Chapter 6.

For creating and transforming the 'meuse' data frame to an 'sf' object, the process is as explained previously.

#12-03

```
# Convert the data frame to an sf object with geometry and crs
meuse_sf = st_as_sf(meuse_df, coords = c("x", "y"), crs = 28992)
```

12.1.3 Reprojecting

For the reprojection of a vector layer in a 'Spatial*' format, one can use the spTransform() function from the 'sp' package. This example uses the previously created 'meuse_sp' object and changes its CRS.

#12-04/01

```
# Create a new projection object
new_proj = CRS("+proj=longlat +datum=WGS84")

# Reproject the data
meuse_sp_reproj = spTransform(meuse_sp, new_proj)

# Check the new projection
proj4string(meuse_sp_reproj)
```

The result is a full description of the crs.

```
[1] "+proj=longlat +datum=WGS84 +no_defs"
```

If the object that we want to reproject is in the 'sf' format, one can use the st_transform() function to reproject vector and raster layers. The following example shows how to transform the 'meuse_sf' object to the WGS84 CRS.

#12-04/02

```
# Define the target CRS (WGS84)
new_crs = st_crs(4326)

# Reproject the sf object to WGS84 using st_transform()
meuse_sf_wgs84 = st_transform(meuse_sf, crs = new_crs)
```

Alternatively, one can directly use the EPSG code for the WGS84 with the parameters 'crs = 4326'.

---

[4] https://github.com/filipkral/meuse

Note that if the object is a raster object, the projectRaster() function from the 'raster' package can be used. If the question is to transform a 'spatRaster' object from the 'terra' package, we can use the project() function to reproject the object to a new CRS (more in Chapter 15, don't rush).

12.2 Buffering

Buffering is another common spatial analysis technique that involves creating a zone or area of influence around a spatial object, by specifying the distance around its elements. The buffer zone can be used to perform various spatial analyses, such as determining if a spatial feature is in the proximity of others, identifying areas within a certain distance of a given location, and determining spatial relationships between objects.

In the example below for the 'meuse' dataset, we create a buffer of 100 metres around the data points using the st_buffer() function. This creates a circular buffer zone around each point, with a radius of 100 metres. The resulting buffer is an 'sf' object that contains the same spatial information as the original 'meuse' dataset but with an additional buffer zone around each point.

Notice that the buffer is created in the original CRS that is 'EPSG:28992' that is projected and therefore in metres.

To correctly plot the data using the 'leaflet' package, it is necessary to transform the CRS to WGS84, done with the st_transform() function.

```
#12-05

# Load the libraries
library(leaflet)

# Load the Meuse dataset
data(meuse)

# Define the url for the icon
ss_icon_url = "https://wiki.openstreetmap.org/w/images/thumb/
e/e3/Volcano-8.svg/8px-Volcano-8.svg.png"

# Create data frame
meuse_df = meuse

# Convert the data frame to an sf object with geometry and crs
meuse_sf = st_as_sf(meuse_df, coords = c("x", "y"), crs = 28992)

# Create a buffer of 100m around the Meuse points
meuse_buf = st_buffer(meuse_sf, dist = 100)

# Transform the coordinates to be plotable in leaflet
meuse_sf_wgs84 = st_transform(meuse_sf, crs = 4326)
meuse_buf_wgs84 = st_transform(meuse_buf, crs = 4326)
```

```
# Plot the buffer using leaflet
leaflet() %>%
  addTiles() %>%
  addPolygons(data = meuse_buf_wgs84, group="Buffer",
fillOpacity = 0.2) %>%
  addMarkers(data = meuse_sf_wgs84,group="Marks",
    icon = makeIcon(ss_icon_url, iconWidth = 10, iconHeight = 10),
    popup = ~paste("Cadmium: ", meuse_buf$cadmium)) %>%
  addLayersControl(overlayGroups = c("Buffer","Marks"))
```

Figure 12.1 shows the result from the 'leaflet' plot.

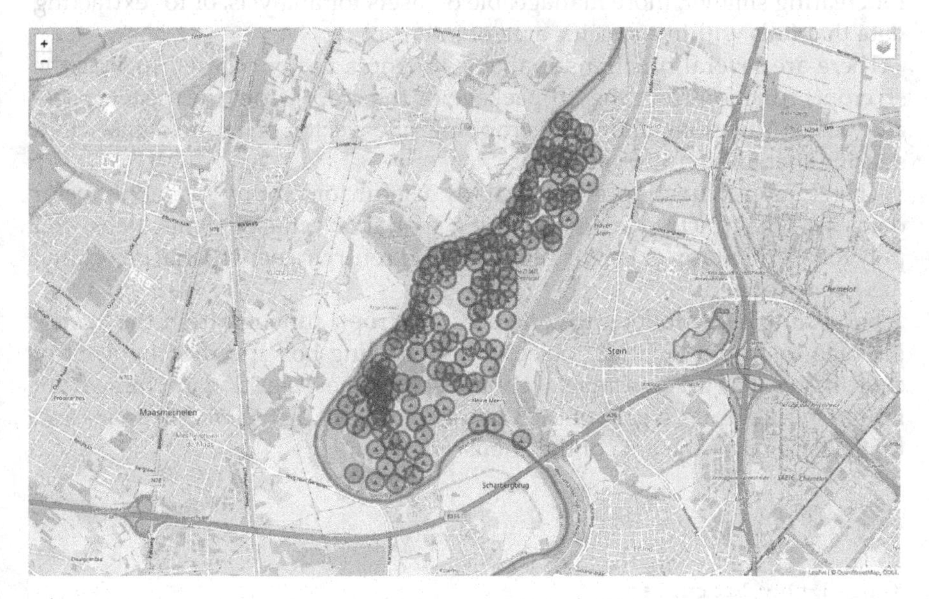

FIGURE 12.1
The result of a buffer of 100 metres for the 'meuse' data.

12.3 Subsetting Spatial Data

In spatial data analysis, 'clip' and 'crop' are two different operations that are used to extract subsets of data.

The main difference between the two operations is in the way they handle the boundary of the extracted subset:

'Crop' operation: With this operation, we extract a subset of the data by defining a rectangular or square window that covers the desired area.

'Clip' operation: With the 'clip' operation is extracted a subset of the data by defining a boundary or mask, and only the parts of the data that fall within the boundary are kept. The boundary may be defined by a polygon or a set of coordinates.

12.3.1 Cropping

Cropping has already been used previously for practical reasons (see Section 11.4.5 in Chapter 11, code snippet #11-13/02). The process of cutting out a portion of a vector dataset based on a spatial extent or geometry can be useful for creating smaller, more manageable datasets for analysis, or for extracting data that falls within a specific area of interest.

There are several functions available to crop a vector dataset, such as the st_crop() function from the 'sf' package or the crop() from the 'raster' package. These functions allow the cropping of a vector or raster dataset to a specific spatial extent or geometry.

The st_crop() function is particularly useful for cropping spatial vector datasets. This function takes as input a spatial object (e.g. points, lines, and polygons) and a crop extent or geometry and returns a new spatial object that has been cropped.

For example, the following code uses the crop() function from the 'raster' package to crop a 'Spatial*' object to a bounding box specified by a set of coordinates. For this example, we will use the 'meuse.grid' data frame from the 'sp' package, i.e. a grid of 3103 points regularly distributed in space and that extends over the area of the 'meuse' dataset.

#12-06

```
# Load the libraries
library(raster)

# Load example dataset
data(meuse.grid)

# Create a copy of meuse.grid
meuse_grid_sp = meuse.grid

# Create the sp object
coordinates(meuse_grid_sp) = c("x", "y")
proj4string(meuse_grid_sp) = CRS("epsg:28992")

# Define the crop extent
crop_extent = c(180000, 330000, 181000, 331000)

# Crop the dataset
meuse_grid_cropped = crop(meuse_grid_sp, crop_extent)
```

```
# Plot the results
par(mfrow=c(1,2))
plot(meuse_grid_sp, main="Meuse Grid Dataset")
rect(crop_extent[1], crop_extent[2], crop_extent[3], crop_
extent[4], border = "red", lwd = 2)
plot(meuse_grid_cropped, main="Meuse Grid Cropped")
```

The 'meuse_grid_sp' data frame when instructed with the coordinates() function is converted to a 'SpatialPolygonsDataFrame' object. The 'crop_extent' is a vector of coordinates defining the extent to be cropped, in the format (xmin, ymin, xmax, and ymax). The crop() function, from the raster' package, is applied to crop the dataset to the defined extent, and the result is stored in a new object called 'meuse_grid_cropped'.

Note that for this example, we are using the 'meuse.grid' dataset and not the most basic 'meuse' dataset that we have been using before.

In this example, the rect() function from the 'graphics' package is used to draw a rectangle in the plot. The resulting plot is presented in Figure 12.2.

Notice that for cropping, the CRS of the coordinates defined for the extent vector must be the same as the CRS of the cropping object; otherwise, the crop will fail.

Cropping an 'sf' object is similar to cropping a 'Spatial*' object, but the syntax and methods differ slightly. In the case of 'sf' objects, the st_crop() function is applied to a specified spatial extent, defined by a second 'sf' object.

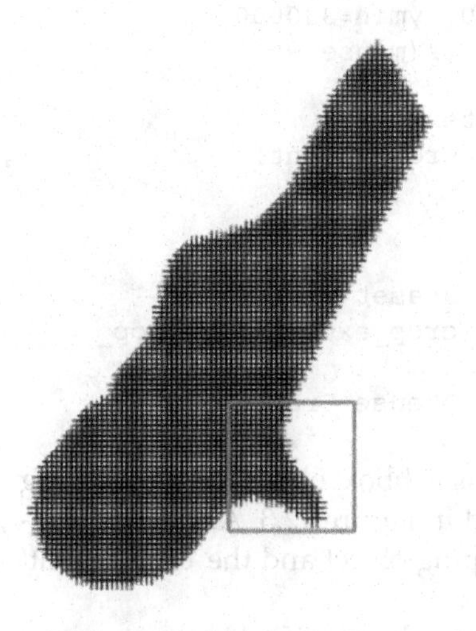

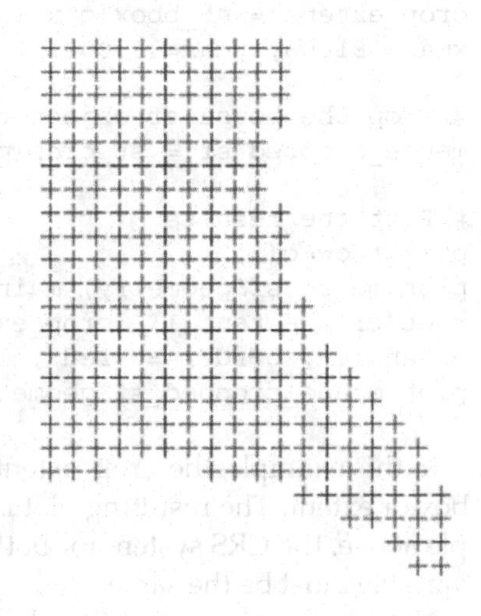

FIGURE 12.2
The result of cropping the 'meuse.grid' dataset.

For example, if we have an 'sf' object named 'my_sf_object' and want to crop it to a smaller extent defined by a second 'sf' object named 'crop_extent', one can use the st_crop() function as follows:

```
my_cropped_sf_object = st_crop(my_sf_object, crop_extent)
```

It's also possible to crop an 'sf' object to the extent of another object, such as a 'Spatial*' object, using the st_as_sf() function to convert the 'Spatial*' object to an 'sf' object and then the st_crop() function to crop the original 'sf' object to the extent of the converted object. For defining the crop extent, this time the variable is created using the st_bbox() function, with the coordinates and the CRS of the coordinates.

In this example, we will crop the 'meuse' data to the same extent. For using this function, it is necessary to convert the 'meuse' variable to an 'sf' type using the st_as_sf() function.

#12-07

```
# Define meuse available data
data(meuse)

# Convert to an sf object
meuse_sf = st_as_sf(meuse, coords = c("x", "y"), crs = 28992)

# Create a second sf object to use as the crop extent
crop_extent = st_bbox(c(xmin = 180000, ymin=330000,
xmax=181000, ymax=331000), crs = st_crs(meuse_sf))

# Crop the meuse sf object to the extent
meuse_cropped_sf = st_crop(meuse_sf, crop_extent)

# Plot the results
par(mfrow=c(1,2))
plot(meuse_sf$geometry, main="Meuse Dataset")
rect(crop_extent[1], crop_extent[2], crop_extent[3], crop_
extent[4], border = "red", lwd = 2)
plot(meuse_cropped_sf$geometry, main="Meuse Cropped")
```

In this example, the 'crop_extent' variable is a 'bbox' object, i.e. a bounding box or extent. The resulting plot is presented in Figure 12.3. As in the example above, the CRS system for both the cropping object and the 'crop_extent' variable must be the same.

Note that there are additional options that can be specified in the st_crop() function, such as the 'drop' option to specify whether or not to drop empty geometries, and the 'crs' option to specify the CRS of the output object.

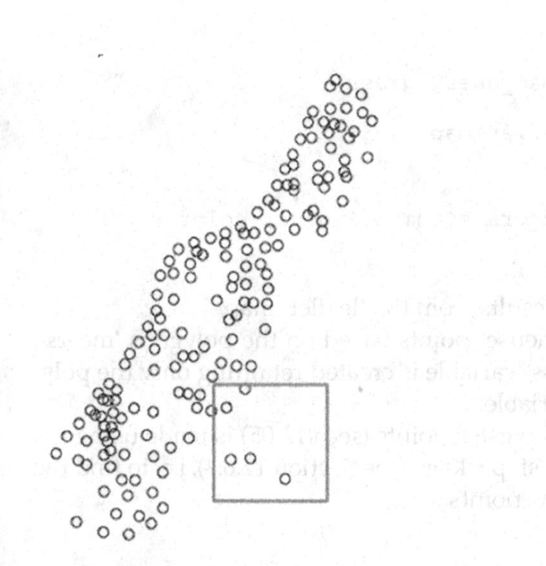

FIGURE 12.3
The crop using the st_crop() in the 'meuse' dataset.

12.3.2 Clipping

Let us suppose that we want to identify the meuse data points that are located in grass terrains. For this, one could use the identifiers 'grass' of the 'landuse' key from data available in the OpenStreetMap. For this, we will clip the points from 'meuse' data that fall inside the grass.

For extracting the grass polygons, the 'osmdata' package is used.

The query is defined to be in the Meers-Stein region in Netherlands, using the opq() function combined with the 'bbox' argument. The features 'key = "landuse"' and 'value = "grass"' are added to the query definition.

The query is executed with the osmdata_sf() function and the variable called 'meers_grass' is created.

In order to verify the results, a 'leaflet' map is plotted (Figure 12.4).

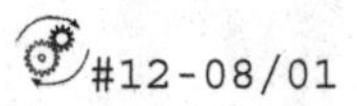

```
#12-08/01

# Load the libraries
library(osmdata)
library(leaflet)
library(sf)

# Use opq to fetch the OSM data for the grass
osm_query = opq(bbox = getbb("Meers, Stein, Netherlands"))
```

```
# Define the feature for the grass
osm_query_grass = add_osm_feature(osm_query, key="landuse",
value = "grass")

# Retrieve the grass
meers_grass = osmdata_sf(osm_query_grass)

# Plot the result on a leaflet map
leaflet() %>%
  addTiles() %>%
  addPolygons(data = meers_grass$osm_polygons, color =
"purple", weight = 2)
```

The Figure 12.4 displays the results from the 'leaflet' map.

The next step is to clip the 'meuse' points based on the polygons 'meers_grass'. For this, the 'meuse_grass' variable is created retaining only the polygons from the 'meers_grass' variable.

The clipping of the 'meuse_sf_wgs84' points (see #12-05) is made using the function st_intersect() from the 'sf' package (see Section 12.6.4), i.e. to find the intersection of polygons with the points.

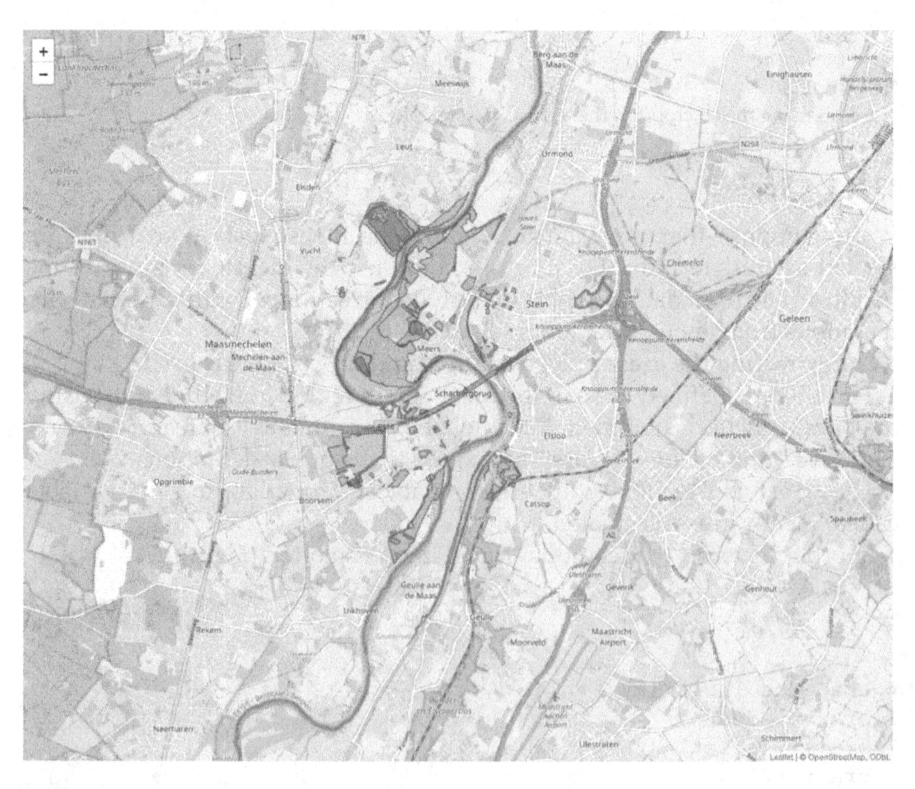

FIGURE 12.4
Map displaying the grass in the Meers-Stein region in Netherlands.

```
#12-08/02

# Create a polygons only variable
meuse_grass = meers_grass$osm_polygons

# Clip the sample points
meuse_smpl_grass = st_intersection(meuse_sf_wgs84, meuse_grass)
```

After these operations, the dataset called 'meuse_smpl_grass' with the points in the grass is created. To visualise the results, a 'leaflet' map is created with a different icon for all the samples (triangle) and in grass sample (dot).

```
#12-08/03

# Url for icons to addMarkers in leaflet
ss_icon_url = "https://wiki.openstreetmap.org/w/images/thumb/
e/e3/Volcano-8.svg/8px-Volcano-8.svg.png"
grass_icon_url = "https://wiki.openstreetmap.org/w/images/a/
ad/Social_amenity_darken_80-16.svg"

# Plot the clipped points on a leaflet map
leaflet() %>%
  addTiles() %>%
  addMarkers(
      data = meuse_sf_wgs84, group="All samples",
      icon = makeIcon(ss_icon_url, iconWidth = 10, iconHeight
= 10),
      popup = ~paste("Cadmium: ", meuse_sf$cadmium)) %>%
  addMarkers(
      data = meuse_smpl_grass, group="Grass",
      icon = makeIcon(grass_icon_url, iconWidth = 20,
iconHeight = 20),
      popup = ~paste("Cadmium: ", meuse_smpl_grass$cadmium)) %>%
  addLayersControl(overlayGroups = c("All samples","Grass"))
```

The resulting map is shown in Figure 12.5.

The sf_intersect() function not only clips the data from an 'sf' object based on the geometry of another but also adds the data table information from the second 'sf' object to the first. Notice the names from the 'meuse_smpl_grass' are the sum of the ones in the 'meuse_df' and the ones in 'meuse_grass'.

```
#12-08/04

# Verify the variable names
names(meuse_df)
names(meuse_grass)
names(meuse_smpl_grass)
```

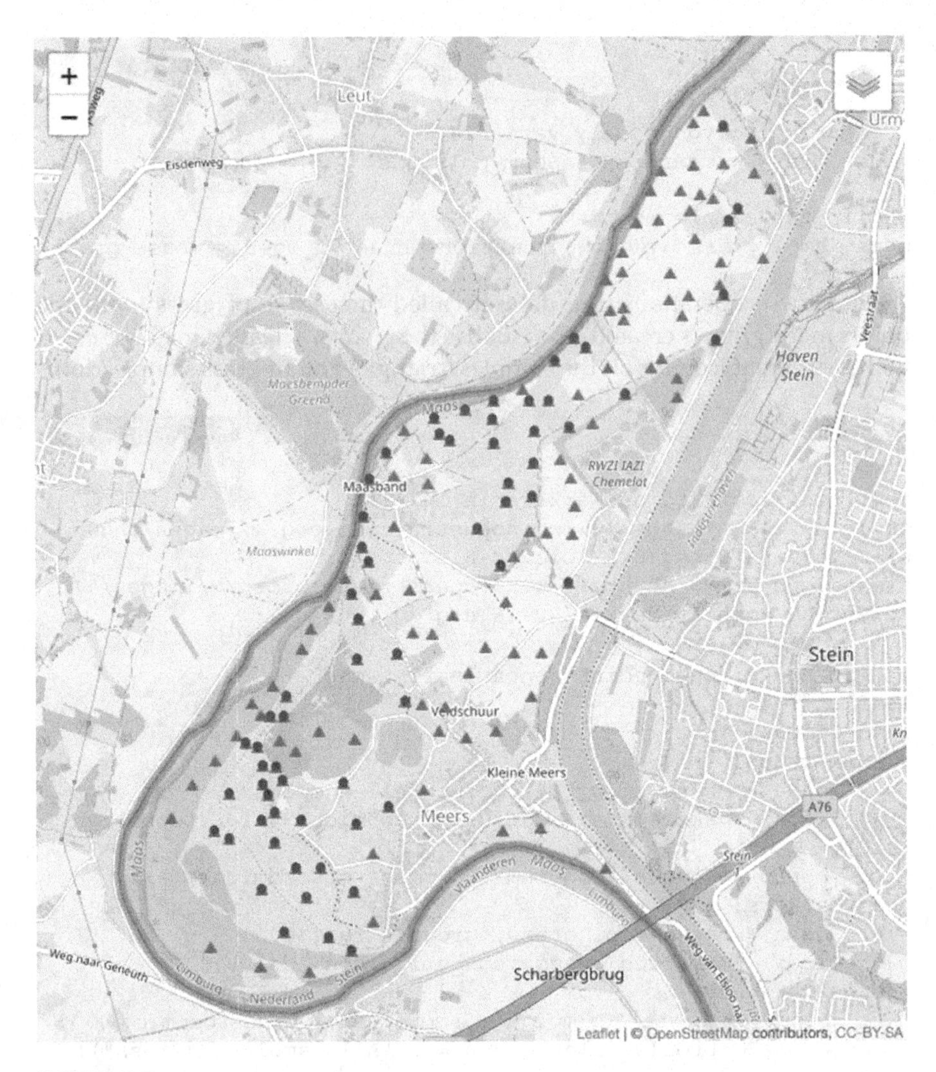

FIGURE 12.5
Leaflet map of all the samples and the ones that are in the grass (with a circle on top).

The output results are as follows:

```
[1] "x" "y" "cadmium" "copper" "lead" "zinc" "elev" "dist"
"om" "ffreq"
[11] "soil"            "lime" "landuse" "dist.m"

[1] "osm_id" "landuse" "natural" "source" "water" "geometry"

 [1] "cadmium" "copper" "lead" "zinc" "elev"
 [6] "dist" "om" "ffreq" "soil" "lime"
```

```
[11] "landuse" "dist.m" "osm_id" "X3dshapes.note" "landuse.1"
[16] "natural" "source" "geometry"
```

Note that as the 'landuse' variable name exists in both objects, the name of the second was added the suffix '.1'.

12.4 Spatial Queries (Logical)

Another important group of spatial operations is the spatial queries that naturally return a logical value. This category compares two or more spatial objects and returns a value that is a logical scalar, i.e. TRUE or FALSE, or a vector of logical.

The following examples are of abstract nature for demonstration purposes; its applications in geological situations are straightforward.

We must emphasise that the cases presented do not cover all the possible cases. It is advisable that, if one wants to consider a specific case, one should test it with an example with simple geometry, as the ones here presented, before applying the tests to a large dataset.

12.4.1 The st_contains() Function

This function tests whether a geometry contains another geometry. In other words, it returns TRUE if all the points of the second geometry are within the first geometry. This example has three squares, one big and two small, that are inside the big one (Figure 12.6).

#12-09/01 1 star

```
# Create two polygons as 'sf' objects
big_square = st_polygon(
  list(rbind(c(0, 0), c(0, 1), c(1, 1), c(1, 0), c(0, 0))))

small_square1 = st_polygon(
  list(rbind(c(0.2, 0.2), c(0.2, 0.4), c(0.4, 0.4), c(0.4,
0.2), c(0.2, 0.2))))

small_square2 = st_polygon(
  list(rbind(c(0.6, 0.6), c(0.6,0.8), c(0.8, 0.8), c(0.8,
0.6), c(0.6, 0.6))))

# Plot
par(mfrow=c(1,1))
plot(big_square, main="Two small squares inside a bigger square")
plot(small_square1, col="red", add=T)
plot(small_square2, col="blue", add=T)
```

Figure 12.6 shows the geometry.

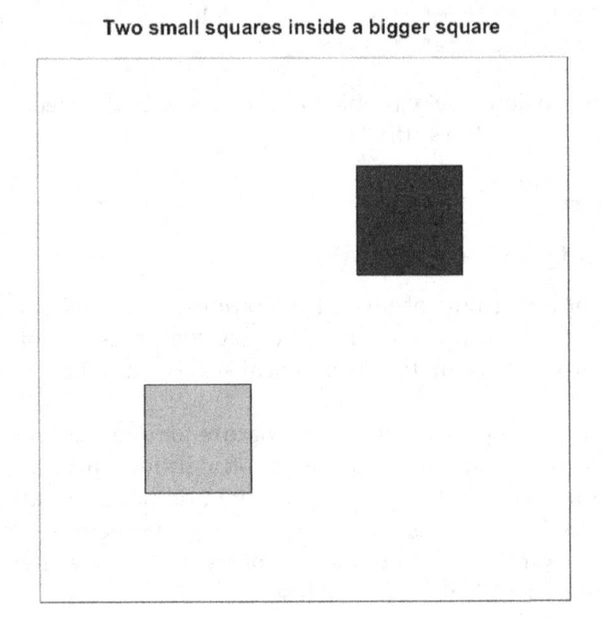

FIGURE 12.6
Two small squares inside a bigger square.

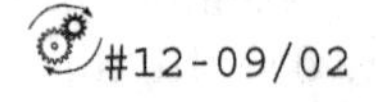#12-09/02

```
# Test whether small_square1 is contained in big_square
contains1 = st_contains(big_square, small_square1)

# Test whether small_square2 is contained in small_square1
contains2 = st_contains(small_square1, small_square2)

# Print the results
contains1
contains2
```

In this example, we use the three square polygons, 'big_square', 'small_square1' and 'small_square2', and then use the st_contains() function to test whether 'small_square1' is contained within 'square' and if 'small_square1' contains 'small_square2'. The result is returned as a logical vector, with one value per row of 'square' and 'small_square1', respectively.

```
Sparse geometry binary predicate list of length 1, where the
predicate was `contains'
 1: 1

Sparse geometry binary predicate list of length 1, where the
predicate was `contains'
 1: (empty)
```

The responses are self-evident: The 'big_square' contains the 'small_square1' and that the 'small_square1' does not contain the 'small_square2'.

12.4.2 The st_touches() Function

This function tests whether two geometries share a boundary. In other words, it returns TRUE if the intersection of the geometries is not empty but does not contain any points in the interior of either geometry.

In this code example, the par() function is used with the 'xpd = TRUE' argument to guarantee that the triangle is correctly drawn as it is necessary that the original plot is expanded when 'triangle' is drawn.

#12-10/01

```
# Create three polygons as 'sf' objects
big_square = st_polygon(
  list(rbind(c(0, 0), c(0, 1), c(1, 1), c(1, 0), c(0, 0))))
triangle = st_polygon(
  list(rbind(c(1, 0), c(1, 1), c(2, 0.5), c(1, 0))))
small_square1 = st_polygon(
  list(rbind(c(0.6, 0.6), c(0.6,0.8), c(0.8, 0.8), c(0.8,
0.6), c(0.6, 0.6))))

# Plot
par(mfrow=c(1,1), xpd = TRUE)
plot(big_square, main="Two squares and a triangle")
plot(triangle, col="red", add=T)
plot(small_square1, col="blue", add=T)
```

Figure 12.7 shows the proposed geometry.

To verify the results, the st_touches() function is used.

#12-10/02

```
# Test whether the geometries touch
touches1 = st_touches(big_square, triangle)
touches2 = st_touches(big_square, small_square1)

# Verify the results
touches1
touches2
```

The results obtained indicate that the 'big_square' touches the 'triangle' and that the 'small_square1' does not touch the 'big-square'.

```
Sparse geometry binary predicate list of length 1, where the
predicate was `touches'
 1: 1
```

```
Sparse geometry binary predicate list of length 1, where the
predicate was `touches'
 1: (empty)
```

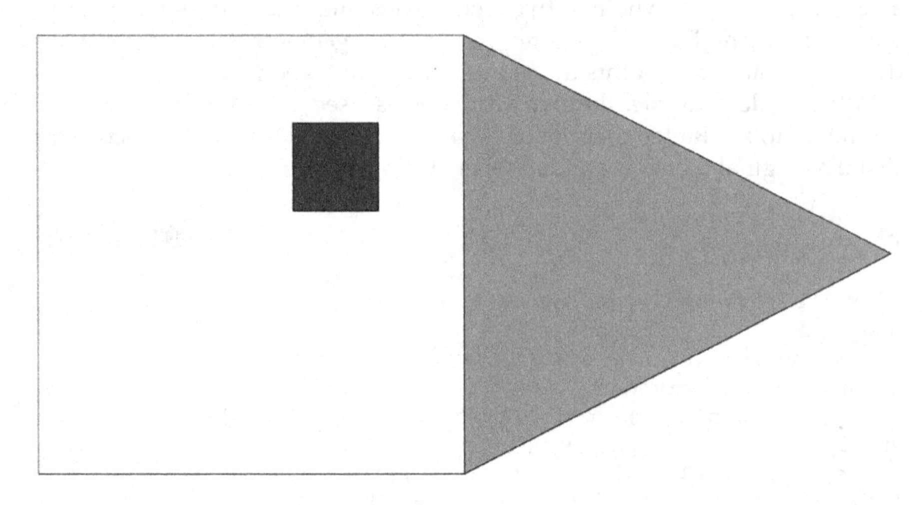

FIGURE 12.7
Two squares and a triangle.

The result is, as above, self-evident.

12.4.3 The st_crosses() Function

This function tests whether two geometries intersect but do not touch or contain each other. In other words, it returns TRUE if the intersection of the geometries is not empty. If more than two features are involved, it returns a logical vector that indicates whether the first set of geometries crosses the second set of geometries elementwise.

#12-11/01 1 star

```
# Create one square and a line
square = st_polygon(
  list(rbind(c(0, 0), c(0, 1), c(1, 1), c(1, 0), c(0, 0))))

line = st_linestring(rbind(c(-.1, 0), c(0.5,0.5), c(0.7,0.9)))

# Plot
plot(square, main="One square and a line",)
plot(line, col="red", add=TRUE)
```

Figure 12.8 shows the test geometry that is composed of a line and a square.

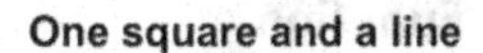

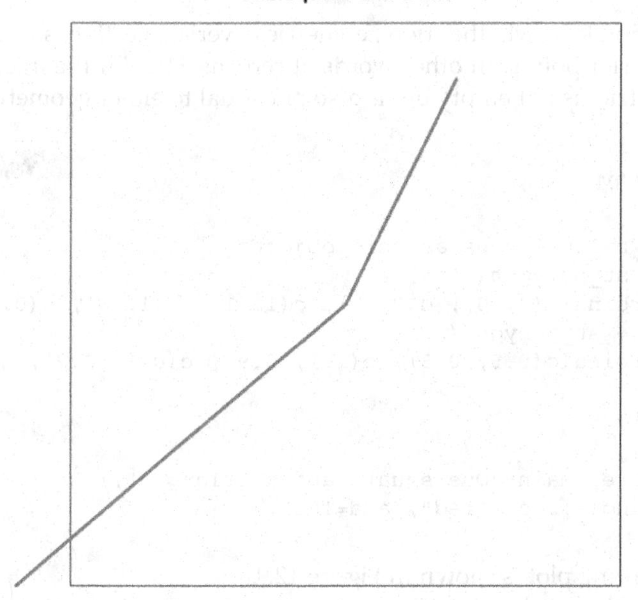

FIGURE 12.8
A square and a line crossing.

To verify the crossing of the line over the square, the st_crosses() function is used.

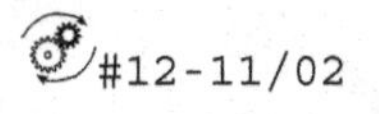
#12-11/02

```
# Test whether square crosses line
crosses = st_crosses(square, line)

# Print the result
crosses
```

The result indicates that the geometries cross.

```
Sparse geometry binary predicate list of length 1, where the
predicate was `crosses'
 1: 1
```

The st_crosses() function can be used to compare two geometry objects and determine whether their intersection corresponds to a spatial cross. This means that the geometries have some, but not all, interior points in common.

Additionally, the intersection of the two geometries must not be equal to either of the source geometries. If any of these conditions are not met, the function returns FALSE.

12.4.4 The st_overlaps() Function

This function tests whether two geometries overlap, i.e. they share some but not all interior points. In other words, it returns TRUE if the intersection of the geometries is not empty but is also not equal to either geometry.

#12-12/01

```
# Create two polygons as 'sf' objects
square = st_polygon(
   list(rbind(c(0, 0), c(0, 1), c(1, 1), c(1, 0), c(0, 0))))
triangle = st_polygon(
   list(rbind(c(0.5, 0.5), c(1.1, 0.9), c(0.5, 0.9), c(0.5,
0.5))))

# Plot
plot(square, main="One square and a triangle",)
plot(triangle, col="red", add=TRUE)
```

The resulting plot is shown in Figure 12.9.

To verify the overlapping, the st_overlaps() function is applied.

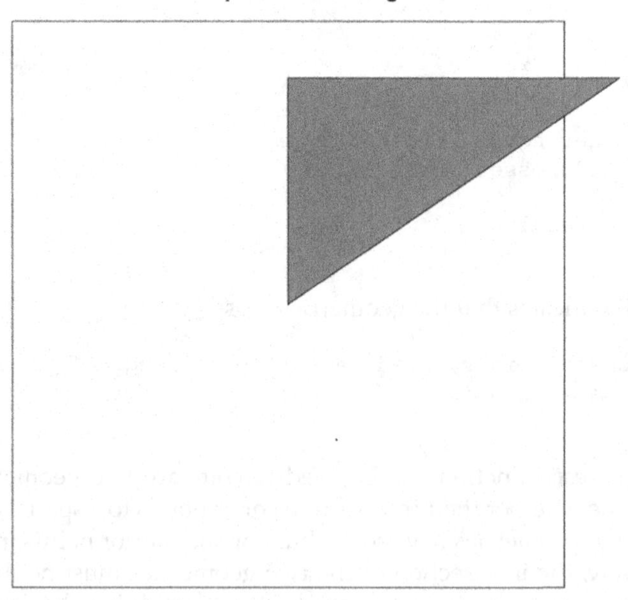

FIGURE 12.9
A triangle that overlaps a square.

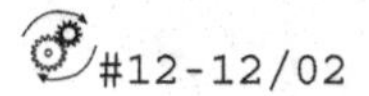#12-12/02

```
# Test whether square overlaps triangle
overlaps = st_overlaps(square, triangle)

# Print the result
overlaps
```

The result is a TRUE logical value.

```
Sparse geometry binary predicate list of length 1, where the
predicate was `overlaps'
 1: 1
```

Notice that while the st_crosses() function checks for a specific type of relationship where boundaries intersect but interiors do not, the st_overlaps() function checks for any partial overlap of areas or boundaries.

12.4.5 The st_within() Function

This function tests whether a geometry is within another geometry. In other words, it returns TRUE if all the points of the first geometry are within the second geometry.

#12-13/01

```
# Create two polygons as 'sf' objects
big_square = st_polygon(
   list(rbind(c(0, 0), c(0, 1), c(1, 1), c(1, 0), c(0, 0))))
small_square = st_polygon(
   list(rbind(c(0.2, 0.2), c(0.2, 0.4), c(0.4, 0.4), c(0.4,
0.2), c(0.2, 0.2))))

# Plot
plot(square, main="One square and another square",)
plot(small_square, col="red", add=TRUE)
```

Figure 12.10 shows the example.

In this example, we create two square polygons, 'square' and 'small_square', then use the st_within() function to test whether 'square' is within 'small_square'. Then test the opposite, i.e. if the 'small_square' is within the 'square'. The result is returned as a logical vector, with one value per row.

#12-13/02

```
# Test whether square is within small_square and the opposite
within1 = st_within(big_square, small_square)
within2 = st_within(small_square, big_square)
```

```
# Print the result
within1
within2
```

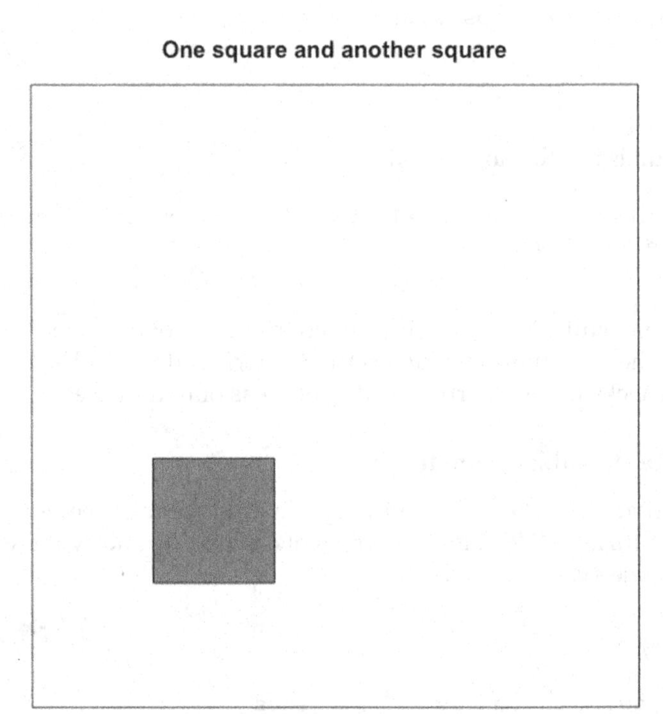

FIGURE 12.10
One square within another square.

The obtained resúlt is as follows:

```
Sparse geometry binary predicate list of length 1, where the
predicate was `within'
 1: (empty)
Sparse geometry binary predicate list of length 1, where the
predicate was `within'
 1: 1
```

The result indicates that the bigger square is not within the smaller one, but the smaller one is within the bigger one.

12.4.6 The st_intersects() Function

This function tests whether two geometries intersect. It means that it returns TRUE if the intersection of the geometries is not empty. In this case if the line is inside the square, it is considered that it intersects it (cf. 'intersects4' variable).

#12-14/01 1 star

```
# Create the 'sf' objects
big_square = st_polygon(
   list(rbind(c(0, 0), c(0, 1), c(1, 1), c(1, 0), c(0, 0))))
line1 = st_linestring(rbind(c(-.1, 0), c(.5, .5), c(.7, .9)))
line2 = st_linestring(rbind(c(.1, 0), c(.5, .5), c(.9, .9)))
line3 = st_linestring(rbind(c(-.2, .2), c(-.1, .5), c(-.2, .6)))
line4 = st_linestring(rbind(c(.1, .9), c(.3, .5), c(.7, .6)))

# Plot
plot(big_square,  main="A square and 4 lines")
plot(line1, col="red", add=TRUE, lwd=3)
plot(line2, col="blue", add=TRUE, lwd=3)
plot(line3, col="green", add=TRUE, lwd=3)
plot(line4, col="black", add=TRUE, lwd=3)
```

Figure 12.11 shows the plot obtained.
To test the intersection, the st_intersects() function is applied.

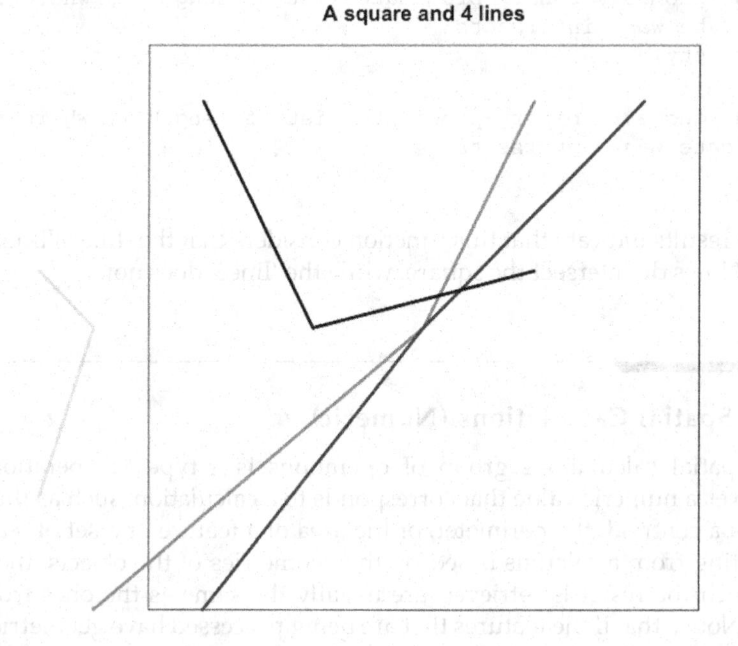

FIGURE 12.11
Four lines and a square.

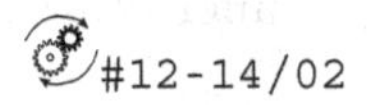

```
#12-14/02

# Test whether square intersects rect
intersects1 = st_intersects(big_square, line1)
intersects2 = st_intersects(big_square, line2)
intersects3 = st_intersects(big_square, line3)
intersects4 = st_intersects(big_square, line4)

# Verify the results
intersects1
intersects2
intersects3
intersects4
```

The results are as follows:

```
Sparse geometry binary predicate list of length 1, where the
predicate was `intersects'
 1: 1
Sparse geometry binary predicate list of length 1, where the
predicate was `intersects'
 1: 1

Sparse geometry binary predicate list of length 1, where the
predicate was `intersects'
 1: (empty)

Sparse geometry binary predicate list of length 1, where the
predicate was `intersects'
 1: 1
```

The results indicate that this function considers that the 'line1', 'line2', and 'line4' lines do intersect the square, while the 'line3' does not.

12.5 Spatial Calculations (Numeric)

The spatial calculations group of operations is a type of operation that retrieves a numeric value that corresponds to a calculation, such as the position of a centroid, the perimeter, or the area of a feature or a set of features. Resulting from a calculus based on the geometries of the objects, the units in which the result is retrieved are usually the same as the ones from the CRS. Notice that if the features that are being processed have geometric CRS, i.e. latitude and longitude, the result will also be in these units, i.e. decimal degrees. If the CRS system used is a projected one, the units will be in metres.

In the case of the examples presented below, the spatial objects do not have a CRS defined; therefore, the units are not defined.

12.5.1 st_centroid() Function

This function calculates the centroid of a 'sf' object, which is the centre point of its spatial extent. The centroid is the weighted average of the x and y coordinates of the features in the 'sf' object.

#12-15

```
# Create a simple square polygon
square = st_polygon(
  list(rbind(c(0, 0), c(0, 1), c(1, 1), c(1, 0), c(0, 0))))

# Compute the centroid of the polygon
centroid = st_centroid(square)

# Print the centroid
centroid

# Plot the square and the centroid
plot(square, main = "A square with its centroid")
points(x=centroid[1], y=centroid[2], pch=2, cex=1, col="red")
```

In this example, we create a simple square polygon as an 'sf' object called 'square'. We then use the st_centroid() function to compute the centroid of the polygon, which is returned as a POINT geometry. The resulting 'sf' object is the centroid, which has a single row with the coordinates of the point.

```
POINT (0.5 0.5)
```

The resulting plot is present in Figure 12.12.

12.5.2 The st_area() Function

This function calculates the area of a 'sf' object. The code example uses a rectangle of sides 2×1 and calculates its area.

#12-16

```
# Create a simple rectangle polygon as an 'sf' object
rect = st_polygon(
  list(rbind(c(0, 0), c(0, 2), c(1, 2), c(1, 0), c(0, 0))))

# Compute the area of the polygon
area = st_area(rect)
```

```
# Print the area
area
```

The result is as follows:

```
[1] 2
```

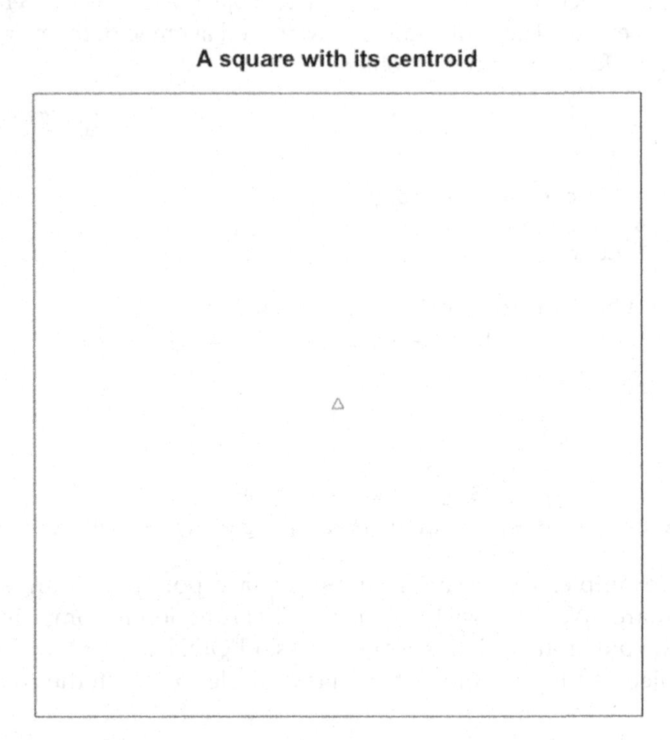

FIGURE 12.12
The centroid of a square.

In this example, we create a simple rectangular polygon as an 'sf' object called 'rect'. We then use the st_area() function to compute the area of the polygon, which is returned as a scalar value.

12.5.3 The st_length() Function

This function calculates the length of a 'sf' object. This might be used to calculate the length of a line or the perimeter of a polygon.

```
#12-17

# Create a simple line as an 'sf' object
line = st_linestring(
   rbind(c(0, 0), c(0, 1), c(1, 1), c(1, 0), c(2, 0)))
```

```
# Compute the length of the line
length = st_length(line)

# Print the length
length
```

In this example, we create a simple line as an 'sf' object called 'line'. We then use the st_length() function to compute the length of the line, which is returned as a scalar value.

```
[1] 4
```

12.5.4 The st_distance() Function

This function calculates the distance between two 'sf' objects, which is the shortest distance between any pair of point coordinates of the two objects.

#12-18

```
# Create two points as 'sf' objects
point1 = st_point(c(0, 0))
point2 = st_point(c(1, 1))

# Compute the distance between the two points
distance = st_distance(point1, point2)

# Print the distance
distance
```

In this example, we create two points as 'sf' objects called 'point1' and 'point2'. We then use the st_distance() function to compute the distance between the two points, which is returned as a scalar value.

The result is as follows:

```
     [,1]
[1,] 1.414214
```

12.6 Spatial Operations (Geometric)

This group of functions perform geometric analysis of the data, might that be geometric difference, geometric join, symmetric difference, or any other possible geometric operation. As before, one must be aware of the importance of having all the objects involved in these operations with the same CRS. If it is not the case, the operation results will be random and naturally invalid.

12.6.1 The st_sym_difference() Function

This function calculates the symmetric difference of two geometries, i.e. the parts of the two geometries that are not in their intersection. In other words, it returns a new geometry that contains all the points that are in one geometry or the other but not in both.

The following code snippet creates a 'square' and a 'triangle' object.

#12-19/01

```
# Create two polygons as 'sf' objects
square = st_polygon(
  list(rbind(c(0, 0), c(0, 1), c(1, 1), c(1, 0), c(0, 0))))
triangle = st_polygon(
  list(rbind(c(0.5, 0.5), c(1.1, 0.9), c(0.5, 0.9), c(0.5, 0.5))))

# Plot
par(mfrow=c(1,2), xpd = TRUE)
plot(square, main="A square and a Triangle")
plot(triangle, col="red", add=TRUE)
```

Note that the 'xpd' parameter in the par() function allows plotting outside of the standard plot region, necessary for this example.

For calculating the symmetric difference, the st_sym_difference() function is used.

#12-19/02

```
# Calculate the symmetric difference of square and rect
sym_diff = st_sym_difference(square, triangle)

# Plot
plot(square, main="The difference")
plot(triangle, col="red", add=TRUE)
plot(sym_diff, add=TRUE, cex=2, col="blue")
```

The final plot is in Figure 12.13. The right image presents the result.

12.6.2 The st_union() Function

This function calculates the union of two geometries, i.e. the geometry that contains all the points that are in either one or both of the input geometries.

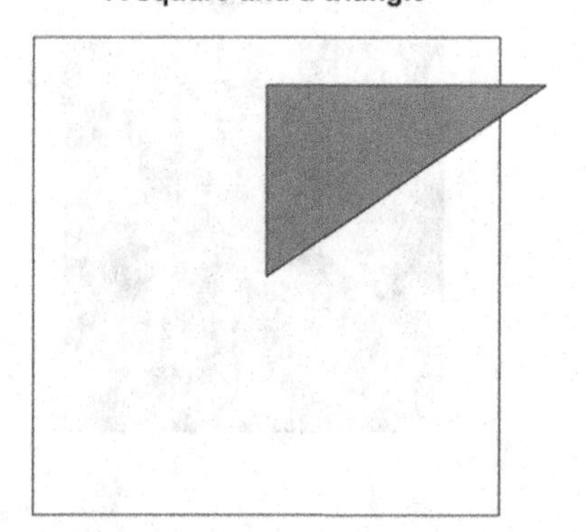

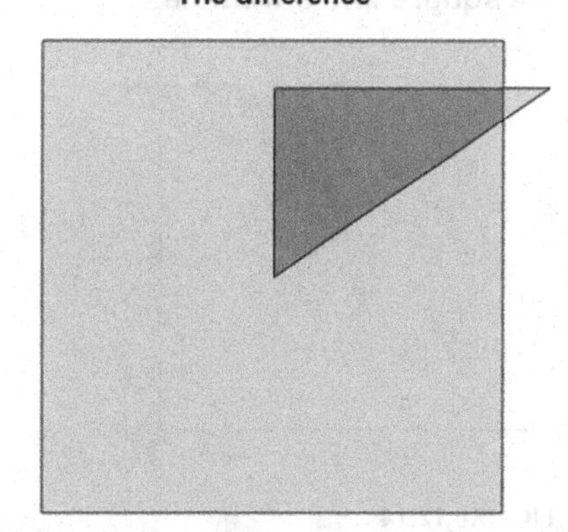

FIGURE 12.13
The symmetric difference between two polygons.

An adaptation of the previous code to use this new function is carried to demonstrate these results.

#12-20

```
# The geometries
par(mfrow=c(1,2), xpd=TRUE)
plot(square,  main="A square and a triangle")
plot(triangle, col="red", add=TRUE)

# Calculate the union of square and rect
union = st_union(square, triangle)

# Plot the result
plot(union, col="blue", main = "The union")
```

In this example, we create two polygons, 'square' and 'triangle', and then use the st_union() function to calculate the union of 'square' and 'triangle' and plot the result.

Figure 12.14 displays the result, on the left the original data and on right the result.

12.6.3 The st_difference() Function

This function calculates the difference between two geometries, i.e. the parts of the first geometry that are not in the second geometry. In other words, it returns a new geometry that contains all the points that are in the first geometry but not in the second geometry.

The code is an adaptation from the previous one.

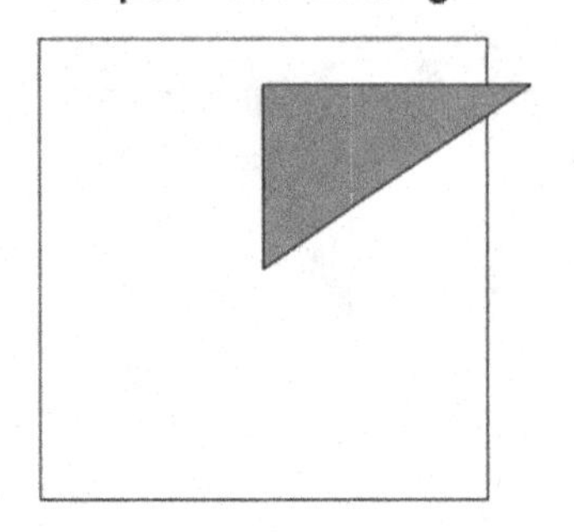

FIGURE 12.14
The union between a square and a triangle.

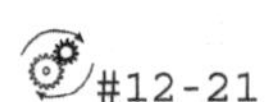
#12-21

```
# The geometries
par(mfrow=c(1,2), xpd=TRUE)
plot(square,  main="A square and a triangle")
plot(triangle, col="red", add=TRUE)

# Calculate the symmetric difference of square and rect
diff = st_difference(square, triangle)

# Plot
plot(diff,  main="The difference", col="blue")
```

Figure 12.15 shows the original objects in the left image, and the resulting geometry in gray shade.

12.6.4 The st_intersection() Function

This function performs a spatial intersection between two 'sf' objects, returning an object that contains the geometric intersection of the input objects. The result will have the same dimensions and CRS as the input objects. Here's an example of how to use st_intersection() in an adaptation of the previous code examples.

#12-22

```
# The geometries
par(mfrow=c(1,2), xpd = TRUE)
plot(square,  main="A square and a triangle")
plot(triangle, col="red", add=TRUE)
```

```
# Perform intersection between the two 'sf' objects
inter = st_intersection(square, triangle)

# Plot the resulting 'sf' object
plot(inter, col = "blue", main="The intersection")
```

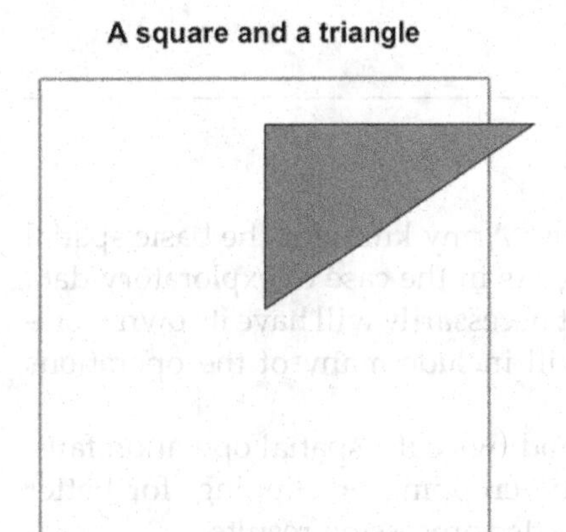

The difference

FIGURE 12.15
The difference between two 'sf' objects.

Figure 12.16 represents the original on the left and the result on the right. Notice that the right image is bigger because the plot is resized to fit the plot area.

In this example, we create two polygons, 'square' and 'triangle', and then use the st_intersection() function to perform an intersection between the two objects and store the result in a new 'sf' object called 'inter'. Finally, we plot the resulting object, which contains the area where the two polygons overlap.

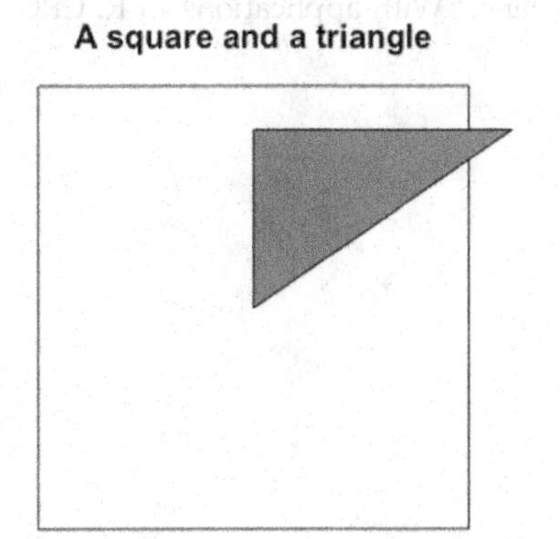

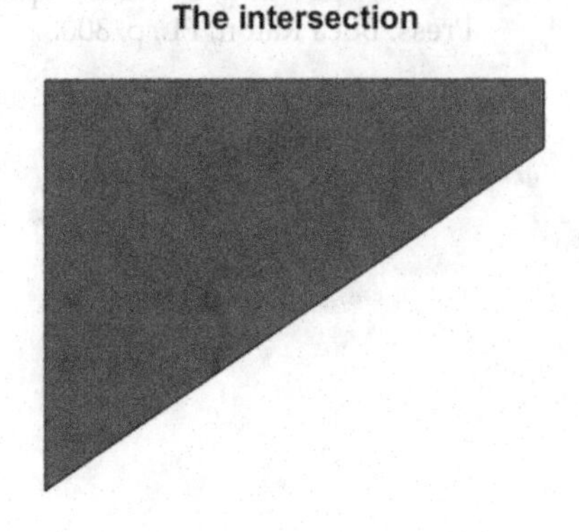

FIGURE 12.16
The result of the intersection of two polygons.

The st_intersection() function can be used with other types of geometries besides polygons, such as points, lines, and multipolygons (see the example used for clipping in Section 12.3.2 in Chapter 12). The function can also be used to perform intersections between multiple 'sf' objects by calling it repeatedly with each pair of objects.

12.7 Concluding Remarks

Consider this chapter as your trusty Swiss Army knife for the basic spatial operations and for spatial data handling. As in the case of exploratory data analysis, each individual research project necessarily will have its own workflow design. But be sure that often it will include many of the operations described.

Remember to check CRSs every time and twice if a spatial operation fails. Use frequently subsetting, might that be cropping or clipping, for better memory management of your data and faster processing results.

Acknowledging the main geometric operations is paramount for any spatial analyst, but remember to not trust the results before checking if the operation you are using provides the results you desire. Geometric operations are often intricate, and therefore, the results need to be double checked. Nevertheless, the spatial operations are an ally of every spatial analyst providing new insights to understand and explore the geology of a region.

References

Pebesma, E., & Bivand, R. (2023). Spatial data science: With applications in R. CRC Press, Boca Raton, FL, p. 300.

13

Putting It All to Work: Part II Vectors

For this wrap-up chapter, we propose a return to the example of the Northern Territory (NT), its geology, its mineral resources, and the data that has been made available by its authorities.[1] In the previous chapter, we selected the data made available by the mining company "Bureau of Mineral Research, Geology and Geophysics" (BMRGG) to start exploring its geological applications using an exploratory approach.

The NT government site contains other geological information[2] and even a geological map of the territory.[3] This information is in various spatial formats and includes mineral titles, petroleum and pipeline titles, geothermal titles, drilling, diamond exploration, geochemistry (from where the data of the stream sediments samples was extracted), geochronology, geology, geophysics, mineral deposits and mines, indexes, graticules, and blocks.

It is a truly remarkable playground for geologists who want to practise spatial data analysis. So, let's continue to play.

13.1 Setting the Data

For better following the text, in this first code snippet (#13-01), is presented the part of the code that was used for reading the data and pre-processing it to the form of having the stream sediments data subsets from the BMRGG data to work with.

It is assumed that you already have downloaded it (why not?) and unzip it (of course!!).

If you are lost, please return to Section 8.1 in Chapter 8 (code snippet #08-01/01), for reference.

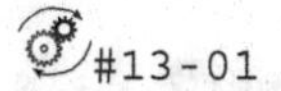#13-01

```
# Read the data
nt_ss = read.csv("NT_SS/GEOCHEM_STREAM_SEDIMENTS.csv")
```

[1] https://geoscience.nt.gov.au/gemis/ntgsjspui/handle/1/82057
[2] https://geoscience.nt.gov.au/downloads/NTWideDownloads.html
[3] https://geoscience.nt.gov.au/gemis/ntgsjspui/bitstream/1/82057/3/NTGeology.pdf

DOI: 10.1201/9781032651880-15

```
# Filter the dataset
company_name = grepl("^Bureau of Mineral", nt_ss$COMPANY)
query = nt_ss$COMPANY == company_name

# Create the subset
nt_ss_bmrgg = nt_ss[company_name,]

# Remove columns that are NA
na_cols = which(base::colSums(is.na(nt_ss_bmrgg)) ==
nrow(nt_ss_bmrgg))
print(paste(length(na_cols), "NA columns rejected"))

# Subset the data frame to exclude the all-NA columns
nt_ss_bmrgg_filtered = nt_ss_bmrgg[,  -na_cols]
print(paste(ncol(nt_ss_bmrgg_filtered),"columns present"))

# Remove -9999 columns
na_cols = which(colSums(nt_ss_bmrgg_filtered  == -9999) ==
nrow(nt_ss_bmrgg_filtered ))

# Verify the results
print(paste(length(na_cols),"-9999 columns rejected"))

# Subset the data frame to exclude the all the NA columns
nt_ss_bmrgg_10 = nt_ss_bmrgg_filtered[,  -na_cols]

# Verify the results
print(paste(ncol(nt_ss_bmrgg_10),"columns present"))

# Replace all the -9999 remaining by NA
nt_ss_bmrgg_10[nt_ss_bmrgg_10 == -9999] = NA

# Verify the results
colnames(nt_ss_bmrgg_10)

# Create two datasets
bmrgg = nt_ss_bmrgg_10 # All the pre-processed data
bmrgg_chem = nt_ss_bmrgg_10[, c(3,21:38)] # The chemical analysis
```

The next step is to load the necessary libraries and to create a 'sf' object from the 'bmrgg' data. Its coordinates (longitude and latitude) and CRS are defined.

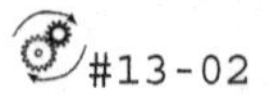
#13-02

```
# The necessary libraries
library(dplyr)
library(sp)
library(sf)
library(raster)
```

```
library(ggplot2)
library(ggspatial)
library(leaflet)

# Define the working folder
setwd("~/Spatial/Data")

# Create a bmrgg stream sediments data
bmrgg_sf = st_as_sf(bmrgg, coords = c("LONGITUDE","LATITUDE"),
crs=4326)
```

In the code snippet (#13-03), it's assumed that the data has already been downloaded, unzipped, and that the shapefiles are ready in the 'NT_Geology_shape/' folder.

The code starts by reading these new datasets, i.e. the geology and the faults, to the R environment.

After reading the shapefiles with the st_read(), they are stored as 'geo' and 'faults' variables that are of 'sf' format.

The next steps involve cropping the geology and the faults to the study region, in this case the area defined by the stream sediments of the BMRGG company that is obtained from the 'bmrgg_sf' variable, by using the st_bbox() function.

#13-03

```
# Read the Geology and the Faults data
geo = st_read("NT_Geology_shape/GEO_INTERP_2500K.shp")
faults = st_read("NT_Geology_shape/GEO_FAULTS_2500K.shp")

# Convert to WGS84
geo = st_transform(geo, crs=4326)
faults = st_transform(faults, crs=4326)

# Create bbox()
geo_bb = st_bbox(bmrgg_sf)
# Crop Geology and Faults
geo_bmrgg = st_crop(geo, geo_bb) # Geology
flt_bmrgg = st_crop(faults, geo_bb) # Faults
```

To verify the result of these operations, a plot is created (Figure 13.1). The geology is drawn as black lines, the stream sediment samples are triangles (pch = 2) with a blue colour (col = "blue"), and the faults are drawn as 3 point thick (cex = 3) and in red lines (col = "red").

#13-04

```
# Plot the results
par(mfrow=c(1, 1))
```

```
plot(geo_bmrgg$geometry, main = "Geology Bureau Mineral Resources")
plot(bmrgg_sf$geometry, pch = 2, cex = 0.5, col = "blue", add
= TRUE)
plot(flt_bmrgg$geometry, cex = 3, col = "red", add = TRUE)
```

Figure 13.1 shows the results of the downloaded and cropped data.

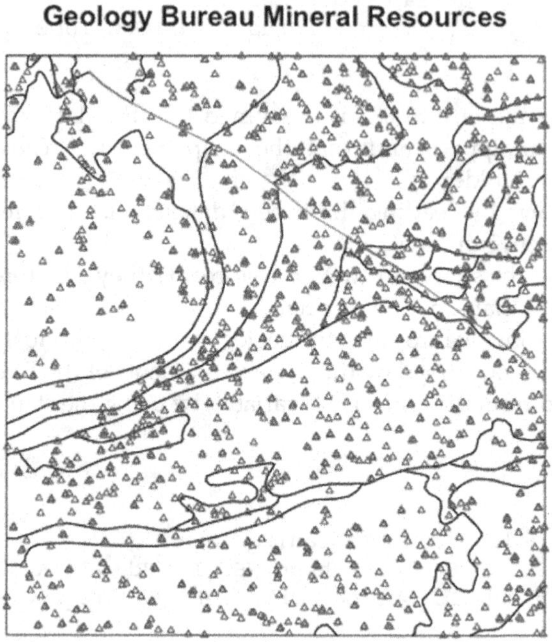

Geology Bureau Mineral Resources

FIGURE 13.1
The data extracted for this example from the Northern Territory repository of the Bureau of Mineral Resources company.

A detailed inspection allows to verify that 4 points (very near the east boundary) fall outside the geological map. This might be due to two different problems: i) decimal places in the coordinates or ii) samples that were collected outside the map boundaries. Up front we will have to deal with these samples.

13.2 Preparing the Maps

The first task is to reproduce, with the better possible quality, the maps and symbology that are available in the literature.[4] For that, it is necessary to verify what are the geological units present in the study area. The unique()

[4] https://geoscience.nt.gov.au/gemis/ntgsjspui/handle/1/82057

function retrieves the values of the variable 'SYMBOLS'. In this case, we opted for sorting them alphabetically with the sort() function.

#13-05

```
# Verify the symbols
sort(unique(geo_bmrgg$SYMBOL))
```

The result retrieved 11 different symbols. The 88 levels present are the factors inherited from the original 'geo' variable.

```
[1] "g5"   "K"    "L6"   "L7"   "L8"   "L9"   "M10" "M6"   "M6b" "Y4"
"Y5"
```

In situations like this, the droplevels() function proves to be invaluable. It's designed to streamline factors in R by eliminating any levels not represented in the dataset. Factors in R are used to handle categorical data and can sometimes include categories (levels) that don't appear in the data, potentially leading to confusion or errors in analysis and visualization. By applying droplevels(), we ensure that only relevant, data-present levels are retained, simplifying subsequent analyses and visualizations.

The next step is to go to the legend of the geological map, find these symbols and get the red, green, and blue (RGB) colour code for each. After that, create a variable 'murphy_colors'[5] with all the RGB codes of each geological unit present.

Yes, it is necessary to cherry-pick the colours from the map if there is not a table with it.

#13-06

```
# Create the colour codes for the geological units
murphy_colors = c(
  rgb(233,31,55, maxColorValue = 255), # g5 Nicholson Granite
Complex
  rgb(232,237,209, maxColorValue = 255), # K Carpentaria Basin
  rgb(213,176,166, maxColorValue = 255), # L6 Wire Creek Sandstone
  rgb(122,179,218, maxColorValue = 255), # L7 Fickling Group
  rgb(195,233,249, maxColorValue = 255), # L8 Doomadgee Formation
  rgb(226,207,194, maxColorValue = 255), # L9 South Nicholson
Group
  rgb(226,227,228, maxColorValue = 255), # M10 Bukalara
Sandstone, Wessel Group
  rgb(213,176,166, maxColorValue = 255), # M6 Parsons Range,
Tawallah, and Katherine River Groups
  rgb(53,189,166, maxColorValue = 255), # M6b Seigal Volcanics
```

[5] The geological map is from the Murphy region; that is why the variable was named like this.

```
  rgb(221,154,97, maxColorValue = 255), # Y4 Murphy Metamorphics
  rgb(255,237,114, maxColorValue = 255) # Y5 Cliffdale Volcanics
)
```

Note that we had only 11 geological units, and therefore, this task was not very time consuming. On occasions where there are hundreds of geological units, this might be a harder task.

13.2.1 The ggplot() Map

After these steps, it is possible to create a plot of the geological map, assigning to each geological symbol the correct colour. For this map created with the 'ggplot2' package, we will use the data variables that are in the 'sf' format.

#13-07

```
# In ggplot() use 'sf' variables with the geom_sf() function
ggplot(data = geo_bmrgg) +
  geom_sf(aes(fill = geo_bmrgg$SYMBOL)) +
  scale_fill_manual(values = murphy_colors, name="Geological
units") +
  geom_sf(data = flt_bmrgg, color = "black") +
  # Add a scale bar in km
  annotation_scale(location = "bl", width_hint = 0.25, text_
cex = 0.75) +
  # Add a title to the plot
  labs(title = "Geology of the Murphy Region")
```

The resulting plot is presented in Figure 13.2.

This plot is somewhat complex, and to understand it, here is the breakdown of the functions and attributes used:

ggplot(): It initialises a new 'ggplot' object with 'geo_bmrgg' as the data source.

geom_sf(): It adds a new layer of data to the plot as simple features 'sf' objects, with fill = geo_bmrgg$SYMBOL mapping the fill colour to the SYMBOL column in geo_bmrgg. This function is used to display the geological units in the Murphy region. The aes() function inside the geom_sf() specifies the aesthetics mapping for the layer.

scale_fill_manual(): It sets a manual colour scale for the fill aesthetic. The values argument specifies the colour values to be used, while 'name' sets the title of the legend.

geom_sf(): It adds another layer of data to the plot, with 'data = flt_bmrgg and 'color = "black"'used to display the faults in the Murphy region. Notice that this 'sf' layer is plotted subsequently to the previous one.

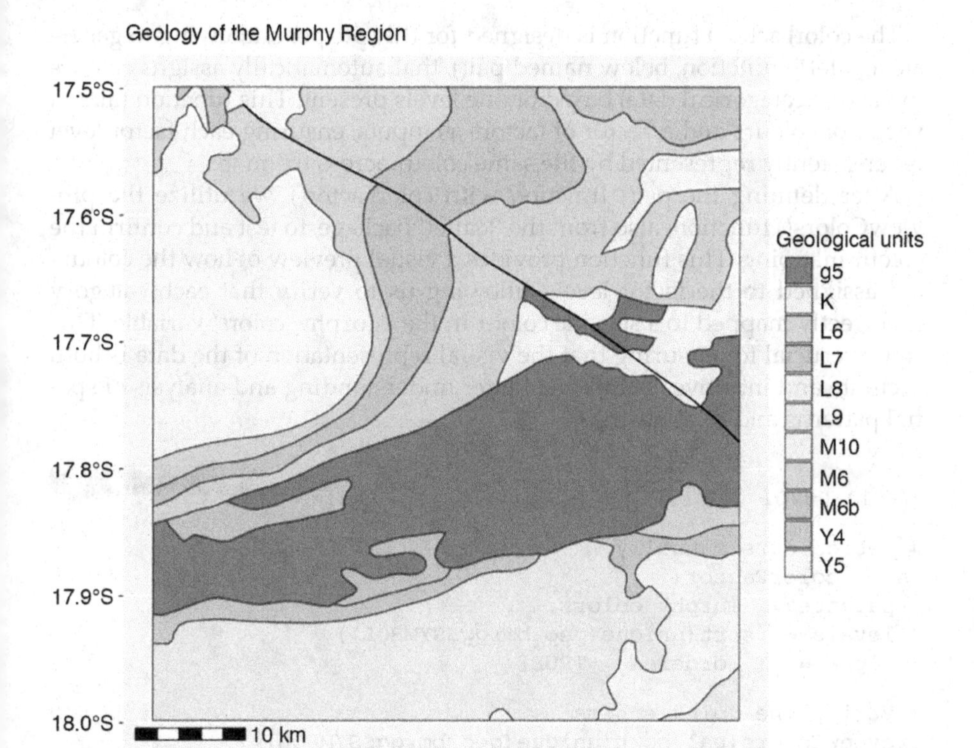

FIGURE 13.2
The geological map of the Murphy Region using the ggplot() function.

annotation_scale(): It adds a scale bar to the plot at the bottom-left (location = "bl") with a width of 0.25 km (width_hint = 0.25) and a text size of 0.75 times the default (text_cex = 0.75).

labs(): It adds a title to the plot with title = "Geology of the Murphy Region". This function is used to provide a descriptive title for the map.

13.2.2 The 'leaflet' Map

The same map can be created using the 'leaflet' package, but it is necessary to make some more tweaking to the data. To effectively visualize spatial data with the 'leaflet' package, it's often useful to assign distinct colours to various categories or values within the dataset. This process involves creating a colour mapping scheme, which can be achieved using the colorFactor() function from the 'leaflet' package. Specifically, in this context, we aim to map a set of predefined colours, referred to as 'murphy_colors', to different categories or values within our data.

The colorFactor() function is designed for this purpose, allowing to generate a palette function, below named pal(), that automatically assigns colours to factors (categorical data) based on the levels present. This function takes a vector of colours and a vector of factors as inputs, ensuring each factor level is consistently represented by the same colour across the map.

After defining the pal() function with colorFactor(), we utilize the previewColors() function, also from the 'leaflet' package, to test and confirm the colour mapping. This function provides a visual preview of how the colours are assigned to the factor levels, allowing us to verify that each category is correctly mapped to a specific colour in the 'murphy_colors' variable. This step is crucial for ensuring that the visual representation of the data is both accurate and intuitive, facilitating better understanding and analysis of spatial patterns and relationships.

```
#13-08/01

# Set colours manually
pal = colorFactor(
  palette =  murphy_colors,
  levels =  sort(unique(geo_bmrgg$SYMBOL)),
  alpha =  T, ordered = TRUE)

# Verify the color scheme
previewColors(pal(sort(unique(geo_bmrgg$SYMBOL))),
sort(unique(geo_bmrgg$SYMBOL)))
```

Here's a breakdown of the colorFactor() function arguments used:

- **'palette = murphy_colors'**: It specifies the palette of colours to use for mapping values. murphy_colors is a vector of named colour values.
- **'levels = sort(unique(geo_bmrgg$SYMBOL))'**: It specifies the levels of the factor to be mapped. This is set to the unique values in the SYMBOL column of the geo_bmrgg data frame, sorted in alphabetical order.
- **'alpha = T'**: It sets the transparency level of the colours to TRUE. This means that the colours can have an alpha channel, i.e. might have a definable degree of transparency, allowing the underlying features to show through.
- **'ordered = TRUE'**: It sets the order of the factor levels to be treated as ordered. This is important for some plot types, such as choropleth maps, where the ordering of the colours is meaningful.

The Figure 13.3 shows the results from the previewColors() function.

The code for displaying the 'leaflet' map is adapted from previous ones but with some juicy options.

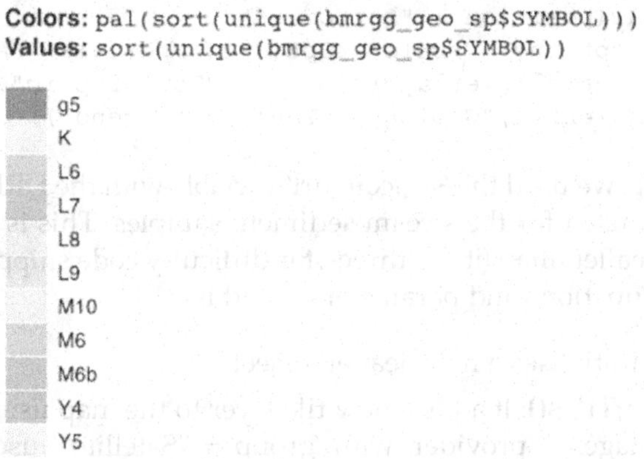

Colors: pal(sort(unique(bmrgg_geo_sp$SYMBOL)))
Values: sort(unique(bmrgg_geo_sp$SYMBOL))

g5
K
L6
L7
L8
L9
M10
M6
M6b
Y4
Y5

FIGURE 13.3
The result of previewColors() function.

#13-08/02 3 stars

```
# Define the icon for the stream sediment samples
ss_icon_url = "https://wiki.openstreetmap.org/w/images/
thumb/e/e3/Volcano-8.svg/8px-Volcano-8.svg.png"

# Create Leaflet map of Murphy geology
leaflet() %>%
 addProviderTiles("Esri.WorldImagery", group = "Satellite")
%>% # satellite map
  addTiles(group = "Topographic") %>% # Base map
  addMarkers(data= bmrgg_sf,
          icon = makeIcon(ss_icon_url, iconWidth = 5,
iconHeight = 5), # icon for the samples
          group = "Stream sediments", # for control
          popup = ~SAMPLEID) %>% # for naming the samples
  addPolygons(data=geo_bmrgg,
          group="Geology", # for control
          fillColor = ~pal(SYMBOL), fillOpacity = 0.7,
# fill colour and opacity
         color="black", weight = 1, # boundary color and width
          popup = ~paste(geo_bmrgg$SYMBOL,"<br>",geo_
bmrgg$STRAT_UNIT)) %>% # For geological units
  addPolylines(data=flt_bmrgg,
          group="Faults",
          color = "black", weight = 2 ) %>%
  addLegend(position = "bottomright",
          group ="Legend",
          colors = pal(sort(unique(geo_bmrgg$SYMBOL))),
          labels = sort(unique(geo_bmrgg$SYMBOL)),
          title = "Symbols") %>%
```

```
  addScaleBar(position = "bottomleft",
            options = scaleBarOptions(imperial = FALSE)) %>%
  addLayersControl(overlayGroups = c("Topographic","Satellite"
,"Stream sediments","Geology","Faults","Legend"))
```

For this map, we used the 'ss_icon_url' variable with the URL that specifies the icon to be used for the stream sediment samples. This is very complete version of a 'leaflet' map (it is a three-star difficulty code snippet). The breakdown of the functions and parameters called is:

leaflet(): It initialises a new 'leaflet' object.

addProviderTiles(): It adds a new tile layer to the map using the "Esri. WorldImagery" provider, with 'group = "Satellite"' used to group the layer for use in the layer control.

addTiles(): It adds a new tile layer to the map using the default provider, with 'group = "Topographic"' used to group the layer for use in the layer control.

addMarkers(): It adds a new marker layer to the map, with 'data = bmrgg_sf' specifying the data source, icon used to specify the icon for the markers, group = "Stream sediments" used to group the markers for use in the layer control, and 'popup = ~SAMPLEID' used to add a popup to each marker displaying the 'SAMPLEID' column from the 'bmrgg_sf' data frame.

addPolygons(): It adds a new polygon layer to the map, with 'data = geo_bmrgg' specifying the data source, 'fillColor = ~pal(SYMBOL)' and 'fillOpacity = 0.7' parameters are used to specify the fill colour and opacity based on the 'SYMBOL' column in 'geo_bmrgg', 'color = "black"' and 'weight = 1' are used to specify the boundary colour and width, and 'popup = ~paste(geo_bmrgg$SYMBOL, "<br>", geo_bmrgg$STRAT_UNIT)' used to add a popup to each polygon displaying the 'SYMBOL' and 'STRAT_UNIT' columns from the 'geo_bmrgg' data frame.

addPolylines(): It adds a new polyline layer to the map, with 'data = flt_bmrgg' specifying the data source, 'color = "black"' and 'weight = 2' used to specify the colour and width of the lines, and 'group = "Faults"' used to group the lines for use in the layer control.

addLegend(): It adds a new legend to the map, with 'position = "bottomright"' specifying the position of the legend, 'group = "Legend"' used to group the legend for use in the layer control, colours and labels used to specify the colours and labels for the legend based on the 'SYMBOL' column in 'geo_bmrgg', and 'title = "Symbols"' used to specify the title of the legend.

addScaleBar(): It adds a new scale bar to the map at the bottom left, with 'options = scaleBarOptions(imperial = FALSE)' used to specify that the scale bar should use metric units.

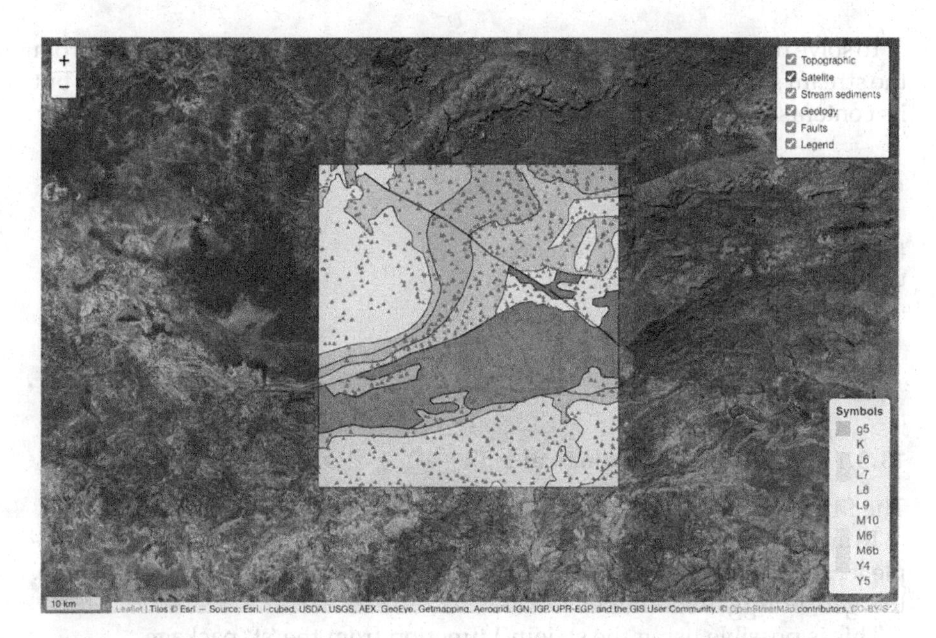

FIGURE 13.4
The geological map of the Murphy region plotted using the 'leaflet' package.

addLayersControl(): It adds a new layer control to the map, with 'overlayGroups' used to specify the groups to be included in the layer control.

Although it's a long sequence of code, it's composed of simple parts that are easy to understand.

Figure 13.4 shows the results from this plot. Note that the control panel is activated and that the background image is the satellite image provided by 'ESRI.WorldImagery'.

This coding provides a greater flexibility for viewing the different information and with different backgrounds. It is also possible to obtain popup information from the plotted layers, in this case 'Stream sediments', 'Geology', the 'Faults', and the 'Legend'.

13.3 Subsetting the Stream Sediments

This example aims to dig further into the analysis of the Northern Territory data, looking to answer to two basic questions:

i. *What is the geological unit that each sample belongs to?*
ii. *How many stream sediment samples each geological unit has?*

To solve these questions, one must verify that in the 'bmrgg_sf' dataset, from the stream sediments campaign, there is a variable named 'LITHOLOGY' but its contents is empty.

#13-09/01

```
# Verify the lithology elements
unique(bmrgg_sf$LITHOLOGY)
```

The result is as follows:

```
[1] ""
```

13.3.1 Subsetting by Geological Unit

The first question is common in any geological campaign.

To answer this question, we can create a cross table that connects each sample location with the variable 'SYMBOL' from the geological unit dataset (i.e. the 'geo_bmrgg' data frame).

This is possible using the st_join() function from the 'sf' package.

With the st_join() function, it is possible to use the nearest polygon to identify the possible geological unit for each stream sediment sample.

#13-09/02

```
# Identify the points that intersect with polygons
bmrgg_sf_join = st_join(bmrgg_sf, geo_bmrgg["SYMBOL"], join =
st_within, left = TRUE)

# How many samples per SYMBOL
count(group_by(as.data.frame(bmrgg_sf_join), SYMBOL))
```

A spatial join is then performed between 'bmrgg_sf' and the 'geo_bmrgg' object based on the nearest feature, and the SYMBOL attribute is added to the resulting 'sf' object using the st_join() function. The 'left = TRUE' argument defines that all rows from the left spatial object will be retained in the result, even if there are no corresponding matches in the right spatial object. When there is no match in the right object, the additional columns coming from the right object will be filled with not available (NA) values. The argument 'join = st_within' indicates that the join is made using the st_within() function to verify the join.

The as.data.frame() function is used to convert the resulting 'sf' object to a data frame, and the 'dplyr' package's count() and group_by() functions are used to count the number of samples for each 'SYMBOL'.

The result is a data frame with two columns, one containing the 'SYMBOL' values and the other containing the count of samples for each 'SYMBOL'.

```
# A tibble: 12 × 2
# Groups:   SYMBOL [12]
   SYMBOL    n
   <chr>  <int>
 1 K       129
 2 L6      10
 3 L7      26
 4 L8      64
 5 L9      203
 6 M10     6
 7 M6      319
 8 M6b     265
 9 Y4      17
10 Y5      125
11 g5      363
12 NA      7
```

Note that there are 7 points that have NA as a result, meaning that these points are not within any geological unit.

As an alternative, one can use the 'st_nearest_feature' parameter. This result attributes the geological unit symbol for every stream sediment based on its nearest feature.

#13-09/03 

```
# Identify the points that intersect with polygons
bmrgg_sf_join = st_join(bmrgg_sf, geo_bmrgg["SYMBOL"], join =
st_nearest_feature, left = TRUE)

# How many samples per SYMBOL
count(group_by(as.data.frame(bmrgg_sf_join), SYMBOL))
```

The resulting join is as follows:

```
# A tibble: 11 × 2
# Groups:   SYMBOL [11]
   SYMBOL    n
   <chr>  <int>
 1 K       129
 2 L6      10
 3 L7      27
 4 L8      65
 5 L9      203
 6 M10     6
 7 M6      321
 8 M6b     266
 9 Y4      17
10 Y5      127
11 g5      363
```

13.3.2 Naming the Geological Units

Another useful function is to aggregate the geological description for each of the 'SYMBOL' variables in the stream sediments datasets. For this, a data frame named 'geo_description' is created with the names for each of the symbols and the corresponding descriptions.

#13-10/01

```
# Create a data frame with symbol and description
geo_description = as.data.frame(cbind( name = c("g5",  "K",
"L6", "L7", "L8", "L9", "M10", "M6","M6b", "Y4", "Y5"),
  description = c("Nicholson Granite Complex","Carpentaria
Basin","Wire Creek Sandstone", "Fickling Group","Doomadgee
Formation","South Nicholson Group", "Bukalara Sandstone,
Wessel Group", "Parsons Range, Tawallah, and Katherine River
Groups", "Seigal Volcanics","Murphy Metamorphics","Cliffdale
Volcanics") ))
```

For aggregating the two datasets, using either 'sp' and 'sf' objects, the merge() function is used.

#13-10/02

```
# Merge the geological description to complete the geology
# data frame
bmrgg_sf_merged = merge(x = bmrgg_sf_join, y = geo_description,
                        by.x = "SYMBOL", by.y = "name")
```

13.3.3 Answering the Questions

The working dataset 'bmrgg_sf_merged' is created as a result of this operation. The first two questions are answered:

i. *What is the geological unit that each sample belongs to?*

To know what geological unit each sample belongs to is straightforward. The following code retrieves 10 random samples of the merged dataset:

#13-10/03

```
# Retrieve 10 samples of the merged dataset
set.seed(123)
bmrgg_sf_merged[sample(1:nrow(bmrgg_sf_merged), 10),
c("SAMPLEID", "description")]
```

The output is as follows:

```
Simple feature collection with 10 features and 2 fields
Geometry type: POINT
Dimension:     XY
Bounding box:  xmin: 137.5244 ymin: -17.89087 xmax: 137.9822
ymax: -17.54048
Geodetic CRS:  WGS 84
        SAMPLEID                          description       geometry
415   7055993      Carpentaria Basin POINT (137.5277 -17.77482)
463   6996357      Carpentaria Basin POINT (137.5244 -17.64066)
179   6994074      Nicholson Granite Complex POINT
                                            (137.6422 -17.88708)
526   6996097      Fickling Group  POINT (137.747 -17.89087)
195   7061273      Nicholson Granite Complex POINT
                                            (137.9822 -17.75346)
938   7058638 Parsons Range, Tawallah, and Katherine River
                                 Groups POINT (137.9123 -17.54564)
1142  6982370      Seigal Volcanics POINT (137.7025 -17.72873)
1323  6988546      Seigal Volcanics POINT (137.7116 -17.59207)
1253  7051577      Seigal Volcanics  POINT (137.799 -17.54048)
1268  6982690      Seigal Volcanics   POINT (137.6231 -17.791)
```

And now let's answer the second question:

ii. *How many stream sediment samples each geological unit has?*

To retrieve this value, simply use a count() function combined with a group_by() function.

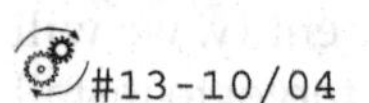

```
#13-10/04

# Count the results
count(group_by(as.data.frame(bmrgg_sf_merged), description))
```

The output is as follows:

```
# A tibble: 11 × 2
# Groups:   description [11]
  description                                          n
  <chr>                                            <int>
 1 Bukalara Sandstone, Wessel Group                    6
 2 Carpentaria Basin                                 129
 3 Cliffdale Volcanics                               127
 4 Doomadgee Formation                                65
 5 Fickling Group                                       27
 6 Murphy Metamorphics                                17
 7 Nicholson Granite Complex                         363
```

```
 8 Parsons Range, Tawallah, and Katherine River Groups      321
 9 Seigal Volcanics                                          266
10 South Nicholson Group                                     203
11 Wire Creek Sandstone                                      10
```

We are now in conditions to further explore these datasets. The next step is to verify the several statistical properties of our data.

13.4 Group Statistics

Group statistics refer to the process of summarising and analysing data within different groups or categories. This approach is especially useful for exploring patterns and differences across subgroups in a dataset, allowing to gain valuable insights and make informed decisions based on the grouped data.

To further develop our spatial data analysis in the study case datasets, we will:

i. Perform the descriptive statistics, by geological unit.
ii. Verify the correlations between chemical elements, based on geological unit.
iii. Find out what are the chemical elements that are dependent or independent from the geological unit.

13.4.1 Descriptive Statistics by Geological Unit

To conduct descriptive statistical analysis for each geological entity, we will use the describeBy() function from the 'psych' package. Don't forget to install it; otherwise, you will receive a friendly error message.

By specifying a grouping variable, such as the geological unit names or abbreviated symbols, one can obtain separate descriptive statistics, including mean, standard deviation, minimum, maximum, and quartiles for each group. This function is particularly valuable when dealing with diverse geological data, as it efficiently generates summary statistics, aiding in the exploration of variations and patterns within different geological units.

Such insights can be essential in geological research and analysis, enabling us to make informed interpretations and comparisons among the distinct geological categories present in the dataset.

#13-11

```
# Load the library
library(psych)
```

```
# Identify the variables present
names(bmrgg_sf_merged)

# Subset to geological symbol and chemical analysis
df = as.data.frame(bmrgg_sf_merged[,c(1,20:37)])

# Describe the data
describeBy(df[,2:19],df$SYMBOL)
```

The result is a very long table for each of the 12 geological units. The results for the 'g5' geological unit are as follows:

```
Descriptive statistics by group
group: g5
       vars n    mean    sd     median trimmed mad     min     max    range   skew  kurtosis se
AG_PPM 1    363   -0.89    0.47  -1.0    -1.00    0.00  -1.00     2.0    3.00  3.98 14.35    0.02
AS_PPM 2    363    3.50    2.23   3.0     3.17    1.48  -1.00    22.0   23.00  3.72 23.79    0.12
BA_PPM 3    363  642.05 177.66 649.0   646.66  183.84 143.00  1170.0 1027.00 -0.19 -0.31    9.32
…
W_PPM  17   363    5.57    9.46   5.0     4.35    2.97  -3.00   113.0  116.00  5.93 54.45    0.50
ZN_PPM 18   363   23.95   15.94  21.0    22.04   11.86   4.00   201.0  197.00  4.76 43.38    0.84
------------------------------------------------------------------------------------------------
```

With these results, it is possible to create correlation diagrams grouped by the geological units that each sample belongs to.

13.4.2 Correlation by Geological Unit

It is general understanding that depending on its mineralogy, geodynamic context, and evolution every geological unit will have different geochemical characteristics. For the data from the Bureau of Mineral Research, this is not surely different, and to understand the relation between the chemical elements for each geological unit, one can use a group statistics approach.

The first step involves some data preparation, subsetting the dataset to remove variables with no data (in this case the 'MO_PPM').

The conversion of a 'sf' object to a data frame using the as.data.frame() function creates an additional column named geometry that contains the geometric information, i.e. usually the coordinates. To avoid this, one can use the st_drop_geometry() function, which removes the geometric information from the 'sf' object.

A working variable named 'data' is created for handling the dataset.

The next operation is to impute zero values to all negatives to perform the correlations.

#13-12/01

```
# Load the library
library(corrgram)

# Only the geochemical data and the SYMBOL
data = as.data.frame( st_drop_geometry( bmrgg_sf_merged[,
c(20:28, 30:37, 1)] ))
```

```
# Convert negatives to NAs
data[data < 0] = NA
```

After the data preparation, it is time to create the correlation diagrams for each unit. This is possible using a for() loop that cycles all the geological units. Inside the loop, a subset of the data based on the geological symbol is created. From this subset, the columns with NAs are removed. In the last step of the loop, the corresponding 'corrgram' of the data is plotted.

The removal of the NAs is made only inside the loop as there are geological units that have NAs for a certain chemical element and not for others. If this operation was made before the loop, the data would have only then elements Ba, Cu, Fe, Ti, and Zn that are the only ones that don't have NAs.

#13-12/02

```
# Cycle all the geological units, sorted
for(var in sort(unique(data$SYMBOL))) {
  # Select a subset that has the geological unit
  df = data[data$SYMBOL == var, ]

  # Remove columns that have NA
  df = df[, colSums(is.na(df)) == 0 ]

  # The corrgram
  corrgram(df, upper.panel=panel.cor,
            main=paste("Correlation for", var, "geological
Unit \n for", ncol(df)-1,"variables and", nrow(df), "rows"),
            order=T)
}
```

This code snippet produces a series of 'corrgrams' for each geological unit. Figure 13.5 displays the examples from the 'g5' and 'M6' geological units.

Note that for creating the title, the paste() function is used to combine several pieces of information, such as the 'var' name (geological unit) and the number of columns (variables) and rows (samples).

It is not the goal of this book to suggest geological interpretations for the data treated in the case studies. Nevertheless, the analysis of the correlation diagrams readily shows that the samples from these geological units have a notoriously different behaviour.

13.4.3 Testing Independency from Geological Unit

Another approach to the analysis of the data is to verify how the variables have independent or dependent behaviours. For this task, we will apply the analysis of variance (ANOVA) test.

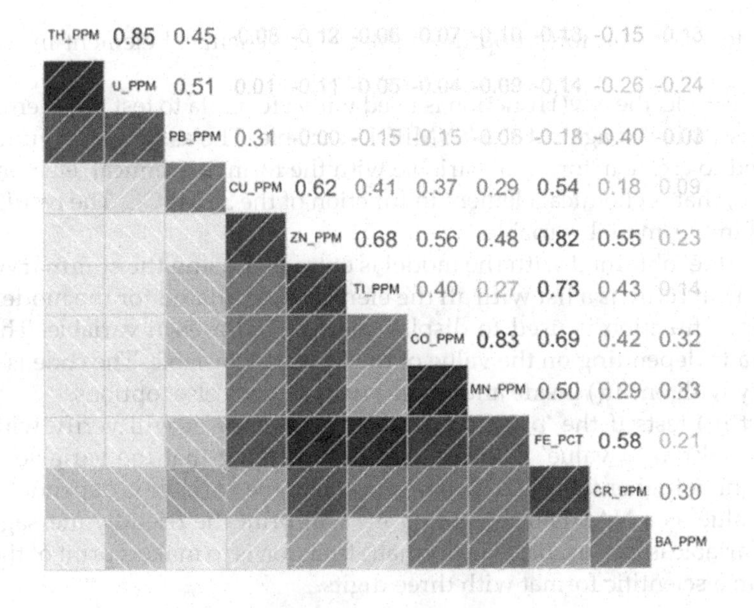

Correlation for g5 geological Unit
for 11 variables and 363 rows
TH_PPM 0.85 0.45 -0.15
U_PPM 0.51 -0.26 -0.24
PB_PPM 0.31 -0.15 -0.15 -0.18 -0.40
CU_PPM 0.62 0.41 0.37 0.29 0.54 0.18 0.09
ZN_PPM 0.68 0.56 0.48 0.82 0.55 0.23
TI_PPM 0.40 0.27 0.73 0.43 0.14
CO_PPM 0.83 0.69 0.42 0.32
MN_PPM 0.50 0.29 0.33
FE_PCT 0.58 0.21
CR_PPM 0.30
BA_PPM
a)

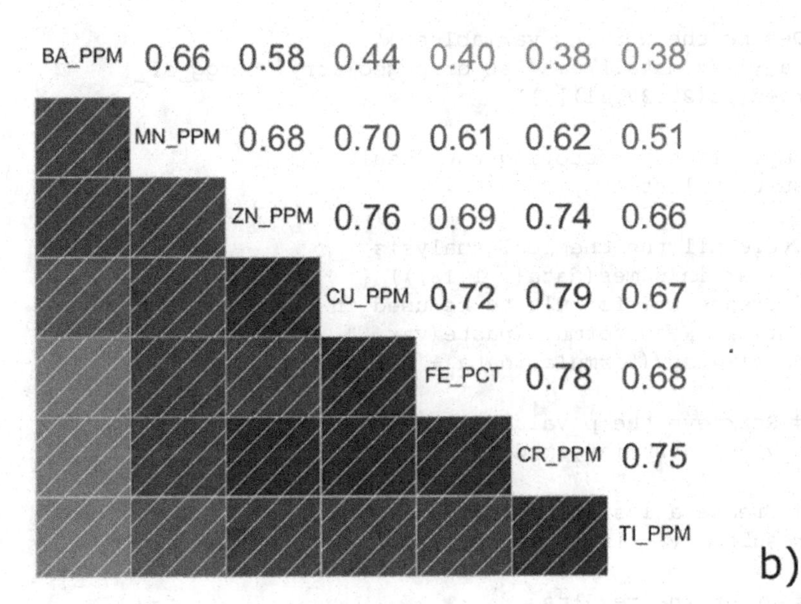

FIGURE 13.5
Correlation for the chemical analysis by geological unit. a) g5 – Nicholson Granite Complex and b) M6 – Parsons Range, Tawallah, and Katherine River Groups.

The geological question is to verify if the chemical analysis of the stream sediment samples is independent from the geological unit in which they were collected.

This is done using a for() loop that cycles every chemical element in the dataset.

Inside the cycle, the aov() function is used with a formula to test the chemical analysis variables against the 'SYMBOL' variable. The as.formula() function is used to create a 'formula' variable with the format 'chemical_element ~ SYMBOL', that is chemical element in function of the SYMBOL. The results are stored in the 'model' variable.

The 'p_value' obtained with the model is extracted using the summary() function. That returns a list with all the elements calculated for the model.

The print() function is used to display the results for each variable. The print is made depending on the value of the 'p_value' variable. The code is a little tricky with one if() inside another if() with several 'else' options.

The first if() tests if the 'p_value' is a NA; if it is not, it will verify with another if()-else if 'p_value' is lower than 0.05 printing that the variable is independent or higher than 0.05 then printing that the variable is dependent. If the 'p_value' is a NA, then the nested if() will print the friendly message that this variable is 'not usable'. The format() function is to make a print of the 'p_value' in a scientific format with three digits.

#13-13/01

```
# Define the working variables
data =  as.data.frame( st_drop_geometry( bmrgg_sf_
merged[,c(20:37, 1)] ))

# Empty list for storing the results
results = list()

# Cycle all the chemical analysis
for (var in names(data[, 1:18])) {
  # Create the formula to be used in the aov() analysis
  formula = as.formula(paste(var, "~ SYMBOL"))
  model = aov(formula, data = data)

  # Retrieve the p_value
  p_value = summary(model)[[1]][["Pr(>F)"]][[1]]

  # Create a list of p_values
  results[[var]] = p_value

  # Print the results
  if (!is.na(p_value)) {
      if (p_value < 0.05) {
```

```
      print(paste(var, "is dependent on geological unit
(p-value =", format(p_value, scientific = TRUE, digits = 3),
")"))
      } else {
      print(paste(var, "is independent of geological unit
(p-value =", format(p_value, scientific = TRUE, digits = 3),
")"))
      }
  } else {
      print(paste(var, "is not usable"))
  }
}
```

The result is as follows:

```
[1] "AG_PPM is dependent on geological unit (p-value = 3.72e-23 )"
[1] "AS_PPM is dependent on geological unit (p-value = 4.22e-37 )"
[1] "BA_PPM is dependent on geological unit (p-value = 1.99e-286 )"
[1] "BI_PPM is dependent on geological unit (p-value = 2.13e-41 )"
[1] "CO_PPM is dependent on geological unit (p-value = 6.44e-252 )"
[1] "CR_PPM is dependent on geological unit (p-value = 2.64e-200 )"
[1] "CU_PPM is dependent on geological unit (p-value = 5.72e-240 )"
[1] "FE_PCT is dependent on geological unit (p-value = 4.11e-234 )"
[1] "MN_PPM is dependent on geological unit (p-value = 5.2e-116 )"
[1] "MO_PPM is independent of geological unit (p-value = 9.76e-01 )"
[1] "NI_PPM is dependent on geological unit (p-value = 5.35e-267 )"
[1] "PB_PPM is dependent on geological unit (p-value = 1.27e-74 )"
[1] "SN_PPM is dependent on geological unit (p-value = 1.24e-23 )"
[1] "TH_PPM is dependent on geological unit (p-value = 1.6e-111 )"
[1] "TI_PPM is dependent on geological unit (p-value = 6.16e-129 )"
[1] "U_PPM is dependent on geological unit (p-value = 7.22e-127 )"
[1] "W_PPM is dependent on geological unit (p-value = 1.15e-53 )"
[1] "ZN_PPM is dependent on geological unit (p-value = 2.49e-167 )"
```

It is noticeable that all the variables, except the molybdenum (yes, we already knew), are dependent on the geological unit where they were collected.

It is possible to retrieve from the data the chemical elements that are more or less independent of the geological unit. For example, one might want to highlight the three more dependent and independent elements.

For this, the created 'results' list from the previous snippet is used. To retrieve the data from this list, the p-values are extracted from the 'results' list using the unlist() function that simplifies the list retrieving the desired values.

After this, the arrange() function, from the 'dplyr' package, is used to order the rows of the 'results_bmrgg' data frame by the values of the 'p_value' column.

#13-13/02

```
# Convert the list to a data frame
results_bmrgg = data.frame(variable = names(results), p_value
= unlist(results))
```

```
# Sort the data frame by ascending p-value
results_bmrgg = arrange(results_bmrgg, p_value)

# Print the top and bottom 3 dependent variables
print("Top 3 dependent variables:")
print(head(results_bmrgg, 3))
print("Top 3 independent variables:")
print(tail(results_bmrgg, 3))
```

The results are as follows:

```
[1] "Top 3 dependent variables:"
       variable     p_value
BA_PPM   BA_PPM 1.994513e-286
NI_PPM   NI_PPM 5.354915e-267
CO_PPM   CO_PPM 6.440951e-252

[1] "Top 3 independent variables:"
       variable     p_value
SN_PPM   SN_PPM 1.241201e-23
AG_PPM   AG_PPM 3.722945e-23
MO_PPM   MO_PPM 9.758559e-01
```

For the Mo, we were already expecting it. What about the others?

To visualise this behaviour, a series of 'boxplot' graphs is created, organised by a geological unit.

13.4.4 Plotting the Independence

The first step is to set the variables and load some helping packages. The 'gridExtra' and 'grid' packages will be used to organise the graphs.

The head() and tail() functions are used to retrieve only the top and bottom values of the dataset. These values are stored in the variable 'vars'.

#13-14/01

```
# Load the library
library(gridExtra)
library(grid)

# The top and bottom 3
vars = c(head(results_bmrgg, 3)$variable, tail(results_bmrgg,
3)$variable)
```

The 'vars' variable will be used to cycle the three most dependent and independent variables (that are chemical elements) and create the boxplots.

The boxplots will be created using the 'ggplot2' package.

Contrary to what we have been doing previously, we will retain the plots in a list named 'plots'. This list is used to organise the final plot as a combination of the different plots.

The for() cycle creates a data frame named 'plot_data' with each dataset to be plotted, after which each individual plot is created using the ggplot() function and stored in a temporary 'p' variable that is added to the 'plots' list.

#13-14/02

2 stars

```
# Create a empty list of plots
plots = list()

# Create a boxplot for the most dependent variable for each symbol
for(i in 1:length(vars)) {
  # Data frame for each geological unit
  plot_data = data.frame(x = data$SYMBOL, y = data[,vars[i]])
  var_name = vars[i]

  # Create the plot
  p = ggplot(plot_data, aes(x = x, y = y, fill=x)) +
      geom_boxplot() +
      scale_fill_manual(values = murphy_colors) +
      guides(fill = "none")+
      labs(x = "Geo Unit", y = var_name) +
      ggtitle(paste(var_name,  "by Geological Unit"))

  # Add the plot to the list of plots
  plots[[i]] = p
}
```

Finally we use the 'plots' list, combined with the functions textgrob() and arrangeGrob(), from the 'grids' package to organise the plots and titles. The textGrob() function is used to create a graphical text grob (graphics object) in the grid system. Grobs are graphical objects in grid graphics, and textGrob() specifically generates a grob for displaying text. The arrangeGrob() is used to arrange multiple grobs in a grid layout. It allows the creation of complex layouts by combining multiple plots or graphical elements into a single display. This function is particularly useful when one wants to arrange multiple plots or graphs together in a custom layout.

To create the final plot, the function 'grid.arrange' from the 'gridExtra' package is used to create a final printable plot.

#13-14/03

```
# Arrange the plots in two rows
# First column of plots (dependent variables)
left_col_plots = list(plots[[1]], plots[[2]], plots[[3]])
```

```
# Second column of plots (independent variables)
right_col_plots = list(plots[[4]], plots[[5]], plots[[6]])

# Title text for both columns
t1 = textGrob("Dependent", gp = gpar(fontsize = 14, fontface =
"bold"))
t2 = textGrob("Independent", gp = gpar(fontsize = 14, fontface
= "bold"))

# Arrange the plots in a grid
grid.arrange(
  arrangeGrob(
      arrangeGrob(grobs = left_col_plots, ncol = 1, top = t1),
      arrangeGrob(grobs = right_col_plots, ncol = 1, top = t2),
      ncol = 2, widths = c(1, 1)
  )
)
```

Figure 13.6 displays the result from the plot obtained with the grid.arrange() function.

It is observable from the boxplots that Mo, Ag, and Sn have very low variability in all the geological units, with exception of some outliers; that is why they are the most independent from the geological unit. Contrarily, Ba, Ni,

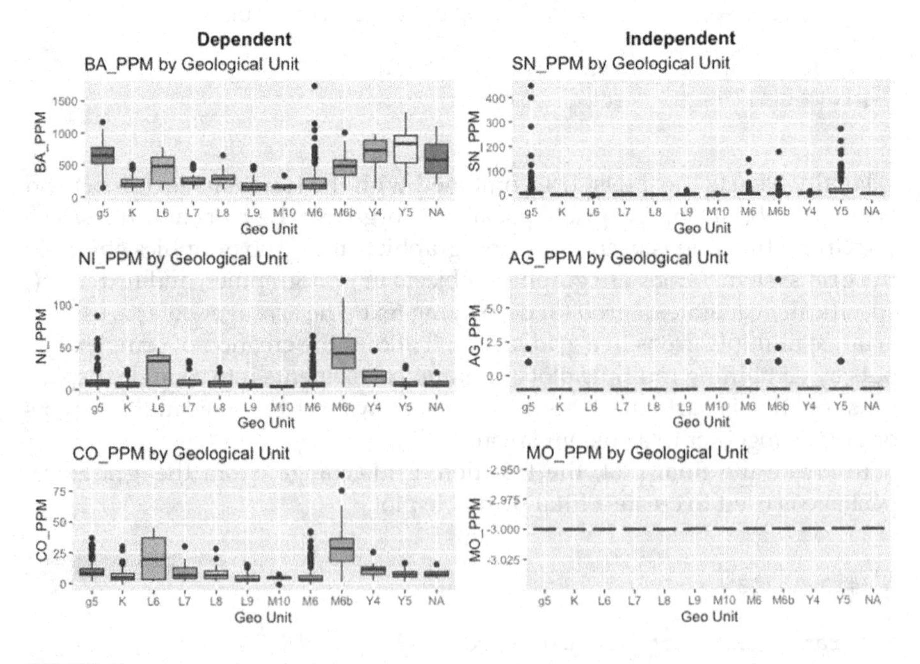

FIGURE 13.6
A plot of the three most dependent and independent variables.

and Co vary significantly between the different geological units, therefore its dependency on the geological unit.

13.5 This Is Not My Fault

Another question that arises from looking at the geological map is as follows:

i. *Is the chemistry of the stream sediments related to the presence of the fault?*

For this analysis, we propose to create a 1000 metres buffer around the fault and select the stream sediments samples inside the buffer. Based on the descriptive statistics from the two datasets, i.e. near the fault and away from the fault, one can discuss the independence of the datasets.

For creating the buffer around the fault, the 'sf' dataset is used with the st_buffer() function.

The function creates two polygon segments that need to be merged to identify the elements inside and outside the polygon. For merging, the st_union() function is used. The merged polygons have to be re-attributed its geometry by using the st_as_sf() function.

#13-15/01

```
# Create the buffer. Use 'sf' for ease of calculating the buffer
buf_flt_sf = st_buffer(flt_bmrgg, 1000)

# Merge the polygons created
buf_flt_sf = st_union(buf_flt_sf)

# Create a new 'sf' object with the merged geometry
buf_flt_sf = st_sf(geometry = buf_flt_sf)
```

Next step is to verify which points are inside and outside the buffer. For this, the st_intersection() and st_difference() functions are used in combination. The intersection identifies points inside the buffer whereas the difference is to retrieve the ones that are not inside, i.e. that are outside. Yes, it is clever!

#13-15/02

```
# Subset the samples inside and outside the buffer
bmrgg_sf_inside = st_intersection(bmrgg_sf_merged, buf_flt_sf)
bmrgg_sf_outside = st_difference(bmrgg_sf_merged, buf_flt_sf)
```

After this, two datasets are created, 'bmrgg_sf_inside' for points inside the buffer and 'bmrgg_sf_outside' for points outside the buffer.

To verify the results, a 'leaflet' map is used where the buffer is plotted and the points inside the buffer are plotted separately in a bigger size.

#13-15/03

1 star

```
# Verify the result
leaflet() %>%
  addTiles(group = "Topographic") %>% # Base map
  addMarkers(data= bmrgg_sf_inside,
            icon = makeIcon(ss_icon_url, iconWidth = 8,
iconHeight = 8), # icon
            group = "Inside",
            popup = ~ SAMPLEID) %>% # for naming the samples
  addMarkers(data= bmrgg_sf_outside,
            icon = makeIcon(ss_icon_url, iconWidth = 5,
iconHeight = 5), # icon
            group = "Outside",
            popup = ~ SAMPLEID) %>% # for naming the samples
  addPolygons(data=geo_bmrgg,
            group="Geology", # for control
            fillColor = ~pal(SYMBOL), fillOpacity = .7,
# color and opacity
            color="black", weight = 1, # boundary color and width
            popup = ~paste(geo_bmrgg$SYMBOL,"<br>",geo_
bmrgg$GEOLREGION, "<br>", geo_bmrgg$description)) %>% # For
geological units
  addPolygons(data=buf_flt_sf,
            group="Buffer", # for control
            fillColor = "grey", fillOpacity = .7,
# fill color and opacity
            color="grey", weight = 1)  %>%
# color and width Buffer Fault
  addPolylines(data=flt_bmrgg,
            group="Faults",
            color = "black", weight = 2 ) %>%
  addLegend(position = "bottomright",
            group ="Legend",
            colors = pal(sort(unique(geo_bmrgg$SYMBOL))),
            labels = sort(unique(geo_bmrgg$SYMBOL)),
            title = "Symbols") %>%
  addScaleBar(position = "bottomleft",
            options = scaleBarOptions(imperial=FALSE)) %>%
  addLayersControl(
       overlayGroups = c("Topographic","Buffer","Inside",
"Outside", "Geology","Faults","Legend"))
```

The resulting map is displayed in Figure 13.7.

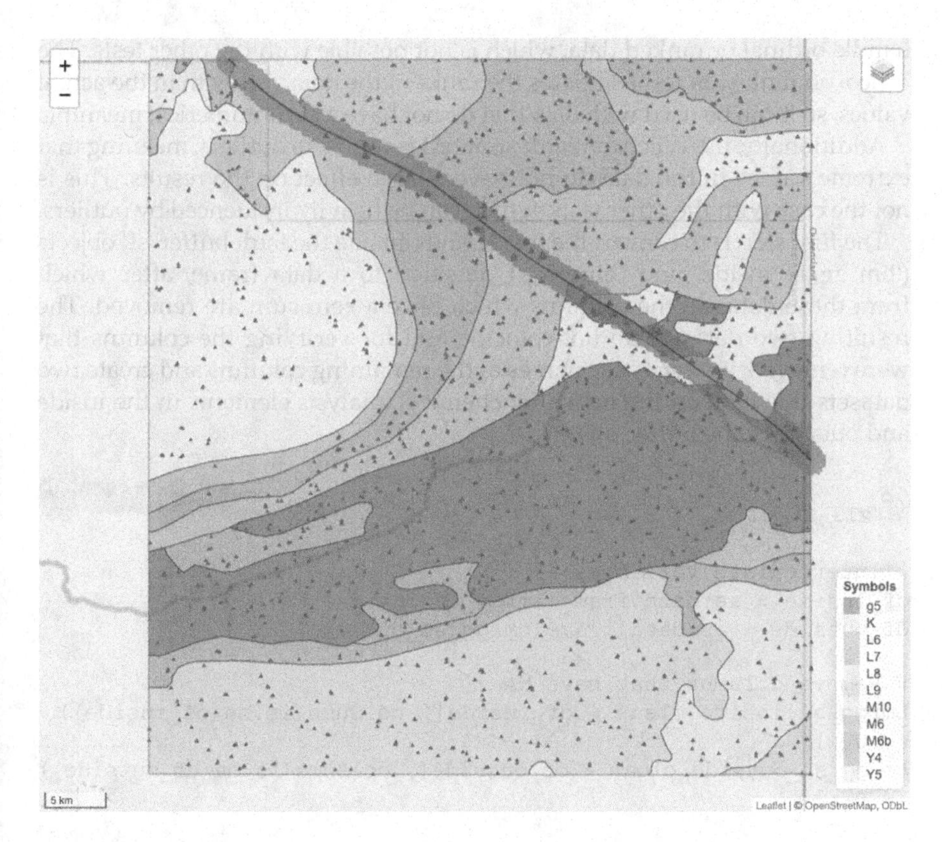

FIGURE 13.7
The map with the buffer of 1000 metres around the fault. The small triangles are the samples outside the buffer, and the bigger triangles are the samples that are inside the buffer.

13.5.1 Where Are You From?

To verify if behaviour of the chemical elements is dependent on its proximity to the fault, it is possible to use independence tests. Previously, the ANOVA and t-tests were used, but their results are influenced by the distribution of the sample and its variance.

For this example, we suggest the use of the Wilcoxon rank-sum test.[6] The Wilcoxon rank-sum test is a non-parametric test that does not make any assumptions about the distribution of the data. This is an advantage over the other tests, which both assume that the data is normally distributed. If the assumptions of the t-test or ANOVA are not met, the results may be unreliable or incorrect. Another advantage of the Wilcoxon rank-sum test is that it can

[6] For more information, consult https://en.wikipedia.org/wiki/Mann%E2%80%93Whitney_U_test

handle ordinal or ranked data, which is not possible with the other tests. The Wilcoxon rank-sum test compares the ranks of the data, rather than the actual values, so it can be used with data that do not have a clear numerical meaning.

Additionally, the Wilcoxon rank-sum test is robust to outliers, meaning that extreme values in the data do not have a large effect on the results. This is not the case with the other tests, which can be heavily influenced by outliers.

The first step is to convert the inside and outside the fault buffer 'sf' objects ('bmrgg_sf_inside' and 'bmrgg_sf_outside') to a data frame, after which from these datasets the columns which have a zero sum are removed. The resulting two datasets do not coincide and for verifying the columns that were removed, we verify the names of the remaining columns and create two datasets that contain the matching chemical analysis elements in the inside and outside of the buffer points.

#13-16/01

```
# Intermediate variables
df_inside = as.data.frame(bmrgg_sf_inside)
df_outside = as.data.frame(bmrgg_sf_outside)

# Remove columns that have NA
bmrgg_sf_inside_clean = df_inside[, colSums(is.na(df_inside))
== 0  ]
bmrgg_sf_outside_clean = df_outside[, colSums(is.na(df_outside))
== 0  ]

# Verify the cleaned data
names(bmrgg_sf_inside_clean)
names(bmrgg_sf_outside_clean)
```

The lists of the column names for the inside data are as follows:

```
 [1] "SYMBOL"          "UNIQ_ID"         "ID"            "SAMPLEID"
     "SAMPLEREF"
 [6] "SAMPLE_TYPE"     "SAMPLE_METHOD"   "LITHOLOGY"     "MINMESH"
     "MAXMESH"
[11] "COMPANY"         "ACCURACY"        "TITLE"
     "MAP_SHEET_100K" "MAP_SHEET_250K"
[16] "REPORT_NO"       "OPEN_FILE"       "JOB_NO"        "COMMENTS"
     "AG_PPM"
[21] "AS_PPM"          "BA_PPM"          "BI_PPM"        "CO_PPM"
     "CR_PPM"
[26] "CU_PPM"          "FE_PCT"          "MN_PPM"        "MO_PPM"
     "NI_PPM"
[31] "PB_PPM"          "SN_PPM"          "TH_PPM"        "TI_PPM"
     "U_PPM"
[36] "W_PPM"           "ZN_PPM"          "description"
     "geometry"
```

And for the outside:

```
 [1] "SYMBOL"           "UNIQ_ID"          "ID"          "SAMPLEID"
     "SAMPLEREF"
 [6] "SAMPLE_TYPE"      "SAMPLE_METHOD"    "LITHOLOGY"   "MINMESH"
     "MAXMESH"
[11] "COMPANY"          "ACCURACY"         "TITLE"
     "MAP_SHEET_100K"   "MAP_SHEET_250K"
[16] "REPORT_NO"        "OPEN_FILE"        "JOB_NO"      "COMMENTS"
     "AG_PPM"
[21] "AS_PPM"           "BA_PPM"           "BI_PPM"      "CO_PPM"
     "CU_PPM"
[26] "FE_PCT"           "MO_PPM"           "NI_PPM"      "PB_PPM"
     "SN_PPM"
[31] "TH_PPM"           "TI_PPM"           "U_PPM"       "W_PPM"
     "ZN_PPM"
[36] "description"      "geometry"
```

The columns of the Cr and Mn do not appear in the outside variable. Therefore, they must be removed from the inside dataset, in order to have both datasets with the same columns.

#13-16/02

```
# The elements Cr [25] and Mn [28] are missing
# from the bmrgg_sf_outside_clean dataset
# Subset the two dataframes with the same variables
df_inside_chem = bmrgg_sf_inside_clean[, c(20:24,26,27,29:37)]
df_outside_chem = bmrgg_sf_outside_clean[, c(20:35)]
```

The datasets created 'df_inside_chem' and 'df_outside_chem' corresponding columns with only the chemical analysis.

To perform the Wilcoxon rank-sum test to all the chemical elements, a for() cycle is used, applying the wilcox.test() function from the 'stats' package. The results of the p-value are stored in a data frame called 'results'.

#13-16/03

```
# Create an empty data frame
results = data.frame(element = character(0), p_value = numeric(0))

# Calculate the Wilcox test for all the chemical elements
for( i in names(df_inside_chem)) {
  results_wilcox =  wilcox.test(as.numeric(df_inside_chem[[i]]),
                                as.numeric(df_outside_chem[[i]]))
```

```
# Print the result from the Wilcox test
  print(paste0(i," p value: ", round(results_wilcox$p.value, 3)))

# Save a variable with the results
  new_data = data.frame(element = i, p_value = results_
wilcox$p.value)
  results = rbind(results, new_data)
}
```

The output of the cycle is the total of all the chemical elements:

```
[1] "AG_PPM p value: 0"
[1] "AS_PPM p value: 0.002"
[1] "BA_PPM p value: 0"
[1] "BI_PPM p value: 0"
[1] "CO_PPM p value: 0.691"
[1] "CU_PPM p value: 0"
[1] "FE_PCT p value: 0.219"
[1] "MO_PPM p value: NaN"
[1] "NI_PPM p value: 0.014"
[1] "PB_PPM p value: 0.004"
[1] "SN_PPM p value: 0"
[1] "TH_PPM p value: 0.056"
[1] "TI_PPM p value: 0.521"
[1] "U_PPM p value: 0.025"
[1] "W_PPM p value: 0.007"
[1] "ZN_PPM p value: 0"
```

To verify the results, one can use the filter() function from the 'dplyr' package.

#13-16/04 1 star

```
# Load the library
library(dplyr)

# Variables with significant Wilcoxon rank-sum test
print(paste("Dependent variables"))
filter(results, p_value < 0.05)

# Variables with significant Wilcoxon rank-sum test
print(paste("Independent variables"))
filter(results, p_value >= 0.05)
```

The outputs are as follows:

```
[1] "Dependent variables"

   element       p_value
1  AG_PPM 7.803035e-10
```

```
2    AS_PPM 2.022294e-03
3    BA_PPM 4.075220e-05
4    BI_PPM 8.452405e-05
5    CU_PPM 2.690277e-04
6    NI_PPM 1.359166e-02
7    PB_PPM 4.326716e-03
8    SN_PPM 4.875605e-06
9     U_PPM 2.502587e-02
10    W_PPM 7.208325e-03
11   ZN_PPM 4.326157e-04
[1] "Independent variables"

  element   p_value
1  CO_PPM 0.69092050
2  FE_PCT 0.21851063
3  TH_PPM 0.05611452
4  TI_PPM 0.52098425
```

In the case of a p-value less than 0.05, there is evidence that the two samples may be from different populations, which means that they are dependent. However, it does not necessarily mean that they are dependent in the sense of a causal relationship between the two variables.

Again, one must notice that for extracting geological and geochemical conclusions, one must consider the specific geological context and have further knowledge of the study area. Nonetheless, one could identify the elements Ag, As, Ba, Bi, Cu, Ni, Pb, Sn, U, W, and Zn as targets to verify their relationship with the proximity to the fault. The remaining elements have an independent behaviour relative to its proximity to the fault. The Mo element does not apply to this case.

13.6 Concluding Remarks

Yes, I know that working with real data to solve everyday quests is far more challenging than the essay tube examples that are presented during the textbook. This chapter is designed to provide some insights of how to solve real but anyway simplified geological questions, presenting the reasoning necessary for solving it.

Notice that sometimes intricate programming approaches such as using nested if()-else expressions or double or multiple for() cycles are necessary to answer the questions posed. Also, the maps and visualisations are becoming more complex to provide the reader with a more usable result. My advice is to always slice the problem in smaller parts and test if each part is delivering what is expected or reasonable, before continuing to the next step.

During this explanation, new statistical and data handling functions were introduced with examples of the application of spatial operations in real-life cases. Anyhow, remember that this is not an introductory textbook on statistics; therefore, not all the possible functions and solutions for statistical problems are presented; mostly I selected the most straightforward approaches to statistical problem solving. Every time a new statistical or spatial analysis problem emerges, remember to read the basic and advanced treatises to identify the solutions that apply to your problem. For most of the problems, I am sure that R will surely have a solution. Naturally, always remember to use internet search wisely to help in identifying what is the better solution to your problem.

14

Into the Grid with Rasters

Raster or gridded data is a key component in spatial analysis. In general terms, raster data is often used to represent spatially continuous surfaces, such as elevation or temperature, as a grid of cells (pixels) with values assigned to each cell. Gridded data is similar but is better used to represent categorical or discrete data, such as land cover or administrative boundaries. However, things can be interchanged, and for most of the cases, their characteristics are somehow similar and can be treated as one unique data type.

Raster data is preferably handled using the 'raster' and 'terra' packages, which possess the most complete set of functions for handling this type of data. Raster data can be created from scratch using functions like raster() and stack(), from 'raster' package, or rast() function from 'terra' package, or if stored in a file, might be imported from the most common raster file formats such as Geographic Tagged Image File Format (GeoTIFF) or Network Common Data Form (NetCDF).

Raster and vector data can often be transformed from one format to the other; for instance, the rasterize() function found in the 'raster' and 'terra' packages facilitates the conversion of vector data, like shapefiles, into raster format using a specified attribute. Nonetheless, it is crucial to recognise that converting vector data into raster format may introduce uncertainty or diminish spatial precision due to the resolution of the resulting raster. This implies that the two formats are not perfectly interchangeable. The process of converting data from vector to raster and then back to vector is inherently lossy, with the degree of precision lost being contingent upon the pixel size of the raster.

Once the data is loaded and pre-processed in the R environment, it can be treated using an ample range of spatial analysis tools from numerous packages, such as the 'gstat' or 'spatstat', that have functions to perform tasks such as calculating summary statistics, interpolations, or creating maps. R packages also provide a variety of visualisation tools, such as the aforementioned 'ggplot2' or 'leaflet', which contain tools for creating plots and interactive maps based on raster data.

From here on, for practical reasons, we will use raster data for meaning raster and gridded data, except when the distinction is necessary.

DOI: 10.1201/9781032651880-16

14.1 Types of Raster Data

There are different types of raster data that we can use, depending on the nature of the data and the application. Some of the common raster types include the following:

Single-layer raster: It contains only one layer of data, such as panchromatic image, elevation, or temperature, and is often used for simple analyses and visualisations.

Multi-layer raster: It contains multiple layers of data, such as the red, green, and blue channels of an image, and is often used for image visualisations or can be applied in complex analyses and modelling.

Time series raster: It contains multiple layers of data representing the same variable at different points in time and is often used for analysing temporal trends and patterns.

Remote sensing raster: It is a specific type of multi-layer data that is collected from satellite or airborne platforms. This data can be used to create images of the earth's surface and to extract information about the environment and land use. Remote sensing data is often stored in raster format, with each cell representing a unique value, such as temperature, vegetation, or land use. Remote sensing data is useful for environmental monitoring, resource management, and land use planning.

As for the case of raster data, there are different types of gridded data, depending on the nature of the data and the application. Some of the common gridded data types include the following:

Categorical grid: This type of gridded data represents discrete categories, such as land cover, administrative boundaries, or land use, assigned to each grid cell.

Continuous grid: This type of gridded data represents continuous variables, such as population density, elevation, or temperature, assigned to each grid cell.

Time series grid: This type of gridded data contains multiple grids representing the same variable at different points in time and is often used for analysing temporal trends and patterns.

14.2 Types of Raster Files

Raster and gridded data types are typically stored in various file formats, each with its own applications, characteristics, advantages, and limitations.

Some of the common file formats include the following:

GeoTIFF: A georeferenced raster file format that is widely used in the geospatial community. GeoTIFF files can store a variety of raster data types, including elevation, land cover, and temperature.

NetCDF: A versatile file format that can be used to store a variety of geospatial data, including raster data, vector data, and time series data. NetCDF files are often used in scientific applications.

HDF (Hierarchical Data Format): It is a file format that is similar to NetCDF, but it is optimised for storing large datasets. HDF files are often used in high-performance computing applications.

Grid ASCII: A simple text file format that is used to store raster data. Grid ASCII files are often used for data exchange.

PNG, JPG, and other image file types can also be used for storing and sharing raster data. While these file types are primarily designed for storing and displaying visual images, they can also be used for storing and sharing raster or gridded data in certain cases.

PNG (Portable Network Graphics): This is a lossless image format that supports transparency and can be compressed without loss of quality. PNG files are commonly used for storing and sharing raster data in web-based applications, such as online maps and interactive dashboards.

JPG (Joint Photographic Experts Group): This is a lossy image format that is widely used for storing and sharing photographic images. JPG files can be compressed to reduce file size, but at the cost of some loss of image quality. While JPG files are not ideal for storing and sharing raster data that requires high accuracy and precision, they can be used for displaying visualisations of data, such as maps and charts.

Other image file types, such as BMP (Bitmap), GIF (Graphics Interchange Format), and TIFF (Tagged Image File Format) can also be used for storing and sharing raster data in specific applications.

It is important to note that while image file types can be used for storing and sharing raster data in certain cases, they may not be the most appropriate or efficient file format for all spatial data types and applications. In general, it is recommended to use file formats specifically designed for handling spatial data, such as GeoTIFF or NetCDF, whenever possible.

In addition to these file formats, there are also various proprietary file formats used by different software and tools for handling raster and gridded data, such as Environmental Systems Research Institute's Arc Geographic Information System(ESRI's ArcGIS) raster format, Environment for Visualizing Images (ENVI's) raster format, and IDRISI's[1] raster format.

[1] To learn more visit: https://clarklabs.org/terrset/idrisi-gis/

14.3 Reading and Writing Raster Data

Reading raster data (or vectorial, for what matters) is usually the first step in many spatial analyses, as it allows access and manipulation of spatially continuous and categorical data. For this, one of the most commonly used packages for manipulating raster data is the 'raster' package (see Section 10.4.3 in Chapter 10). This package provides functions for reading and writing raster data in a variety of file formats, such as GeoTIFF, NetCDF, and HDF.

For example, to read a GeoTIFF file, one can use the raster() function, like this:

```
# Load the Libraries
library(raster)

# Load the file
myRaster = raster("myRaster.tif")
```

This will create a 'RasterLayer' object, representing the raster data stored in the GeoTIFF file. The raster() function can also read multi-layer rasters, such as those stored in NetCDF files, by specifying the desired layer using the 'varname' argument.

Alternatively, the brick() function, from the 'raster' package, reads a multilayer raster file from various formats, including GeoTIFF, NetCDF, and HDF5.

For example, to read a NetCDF file named 'myStack.nc' containing multiple layers as a 'RasterBrick' object, one can use:

```
# Load the file
myStack = brick("myStack.nc")
```

The 'raster' package also provides functions for writing raster data to file formats such as GeoTIFF, NetCDF, and HDF. For example, to write a raster to a GeoTIFF file, one can use the writeRaster() function.

```
# Write the file
writeRaster(myRaster, "myRaster_output.tif", format = "GTiff")
```

This will write the 'RasterLayer' object 'myRaster' to a new GeoTIFF file named 'myRaster_output.tif'.

The 'terra' package is becoming the new standard for dealing with raster data, and it also provides methods for reading and writing rasters and raster stacks where the rast() function reads a single-layer raster file from various formats, including GeoTIFF, NetCDF, and HDF5.

For example, to read a GeoTIFF file named 'myRaster.tif' as a 'SpatRaster' object, into a variable called 'my_spatRaster', one can use the following code:

```
# Load the Libraries
library(terra)

# Read the file
my_spatRaster = rast("myRaster.tif")
```

To save the data, the similarly named 'terra' package function named writeRaster() writes a 'spatRaster*' object to a file in various formats, including GeoTIFF, NetCDF, and HDF5.

For example, to write a 'spatRaster*' object named 'myRaster' to a GeoTIFF file named 'myRaster.tif, one can use:

```
# Write the file
writeRaster(my_spatRaster, "myRaster.tif", format = "GTiff")
```

It is worth noting that the 'terra' package has several functions for handling raster data, including functions for summarising and manipulating raster data, extracting values from raster data, and performing spatial analysis on raster data.

Although both the 'raster' package and the 'terra' package provide functionality for working with raster data, there are some key differences between the raster objects created by these two packages.

One of the main differences is in the data structure of the raster objects. The 'raster' package uses a hierarchical data structure, with a 'RasterLayer' object representing a single layer of raster data, and a 'RasterStack' or 'RasterBrick' object representing multiple layers of raster data. On the other hand, the 'terra' package uses a flat data structure, with a 'SpatRaster' object representing a single layer of raster data, and a 'SpatRasterStack' or 'SpatRasterList' object representing multiple layers of raster data.

Another difference is in the syntax and functionality of the two packages. While both packages provide similar functionality for working with raster data, the syntax and function names can be slightly different. For example, in the 'raster' package, functions such as raster() and stack() are used to create 'RasterLayer' and 'RasterStack' objects, respectively. In the 'terra' package, the function rast() is used to create 'SpatRaster' objects.

Furthermore, the 'terra' package provides some additional functionality that is not available in the 'raster' package. For example, the 'terra' package provides support for processing very large raster datasets that cannot fit into memory, by using a chunk-based processing approach. The 'terra' package also provides advanced functionality for working with raster time series data, including support for irregular time steps and calendar systems.

14.4 Converting between Raster Types

Converting between 'terra' and 'raster' objects can be useful when working with R packages that expect some specific format or when combining raster data from different sources. The 'terra' package provides the function rast() for converting 'RasterLayer' and 'RasterBrick' objects from the 'raster' package to 'SpatRaster' and 'SpatRasterStack' objects in the 'terra' package. Similarly, the 'raster' package provides the raster() function. Additionally, an as.raster() function might be available for similar conversions, facilitating flexibility in data handling and analysis.

Here is an example of converting a 'RasterLayer' object from the 'raster' package to a 'SpatRaster' object in the 'terra' package.

```
# Load the Libraries
library(raster)
library(terra)

# Reading a raster file using the 'raster' package
myRaster_raster = raster("myRaster.tif")

# Convert the RasterLayer object to a SpatRaster object using
# the 'terra' package
myRaster_terra = rast(myRaster_raster)
```

Here is an example of converting a 'SpatRaster' object from the 'terra' package to a 'RasterLayer' object in the 'raster' package:

```
# Load the Libraries
library(raster)
library(terra)

# Read a raster file using the 'terra' package
myRaster_terra = rast("myRaster.tif")
```

```
# Convert the SpatRaster object to a RasterLayer object using
# the 'raster' package
myRaster_raster = as.raster(myRaster_terra)
```

Note that when converting between 'terra' and 'raster' objects, some of the metadata associated with the raster data may be lost or modified. It is important to carefully review the output object and ensure that it retains the necessary metadata and properties for the analysis or task being carried.

14.5 Retrieving Elevation

The 'elevatr' package[2] is a package for retrieving elevation data from various online sources. One of the primary sources of elevation data is the Shuttle Radar Topography Mission (SRTM), a NASA mission that collected high-resolution elevation data using radar technology.

The SRTM data consists of a global digital elevation model (DEM) with a spatial resolution of approximately 90 metres. The SRTM data is available in two versions: SRTM1, which has a spatial resolution of approximately 30 metres, and SRTM3, which has a spatial resolution of approximately 90 metres. The SRTM3 dataset is freely available and can be accessed through various online sources, including the United States Geological Survey (USGS) EarthExplorer website. The SRTM1 has 1-arc second grid spacing and is available only for the United States, whereas the SRTM3 has 3-arc second grid spacing and is available for the entire planet between latitudes of –56 and 60.

The 'elevatr' package provides a convenient way to download and process SRTM elevation data. The package includes functions for retrieving elevation data for specific locations or regions, as well as for dealing with the data.

For example, to retrieve SRTM elevation data for a specific location using the 'elevatr' package, one can use the get_elev_point() function.

#14-01

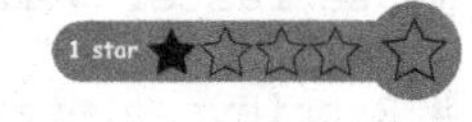


```
# Load the libraries
library(elevatr)
library(sf)

# Define a location
location = data.frame(lng = -7.9072, lat = 38.5725)
location_sf = st_as_sf(location, coords = c("lng","lat"),
crs=4326)
```

[2] https://cran.r-project.org/web/packages/elevatr/vignettes/introduction_to_elevatr.html

```
# Retrieve elevation data for the location
elevation = get_elev_point(locations = location_sf, units=
"meters",  src = "aws")

# View the elevation in metres
elevation$elevation
```

This will retrieve the elevation data for the location with longitude –7.9072 and latitude 38.5725 and store it in a 'data.frame' object named 'elevation'. The result for this data is 276 metres.

Alternatively, to retrieve SRTM elevation data for a larger region using the 'elevatr' package, one can use the get_elev_raster() function. This will retrieve a raster object with the elevation corresponding for each cell or pixel.

In this example, we use the 'meuse.grid' data to create a raster that is provided to the get_elev_raster() function to define the study area.

#14-02

2 stars

```
# Load the library
library(elevatr)
library(raster)
library(sf)

# Retrieve the data
data(meuse.grid)
data(meuse)

# Convert the meuse.grid data to an 'sf' object
sf_meuse.grid = st_as_sf(meuse.grid, coords = c("x", "y"))

# Assign the CRS to the 'sf' object
st_crs(sf_meuse.grid) = st_crs(28992)

# Create a raster from the meuse grid data
meuse_raster = raster(sf_meuse.grid)

# Retrieve elevation data for a larger region
elevation = get_elev_raster(meuse_raster, units= "meters",
src = "aws", z = 14)

# Verify the data type
class(elevation)

# View the results
plot(elevation, main = "Elevation for the Meuse area")
points(meuse, pch = 2, cex = 0.5, col = "red")
```

The parameters for the get_elev_raster() function allow to fine-tune the data obtained. The list of parameters include the following:

'meuse_raster': It in this case represents the study area or region of interest for which we want to obtain elevation data. It typically contains information about the extent, resolution, and coordinate reference system of the area.

'units': It is a parameter indicating the desired units for the elevation data. In this case, "meters" is specified, meaning that the elevation values retrieved will be in metres.

'src': It specifies the data source from which the elevation data will be obtained. In this example, "aws" is specified, indicating that the data will be fetched from an Amazon Web Services server that hosts elevation data.

'z': It is the parameter that is used to specify the desired resolution, or level of detail, of the elevation data. It represents the level of zoom, with higher values indicating a higher level of detail.

When it is not specified, the function automatically determines the appropriate level of zoom based on the spatial extent and desired resolution of the output raster. However, in some cases, it may be useful to manually specify the 'z' parameter to control the level of detail of the elevation data.

The 'z' parameter ranges from 0 to 18, with larger values indicating higher levels of detail. The actual resolution of the elevation data corresponding to a given 'z' value depends on the source data and the spatial extent of the query. In general, a higher 'z' value results in more detailed elevation data but may also increase the processing time and file size of the output raster. It is worth noting that the appropriate value for the 'z' parameter depends on the specific analysis or visualisation task as well as the computational resources available. In general, it is recommended to test different values and compare the resulting elevation data to ensure that the desired level of detail is achieved without excessive processing time or file size.

This outputs the class of elevation, i.e. a raster,

```
[1] "RasterLayer"
attr(,"package")
[1] "raster"
```

and plots the Meuse points in a background with the elevation downloaded (Figure 14.1).

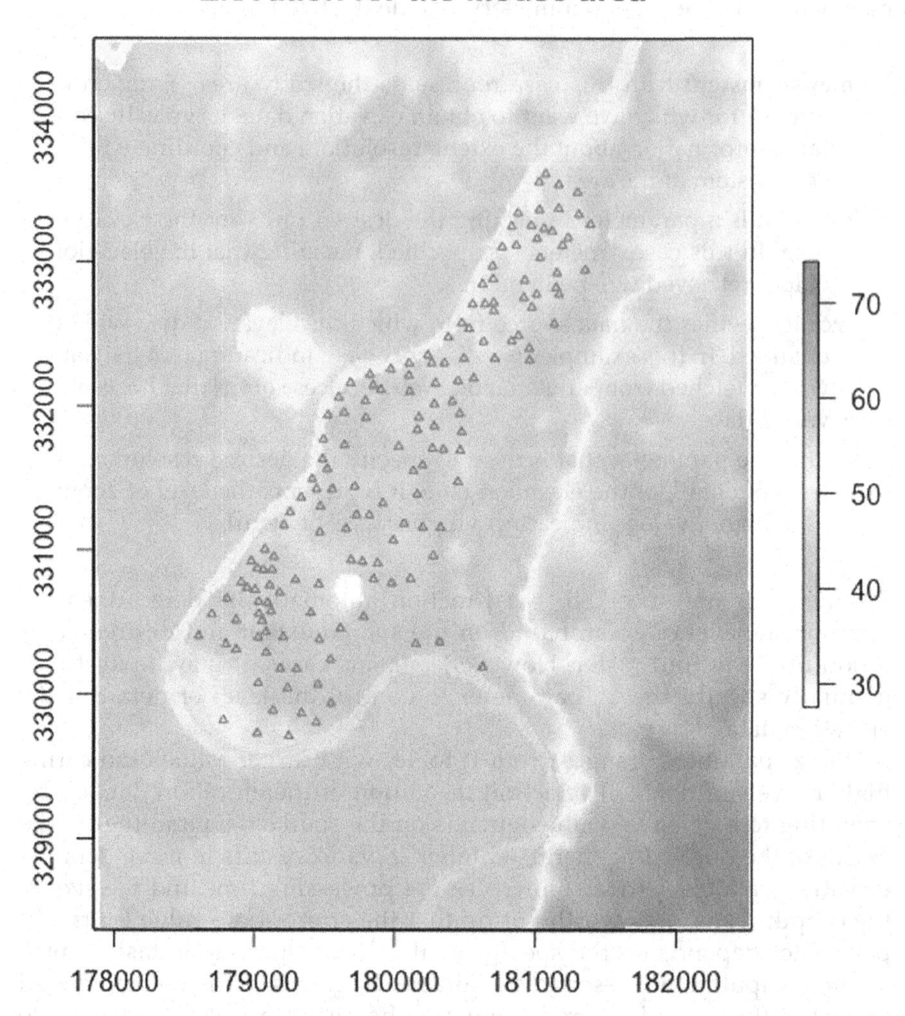

FIGURE 14.1
The elevation of the Meuse dataset area. The triangles correspond to the sample points.

14.6 Retrieving Bathymetry

The European Marine Observation and Data Network (EMODnet) is a consortium of organisations that has developed an infrastructure to provide access to a range of marine data, including bathymetry, geology, chemistry, biology, and physical parameters. The EMODnet Bathymetry project provides bathymetric data at a high resolution for European waters, making it an important resource for marine geologists, oceanographers, and other researchers interested in understanding the seafloor topography and its relationship to the marine environment.

One way to access the EMODnet Bathymetry data[3] is through the EMODnet Web Coverage Service (WCS), which allows users to retrieve bathymetry data in a variety of formats, including netCDF, GeoTIFF, and HDF5. Using the provided package, it is possible to connect to the EMODnet server and retrieve bathymetry data for a specific area of interest.

The benefits of using the EMODnet WCS[4] include the ability to automate data retrieval and analysis and to integrate the bathymetry data with other data sources, such as oceanographic and geologic data. This can enable researchers to gain a more comprehensive understanding of the marine environment and the processes that shape it.

Applications of EMODnet Bathymetry data in marine geology include mapping of seafloor features such as submarine canyons, seamounts, and ridges as well as characterising the geomorphology and geology of the seafloor. This information can be used to better understand the geological processes that shape the seafloor, such as tectonic activity, volcanic activity, and sediment transport.

Remember to install the package using install.packages() function.

#14-03/01

```
# Load the libraries
library(EMODnetWCS)
```

The 'EMODnetWCS' package provides access to a series of services to retrieve specific data available in the repository. To verify what data is available, use the emdn_wcs() function.

#14-03/02

```
# Verify the WCS server data availability
emdn_wcs()
```

The output is as follows:

```
# A tibble: 5 × 2
  service_name     service_url
```

[3] https://github.com/EMODnet/EMODnetWCS

[4] WCS stands for 'Web Coverage Service' that is a standard protocol developed by the Open Geospatial Consortium (OGC) for serving geospatial data that represents coverages, such as raster datasets, over the web. The WCS allows users to access and request coverage data, including satellite images, aerial photographs, and other gridded or raster data, from a remote server over the internet.

```
  <chr>             <chr>
1 bathymetry        https://ows.emodnet-bathymetry.eu/wcs
2 biology           https://geo.vliz.be/geoserver/Emodnetbio/wcs
3 human_activities https://ows.emodnet-humanactivities.eu/wcs
4 physics           https://geoserver.emodnet-physics.eu/
geoserver/wcs
5 seabed_habitats  https://ows.emodnet-seabedhabitats.eu/
geoserver/emodnet_open_maplibrary/wcs
```

This lists the services available from the EMODnet WCS server at the moment. It might be updated by the EMODnet consortium at any time.

For working with the bathymetry WCS information, it is necessary to initiate a client to the service with the emdn_init_wcs_client() function. With the service open, it is possible to retrieve the information that the WCS client has access by using the function emdn_get_wcs_info().

#14-03/03 2 stars

```
# Initiate the client
wcs = emdn_init_wcs_client(service = "bathymetry")

# Retrieve the client information
emdn_get_wcs_info(wcs)
```

The result is as follows:

```
✔ WCS client created succesfully
i Service: <https://ows.emodnet-bathymetry.eu/wcs>
i Service: "2.0.1"

$data_source
[1] "emodnet_wcs"

$service_name
[1] "bathymetry"

$service_url
[1] "https://ows.emodnet-bathymetry.eu/wcs"

$service_title
[1] "EMODnet Bathymetry WCS"

$service_abstract
[1] ""

$service_access_constraits
[1] "NONE"

$service_fees
[1] "NONE"
```

```
$service_type
[1] "urn:ogc:service:wcs"

$coverage_details
# A tibble: 6 × 9
  coverage_id                  dim_n dim_names          extent crs   wgs84...¹ tempo...² verti...³ subtype
  <chr>                        <int> <chr>              <chr>  <chr> <chr>     <chr>    <chr>     <chr>
1 emodnet__mean                0 lat(deg):geograph... -36, ... EPSG... -36, 1... NA    NA    Rectif...
2 emodnet__mean_2016           2 lat(deg):geograph... -36, ... EPSG... -36, 2... NA    NA    Rectif...
3 emodnet__mean_2018           2 lat(deg):geograph... -36, ... EPSG... -36, 1... NA    NA    Rectif...
4 emodnet__mean_atlas_land     2 lat(deg):geograph... -36, ... EPSG... -36, 1... NA    NA    Rectif...
5 emodnet__mean_multicolour    2 lat(deg):geograph... -36, ... EPSG... -36, 1... NA    NA    Rectif...
6 emodnet__mean_rainbowcolour 2 lat(deg):geograph... -36, ... EPSG... -36, 1... NA    NA    Rectif...
# ... with abbreviated variable names ¹wgs84_bbox, ²temporal_extent, ³vertical_extent
```

This metadata information helps to verify the services available.

To verify the coverage details, use the emdn_get_coverage_ids() function.

#14-03/04

```
# Retrieve the coverage information
emdn_get_coverage_ids(wcs)
```

The results are as follows:

```
[1] "emodnet__mean" "emodnet__mean_2016" "emodnet__mean_2018"
[4] "emodnet__mean_atlas_land" "emodnet__mean_multicolour"
"emodnet__mean_rainbowcolour"
```

To retrieve the coverage from one of the 'ids', use the function emdn_get_coverage() function.

This example retrieves the bathymetric data from the Pico Island region defined by its longitude and latitude coordinates (–29, –27, 38, and 39) using the 'emodnet_mean' coverage.

#14-03/05

```
# Extract bathymetric data from the Pico island
pico_bath = emdn_get_coverage(wcs,
            coverage_id = "emodnet__mean",
            bbox = c(xmin = -29,
                  ymin = 38,
                  xmax = -27.5,
                  ymax = 39),
            nil_values_as_na = TRUE )
```

It is possible to view the obtained data with the plot function. The 'pico_bath' variable is a 'spatRaster' object from the 'terra' package.

#14-03/06

```
# Verify the data type
class(pico_bath)
```

```
# Plot the results
plot(pico_bath, main = "Bathymetry from the Pico Island region")
```

The output result is as follows:

```
[1] "SpatRaster" attr(,"package")
[1] "terra"
```

The resulting plot is presented in Figure 14.2.

FIGURE 14.2
Bathymetry of the Pico Island region.

14.6.1 WFS Service (More Than Bathymetry)

The EMODnet service also provides WFS[5] services that retrieve features available in their database. These features can be consulted and downloaded

[5] WFS stands for 'Web Feature Service' that is a standard protocol developed by the OGC for serving geospatial vector data over the web. The WFS allows users to access and query geographic feature data, such as points, lines, and polygons, from a remote server via the internet. With WFS, users can request specific feature types, attributes, and spatial extents from the server, and the service will return the requested data in a standardised format, typically in XML or GeoJSON.

as 'sf' objects using the 'EMODnetWFS' package.[6] For this example, remember to install the 'EMODnetWFS' package as well the 'tidyterra' package for data visualisation.

The emodnet_wfs() function retrieves the services. In the 'service_name' variable, you get the service names available.

#14-04/01

```
# Load the libraries
library(EMODnetWFS)
library(terra)
library(tidyterra)
library(sf)
library(ggplot2)

# List the services
wfs_services = emodnet_wfs()
wfs_services$service_name
```

The result is as follows:

```
 [1] "bathymetry"
 [2] "biology"
 [3] "biology_occurrence_data"
 [4] "chemistry_cdi_data_discovery_and_access_service"
 [5] "chemistry_cdi_distribution_observations_per_category_
     and_region"
 [6] "chemistry_contaminants"
 [7] "chemistry_marine_litter"
 [8] "geology_coastal_behavior"
 [9] "geology_events_and_probabilities"
[10] "geology_marine_minerals"
[11] "geology_sea_floor_bedrock"
[12] "geology_seabed_substrate_maps"
[13] "geology_submerged_landscapes"
[14] "human_activities"
[15] "physics"
[16] "seabed_habitats_general_datasets_and_products"
[17] "seabed_habitats_individual_habitat_map_and_model_datasets"
```

To retrieve the WFS features, it is necessary to apply as a client for the service with the emodnet_init_wfs_client() function.

In this case, we selected the "geology_marine_minerals" service. To verify what this service provides, the function emodnet_get_wfs_info() retrieves the information.

[6] https://github.com/EMODnet/EMODnetWFS

#14-04/02

```
# Create new WFS Client for geology marine minerals
wfs_geo = emodnet_init_wfs_client(service = "geology_marine_
minerals")

# Retrieve the layers names
wfs_info = emodnet_get_wfs_info(wfs_geo)
wfs_info$layer_name
```

The result is as follows:

```
 [1] "MarineAggregatesArea"                  "MarineAggregatesPoint"
 [3] "CobaltRichFerromanganeseCrustsArea"    "CobaltRichFerromanganeseCrustsPoint"
 [5] "EvaporitesArea"                        "EvaporitesPoint"
 [7] "GasHydratesArea"                       "GasHydratesPoint"
 [9] "MarineHydrocarbonsArea"                "MarineHydrocarbonsPoint"
[11] "MarinePlacerDepositsArea"              "MarinePlacerDepositsPoint"
[13] "MarineSapropelArea"                    "MarineSapropelPoint"
[15] "MetalRichSedimentsArea"                "MetalRichSedimentsPoint"
[17] "PhosphoritesArea"                      "PhosphoritesPoint"
[19] "PolymetallicNodulesArea"               "PolymetallicNodulesPoint"
[21] "PolymetallicSulphidesPoint"            "RockPegVeinArea"
[23] "RockPegVeinPoint"
```

It is possible to retrieve one or more layers from each of this data. Verify that some are points and others are polygons. For this example, let's retrieve the 'PolymetallicSulphidesPoint' layer.

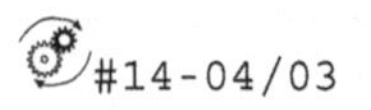

#14-04/03

```
# Define the working layer (or layers)
layers = c("PolymetallicSulphidesPoint")

# Get layer info
emodnet_get_layer_info(wfs = wfs_geo, layers = layers)

# Retrieve the layer
MAR_minerals = emodnet_get_layers(wfs = wfs_geo, layers =
layers, crs=4326)

mineral = MAR_minerals[[1]]
```

The 'MAR_minerals' variable is a list. To retrieve the first (and in this case only) element in the list, use the subsetting list option, i.e. mineral = MAR_minerals[[1]].

For this example, the Mid-Atlantic Ridge southwest from Azores islands will be displayed. The geology of this area and its polymetallic minerals can be found at Klischies et al. (2019). So, the bathymetric information for the area is fetched.

#14-04/04

```
# Retrieve the Madeira-Tore study area
MAR_bath = emdn_get_coverage(wcs,
                             coverage_id = "emodnet__mean",
                             bbox = c(xmin = -34,
                                      ymin = 36,
                                      xmax = -30,
                                      ymax = 39),
                             nil_values_as_na = TRUE)

# Give name for the pixel value = depth
names(MAR_bath) = "depth"
```

One must remember that the files obtained from the coverage of a region are big. The retrieval of large areas might not be possible due to net constraints. When retrieving large areas, it is possible to have a message like this:

```
— Downloading coverage "emodnet__mean_2020"
____________________________________________________________

<GMLEnvelope>
....|-- lowerCorner: 36 -16
....|-- upperCorner: 38 -14Error: [rast] cannot open this file
as a SpatRaster: /private/var/folders/hj/wj29hm4s0fg1h2njpg2vf
zkr0000gn/T/RtmpcIcrRH/emodnet__mean_2020_36,-16,38,-14.tif
In addition: Warning message:
`/private/var/folders/hj/wj29hm4s0fg1h2njpg2vfzkr0000gn/T/
RtmpcIcrRH/emodnet__mean_2020_36,-16,38,-14.tif' not
recognized as a supported file format. (GDAL error 4)
```

If this happens, just define a smaller region and try again.

Notice that for practical purposes, the name for the pixel values is renamed 'depth'.

The next step is to crop the 'mineral' points object to the study area.

#14-04/05

```
# Select and crop the layer
mineral_poly = st_crop(mineral, st_bbox(MAR_bath))
```

This time, to visualise the data we will use the 'ggplot2' package combined with the function geom_spatraster() from the 'tidyterra' package that is used to visualise 'SpatRaster' objects. Also the parameter 'show. legend = FALSE' was defined to avoid showing the bathymetric legend that would clutter the plot.

#14-04/06

```
# Plot the bathymetry data using ggplot2 and tidyterra
ggplot() +
  geom_spatraster(data = MAR_bath, aes(fill = depth), show.
legend = FALSE) +
  geom_sf(data=mineral_poly, aes(color=all_other), size = 2) +
  scale_fill_gradientn(colors = rev(terrain.colors(255)),
                       na.value = "black") +
  scale_size_identity() +
  coord_sf(datum = 4326) +
  labs(title = "Mid-Atlantic Ridge Polymetallic Minerals") +
  theme(legend.position = "bottom",
      legend.box = "horizontal", legend.justification =
"center") +
  guides(color = guide_legend(title = "Deposit Type", nrow = 2))
```

Notice that for the crust's legend, the field 'all_other' was used. The resulting plot is shown in Figure 14.3.

14.7 Know Your Data, Again...

Both the 'raster' and 'terra' packages provide functions to retrieve information about the parameters and metadata of a raster object. Here are some examples of functions for getting information about raster objects in both packages (Table 14.1).

For example, to retrieve the extent, resolution, and number of cells of a 'raster' object named 'myRaster' using the 'raster' package, one can use the following:

```
# Load the library
library(raster)

# Retrieve information about the raster object
myExtent = extent(myRaster)
myRes = res(myRaster)
myNcell = ncell(myRaster)
```

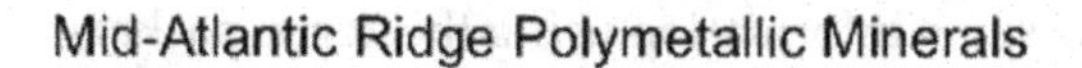

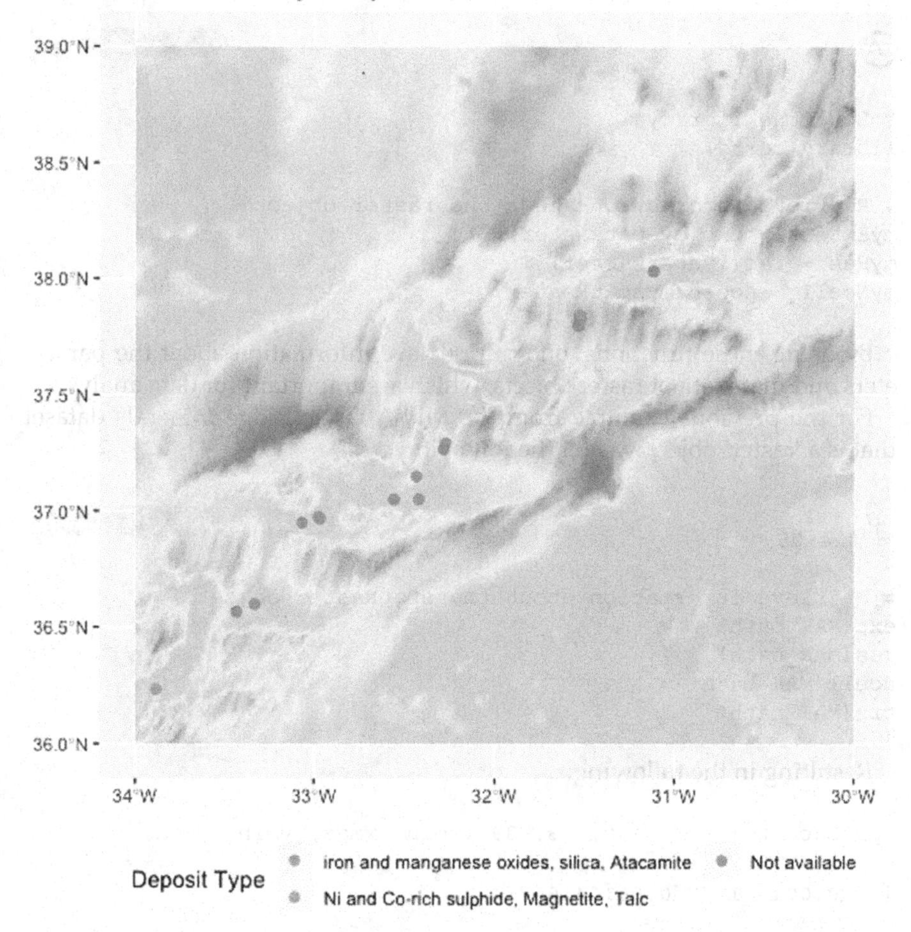

FIGURE 14.3
The polymetallic minerals in the Mid-Atlantic ridge SW from the Azores islands.

TABLE 14.1
Comparison of Basic Function from the 'raster' and 'terra' Packages

Package		
'raster'	'terra'	Description
extent()	ext()	Returns the spatial extent of the raster object.
res()	res()	Returns the resolution (i.e. the size of each pixel) of the raster object.
ncell()	ncell()	Returns the number of cells in the raster object.
values()	values()	Returns the values of the cells in the raster object.
projection()	crs()	Returns the coordinate reference system (CRS) of the raster object.

Similarly, to retrieve the same information using the 'terra' package:

```
# Load the library
library(terra)

# Retrieve information about the raster object
myExtent = ext(myspatRaster)
myRes = res(myspatRaster)
myNcell = ncell(myspatRaster)
```

By using these functions, one can retrieve information about the parameters and metadata of raster objects, which are important for data analysis.

For the previous example from the Mid-Atlantic Ridge minerals dataset that is a 'raster' object, we get the following:

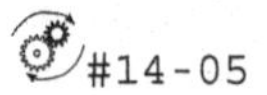

```
# Retrieve information about the spatRaster object
ext(MAR_bath)
res(MAR_bath)
ncell(MAR_bath)
crs(MAR_bath)
```

Resulting in the following:

```
SpatExtent : -34, -30, 36, 39 (xmin, xmax, ymin, ymax)

[1] 0.001041667 0.001041667

[1] 11059200

[1] "GEOGCRS[\"WGS 84\",\n
     ENSEMBLE[\"World Geodetic System 1984 ensemble\",\n
     MEMBER[\"World Geodetic System 1984 (Transit)\"],\n
     MEMBER[\"World Geodetic System 1984 (G730)\"],\n
     MEMBER[\"World Geodetic System 1984 (G873)\"],\n
     MEMBER[\"World Geodetic System 1984 (G1150)\"],\n
     MEMBER[\"World Geodetic System 1984 (G1674)\"],\n
     MEMBER[\"World Geodetic System 1984 (G1762)\"],\n
     MEMBER[\"World Geodetic System 1984 (G2139)\"],\n
     ELLIPSOID[\"WGS 84\",6378137,298.257223563,\n
     LENGTHUNIT[\"metre\",1]],\n
     ENSEMBLEACCURACY[2.0]],\n
     PRIMEM[\"Greenwich\",0,\n
     ANGLEUNIT[\"degree\",0.0174532925199433]],\n
```

```
CS[ellipsoidal,2],\n
AXIS[\"geodetic latitude (Lat)\",north,\n
ORDER[1],\n
ANGLEUNIT[\"degree\",0.0174532925199433]],\n
AXIS[\"geodetic longitude (Lon)\",east,\n
ORDER[2],\n
ANGLEUNIT[\"degree\",0.0174532925199433]],\n
USAGE[\n
SCOPE[\"Horizontal component of 3D system.\"],\n
AREA[\"World.\"],\n
BBOX[-90,-180,90,180]],\n
ID[\"EPSG\",4326]]"
```

14.8 Basic Descriptive Statistics of a Raster

To analyse some basic descriptive statistics on a raster, it is better to convert it to a matrix, after which it is possible to handle the data as a numeric object.

To convert an elevation raster to a matrix and calculate descriptive statistics, one can use the 'raster' package.

Returning to the example from the Meuse River elevation in the code snippet #14-02, and using the elevation raster, one can calculate its basic statistics.

#14-06

```
# Convert the raster to a matrix
meuse_matrix = as.matrix(elevation)

# Calculate descriptive statistics
meuse_mean = mean(meuse_matrix, na.rm = TRUE)
meuse_sd = sd(meuse_matrix, na.rm = TRUE)
meuse_min = min(meuse_matrix, na.rm = TRUE)
meuse_max = max(meuse_matrix, na.rm = TRUE)

# Print the results
cat("Mean:", meuse_mean, "\n")
cat("Standard deviation:", meuse_sd, "\n")
cat("Minimum value:", meuse_min, "\n")
cat("Maximum value:", meuse_max, "\n")
```

In this example, the as.matrix() function is then used to convert the 'raster' object to a matrix. Finally, the mean(), sd(), min(), and max() functions are used to calculate descriptive statistics for the matrix, and the cat() function is used to print the results.

Note that the 'na.rm = TRUE' argument is used in the descriptive statistics functions to remove any missing or not applicable (NA) values in the matrix.

The results are as follows:

```
Mean: 42.17486

Standard deviation: 10.60517

Minimum value: 27.4819

Maximum value: 74.40229
```

14.9 Concluding Remarks

Rasters or gridded data are one of the pillars of spatial data analysis. Its simple format in a matrix-like structure implies that the calculations performed are familiar to many other mathematical and statistical methods. As in most of the cases, there are several types of data and file formats and it is paramount to a data analyst to understand this multiplicity of options, to better decide what tools to use, and how to travel between different data types and formats.

For exemplifying, the basic operations of raster management to web services and the corresponding R packages are introduced. The elevation and bathymetry are important elements of geomorphological analysis, and the 'elevatr' and 'EMODnet' packages are elegant tools to retrieve this type of data. The possibilities are immense for studying the morphological features of onshore and offshore regions.

However, one advertisement must be made; remember that often the available databases on the internet are updated and what is perfectly working in one moment might need to be adapted in another. During the writing of this book, some examples had to be adapted due to these changes. Please refer to the web page of the book to report such changes and to verify the changes necessary to have fully running examples.

Reference

Klischies, M., Petersen, S., & Devey, C. W. (2019). Geological mapping of the Menez Gwen segment at 37 50′ N on the Mid-Atlantic Ridge: Implications for accretion mechanisms and associated hydrothermal activity at slow-spreading mid-ocean ridges. Marine Geology, 412, 107–122.

15

Basic Raster Operations

Raster data, including elevation and bathymetric data, as much as any type of data, often requires various pre-processing and manipulation steps before analysis and visualisation. Along this chapter, we present some of the basic raster operations that can be performed using the elevation data raster as an example for applying the 'raster' and 'terra' packages.

Once more, the 'meuse' elevation data is used as a departure example (Pebesma, 2022).

15.1 Reprojecting Rasters

Reprojection is the process of transforming raster data from one coordinate reference system (CRS) to another. This is often a necessary operation when different datasets are collected and need to be compatibilised for further analysis.

Again, we return to the elevation from the 'meuse' area. In this example code, the elevation data is retrieved in the original projected CRS (European Petroleum Survey Group [EPSG] = 28992) of the 'meuse' dataset. This system is a regional projected CRS and the data is in metres.

If necessary, this can be converted to any CRS, using the projectRaster() function, from the 'raster' package. In this case, it will be converted to a geographic CRS, the WGS84.

The information from the original and reprojected rasters is shown at the end of the code.

#15-01

```
# Load the library
library(elevatr)
library(raster)
library(sf)

# Retrieve the data
data(meuse.grid)
data(meuse)

# Convert the meuse.grid data to an sf object
sf_meuse.grid = st_as_sf(meuse.grid, coords = c("x", "y"))
```

DOI: 10.1201/9781032651880-17

```
# Assign the CRS to the sf object
st_crs(sf_meuse.grid) = 28992

# Create a raster from the meuse grid data
meuse_raster = raster(sf_meuse.grid)

# Retrieve elevation data for a larger region
elevation = get_elev_raster(meuse_raster, units= "meters",
src = "aws", z = 14)

# Reproject the elevation raster to the original projected CRS
elevation_wgs84 = projectRaster(elevation, crs = 4326)

# Show the information of the elevation raster in CRS=28992 format
format(res(elevation), scientific = FALSE)
ncell(elevation)
projection(elevation)

# Show the information of the elevation raster in WGS84
format(res(elevation_wgs84), scientific = FALSE)
ncell(elevation_wgs84)
projection(elevation_wgs84)
```

In this example, the projectRaster() function is used to reproject the 'elevation' raster from the original projected CRS of the 'meuse' dataset to the WGS84 system.

The format() function is used to show the resolution of the rasters in a decimal notification to view all the decimal places. In the case of the original CRS = 28992, the resolution is about 1 pixel = 3 metres. The values for the WGS84 data are in decimal degrees, therefore with many decimal places. This is because 1 metre in decimal degrees at the equator region (used for reference) is about 0.000008 decimal degrees.

The results indicate the following:

```
[1] "3.009277" "3.009277"

[1] 3181646

[1] "+proj=sterea +lat_0=52.1561605555556
+lon_0=5.38763888888889 +k=0.9999079 +x_0=155000 +y_0=463000
+ellps=bessel +units=m +no_defs"

[1] "0.0000428" "0.0000271"

[1] 3258254

[1] "+proj=longlat +datum=WGS84 +no_defs"
```

The number of cells changes slightly due to adjustments between both rasters. The projection system is shown accordingly to the changes.

15.2 Cropping Rasters

Cropping and clipping are the processes of selecting a subset of the original raster data based on a specific spatial extent or polygon boundary. To crop the elevation data from the Meuse region, one can use the crop() function (cf. Section 12.3.1 in Chapter 12).

#15-02

```
# Define the meuse 'sf' dats
meuse_sf = st_as_sf(meuse, coords = c("x","y"), crs = 28992)

# Convert to wgs84
meuse_sf = st_transform(meuse_sf, crs = 4326)

# Crop the elevation raster to the extent of the shapefile
elevation_wgs84_crop = crop(elevation_wgs84, extent(meuse_sf))

# Plot the cropped and clipped elevation rasters
plot(elevation_wgs84_crop, main = "Elevation of the Meuse area
(cropped)")
plot(meuse_sf,pch=2, cex=.5, col="red", add = TRUE)
```

In this example, the crop() function is used to crop the 'elevation_wgs84' object to the extent of the 'meuse_sf' dataset. The result of the plot is presented in Figure 15.1.

The first plot() draws the elevation cropped, and the second plot() used with the 'add = TRUE' argument is used to plot the sample points as triangles.

For clipping a raster, it will be necessary to combine crop(), mask(), and rasterize() functions (see Section 15.6).

15.3 Resampling Raster Data

Resampling is the process of changing the resolution or cell size of a raster.

For this example, we will resample the 3 × 3 metres original elevation raster to a lower resolution 100 × 100 one. This change, lowering the resolution, will be visible when comparing the original plot with the resampled one. Resampling is done when necessary to compare rasters with different resolutions or when it is necessary to harmonise a multi-band raster to a unique resolution.

The first step is to create a blank raster with the intended extent and resolution.[1] After that, the resample() function is used to create the new raster. The

[1] The blank raster is created in order to assure that the pixels in corresponding images are all aligned according to the previously defined raster extent.

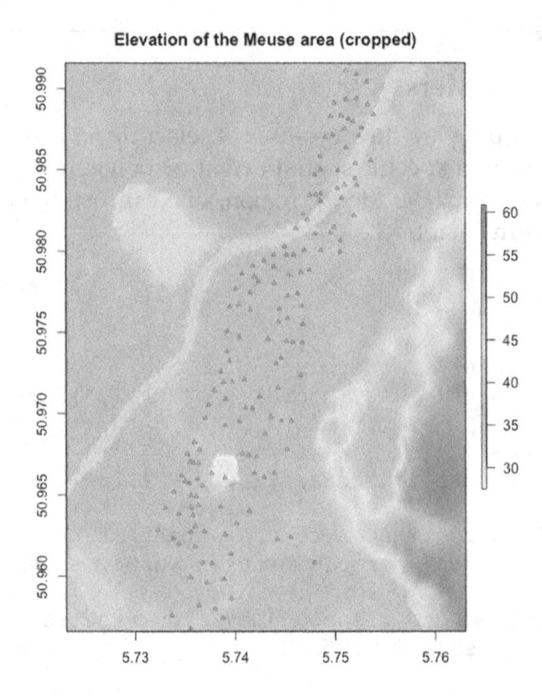

FIGURE 15.1
The elevation of the Meuse area cropped to the extent of the 'meuse' dataset. The triangles represent the sample points.

resample() function maps pixels (or cells from a grid), according to a defined transformation, from the original image to the resampled one.

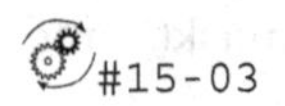
#15-03

```
# Define the new resolution and extent for the resampled raster
resolution = c(100, 100)
extent_data = extent(elevation)

# Create a new Raster object with the desired properties
raster_new = raster(ext = extent_data, res = resolution, crs =
28992, vals = NA)

# Resample elevation to the desired resolution and extent
elevation_resampled = resample(elevation, raster_new, method =
"bilinear")

# Plot the resampled elevation raster
par(mfrow=c(1,2))
plot(elevation, main=paste("res:",round(res(elevation),0)))
plot(elevation_resampled,
main=paste("res:",round(res(elevation_resampled),0)))
```

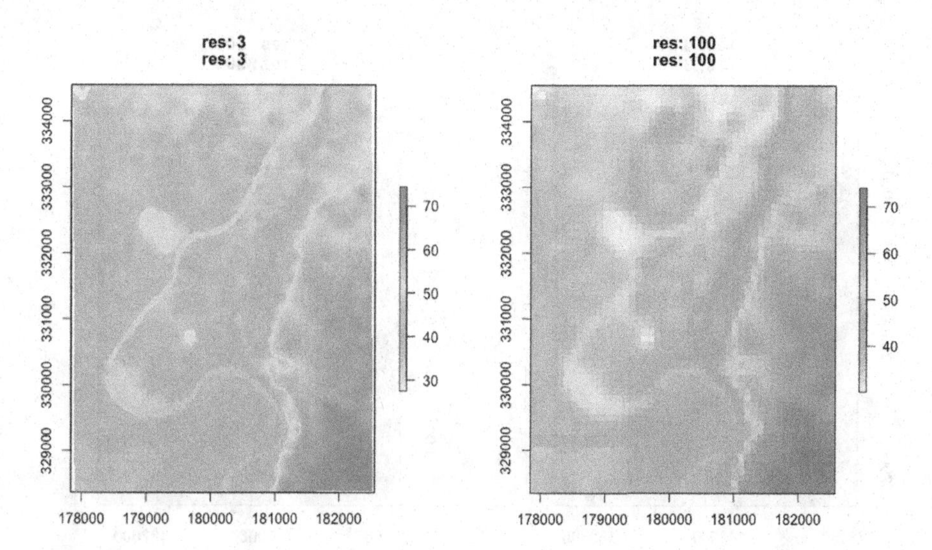

FIGURE 15.2
Resampling of a raster image.

The resulting plot is presented in Figure 15.2. Notice the difference in resolution between the original image in the left and the resample one in the right.

15.4 Aggregating and Disaggregating

As an alternative to resample the raster data, the functions aggregate() and disaggregate() can be utilised. In this example, we use the cell size of 3 (see above) to divide it by 300 and get a pixel size of approximately 300 in a resampling process.

#15-04

```
# Coarsen the resolution of the elevation raster to approx. 300m
elevation_300m = aggregate(elevation, fact=300/3)

# Plot the resampled elevation raster
par(mfrow=c(1,2))
plot(elevation, main=paste("res:", round(res(elevation),0)))
plot(elevation_300m, main=paste("res:", round(res(elevation_
300m),0)))
```

In this example, the aggregate() function is used to resample the elevation raster to a new resolution with a factor of 300/3; it returns 301 as can be seen in Figure 15.3.

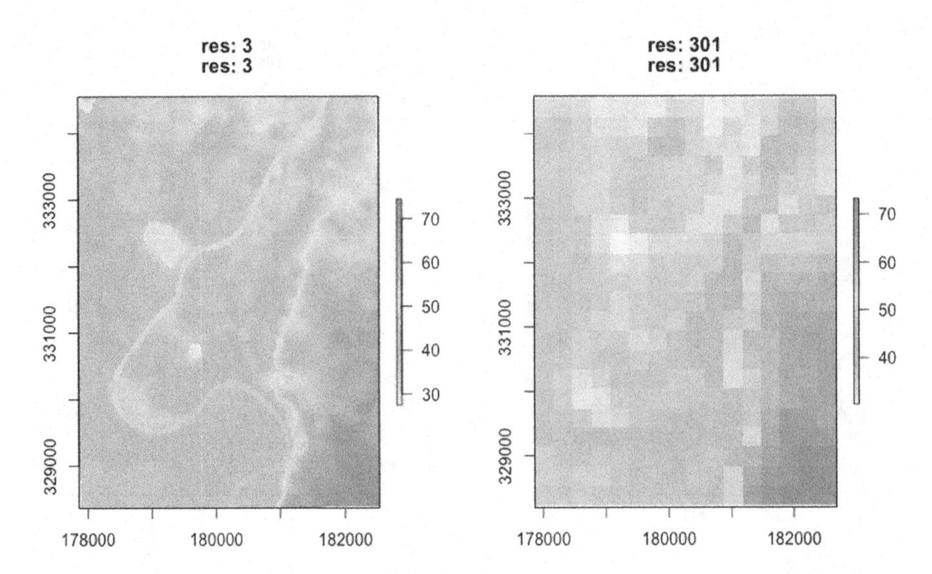

FIGURE 15.3
A resampling using the aggregate() function.

Naturally, the disaggregate() function works in the reverse sense. The resolution is equal to 301 because the original resolution is 3.009277 that multiplied by 100 is rounded to 301.

15.5 Filtering Raster Data

As in the case of vectorial or numerical data, the raster data can be filtered based on its characteristics.

Suppose that it is necessary to identify the areas where the elevation is higher than 35 metres. This is easily filtered with a condition.

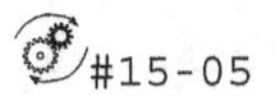#15-05 

```
# Filter the elevation raster based on the elevation
elevation_filtered = elevation > 35

# View the Results
par(mfrow=c(1,1))
plot(elevation_filtered, main = "Meuse elevation > 35m", col=
c("white", "blue"), legend = FALSE)
```

With this code, we demonstrate how to filter an elevation raster dataset to identify areas that exceed a certain height threshold, specifically 35 metres.

Using a simple comparison operation (elevation > 35), a binary raster is created where the pixels representing elevations above 35 metres are marked with a value of 1 (true), and those below or equal to 35 metres are marked with a value of 0 (false). The resulting binary raster is then visualised, highlighting areas higher than 35 metres in blue (dark grey in the image), as shown in Figure 15.4.

This technique is not only valuable for filtering data but also serves as a foundational method for creating raster masks. To illustrate, if we wish to isolate and preserve pixel values from a specific area of a raster, we can generate a mask by setting the desired pixels to 1 and the others to 0. Multiplying the original raster by this mask will retain the values in the areas of interest (where the mask is 1) and eliminate the rest (where the mask is 0), effectively applying the mask. This approach is widely used for selectively focusing on or excluding particular regions of interest, within a raster dataset.

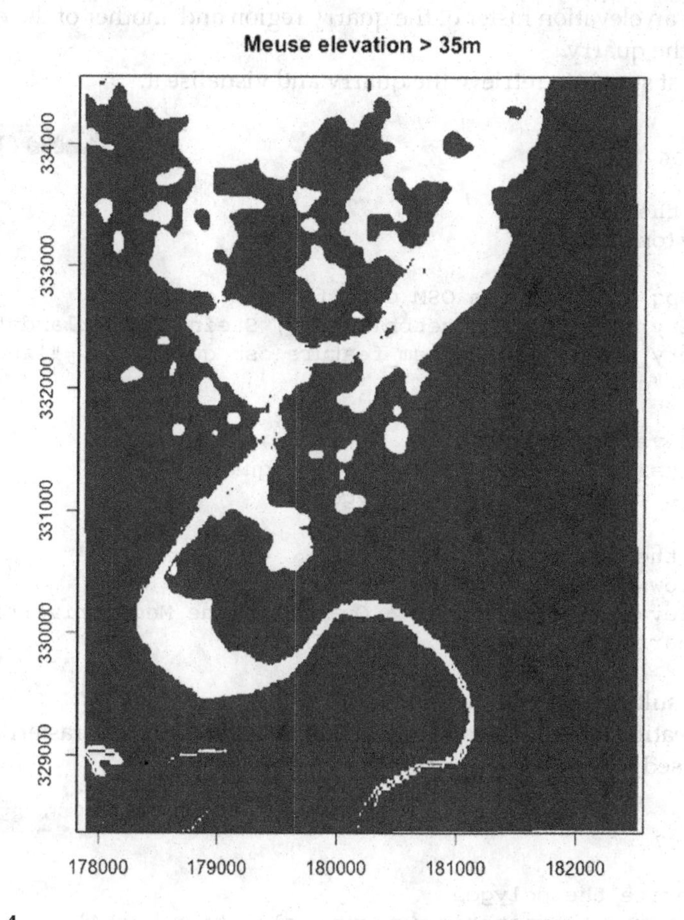

FIGURE 15.4
The elevation filtered to >35 metres. The filled places where the elevation is higher than 35 metres are shown in dark grey.

15.6 Masking Raster Data

As stated above, masking is the process of selecting or retaining only the cells of a raster within a specific region, e.g. a polygon or defined spatial extent.

In the Meuse area using the OpenStreetMap, it is possible to verify that there is a quarry. Its contours can be extracted as a polygon, using the 'osmdata' package.

In this example, we will manage the elevation raster and create one with only the elevation inside the quarry and another with the elevation outside the quarry.

The example starts by extracting the quarry polygon using the 'osmdata', after which the rasterize() function is used to transform the quarry polygon to a raster. Finally, this raster is used as a mask, applied to the elevation raster to create an elevation raster of the quarry region and another of the elevation outside the quarry.

The first step is to retrieve the quarry and visualise it.

#15-06 (3 stars)

```
# Load the library
library(osmdata)

# Use opq to fetch the OSM data for the quarry
osm_query = opq(bbox = getbb("Meers, Stein, Netherlands"))
osm_query_quarry =  add_osm_feature(osm_query, key="landuse",
value = "quarry")

# Retrieve the quarry
meers_quarry = osmdata_sf(osm_query_quarry)
quarry = meers_quarry$osm_polygons

# View the Quarry
par(mfrow=c(1,1))
plot(elevation_wgs84, main=" Quarry in the Meuse river")
plot(quarry, col = "red", add = TRUE)
```

The resulting plot is shown in Figure 15.5.

For creating a mask, the polygon of the quarry is used. The rasterize() function is used to transform a spatial object into a raster one.

#15-07 (1 star)

```
# Rasterize the polygon
quarry_raster = rasterize(quarry, elevation_wgs84)

# View the result
plot(quarry_raster, main="The quarry area", legend = FALSE)
```

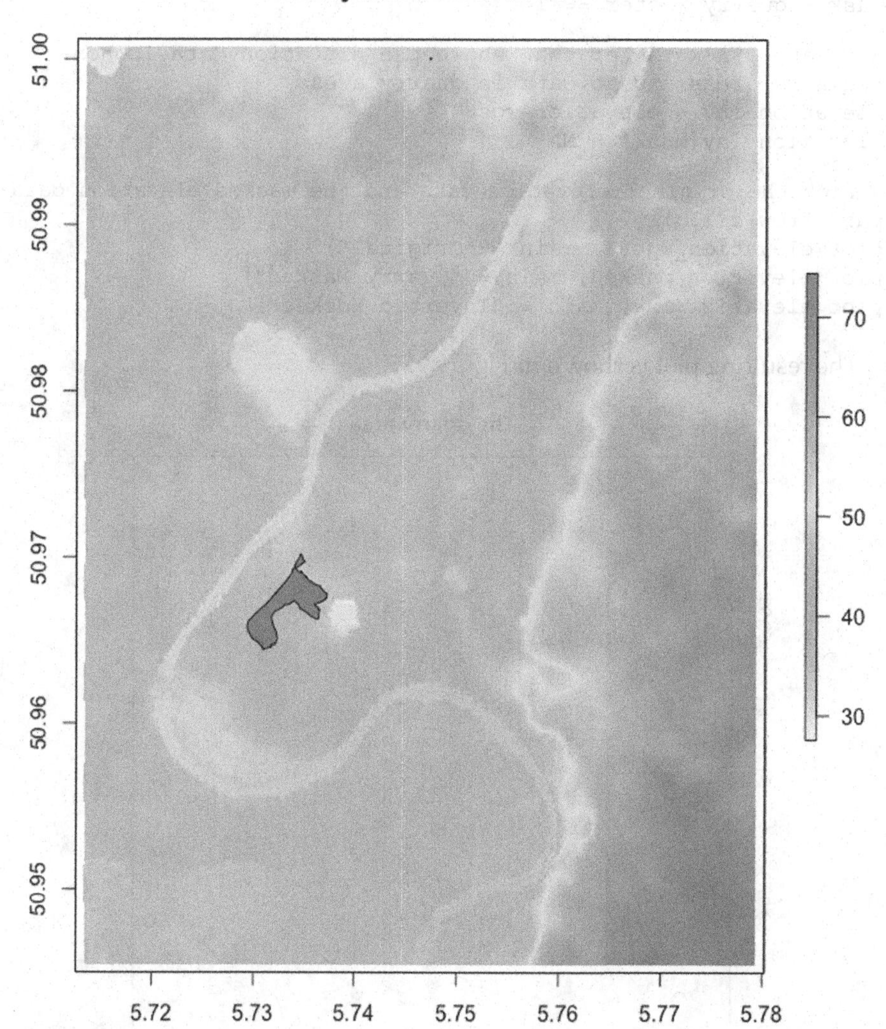

FIGURE 15.5
The quarry in the Meuse data.

The resulting raster is a [0,1] where 1 is the polygon and 0 is the outside of the polygon (Figure 15.6).

Finally, the mask is applied to the elevation data and an example of the inverse of the mask is also calculated.

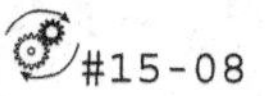#15-08

```
# Only the Quarry masked
elevation_masked = mask(elevation_wgs84, quarry_raster)
```

```
# Invert the mask to only retain the cells outside of the quarry
mask = quarry_raster == 1

# Create a new raster that shows the elevation data in non-
# quarry areas and no data in quarry areas
elevation_inv = elevation_wgs84
elevation_inv[mask] = NA

# Plot the original elevation data and the masked elevation data
par(mfrow=c(1,3))
plot(elevation_wgs84, main = "Original")
plot(elevation_masked, main = "Quarry Masked")
plot(elevation_inv, main = "Inverted Masked")
```

The resulting plot is shown in Figure 15.7.

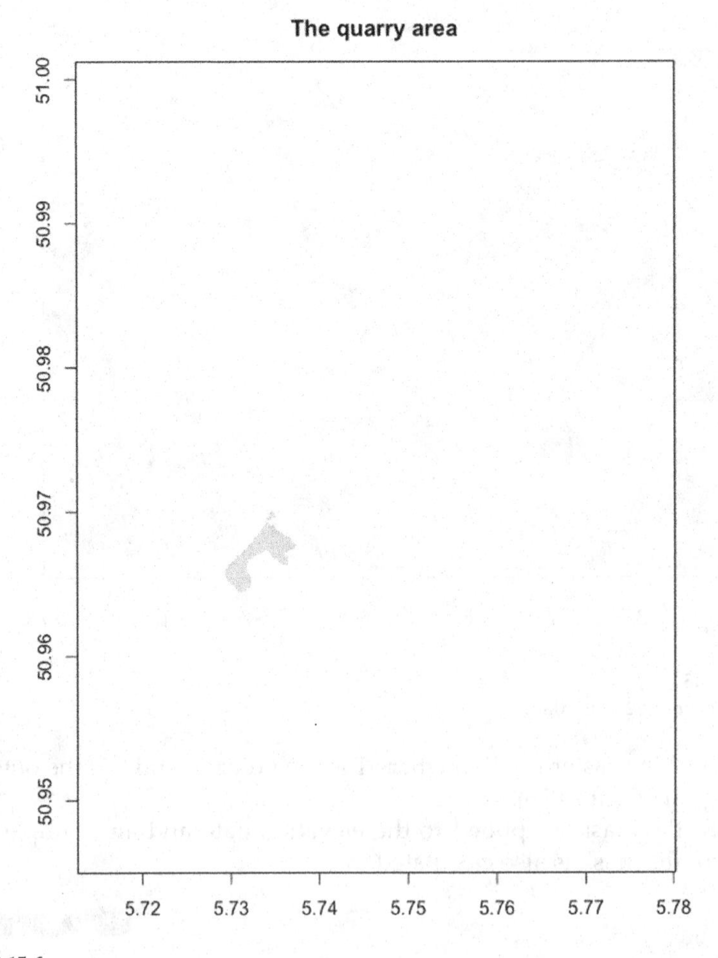

FIGURE 15.6
The resulting raster with the quarry.

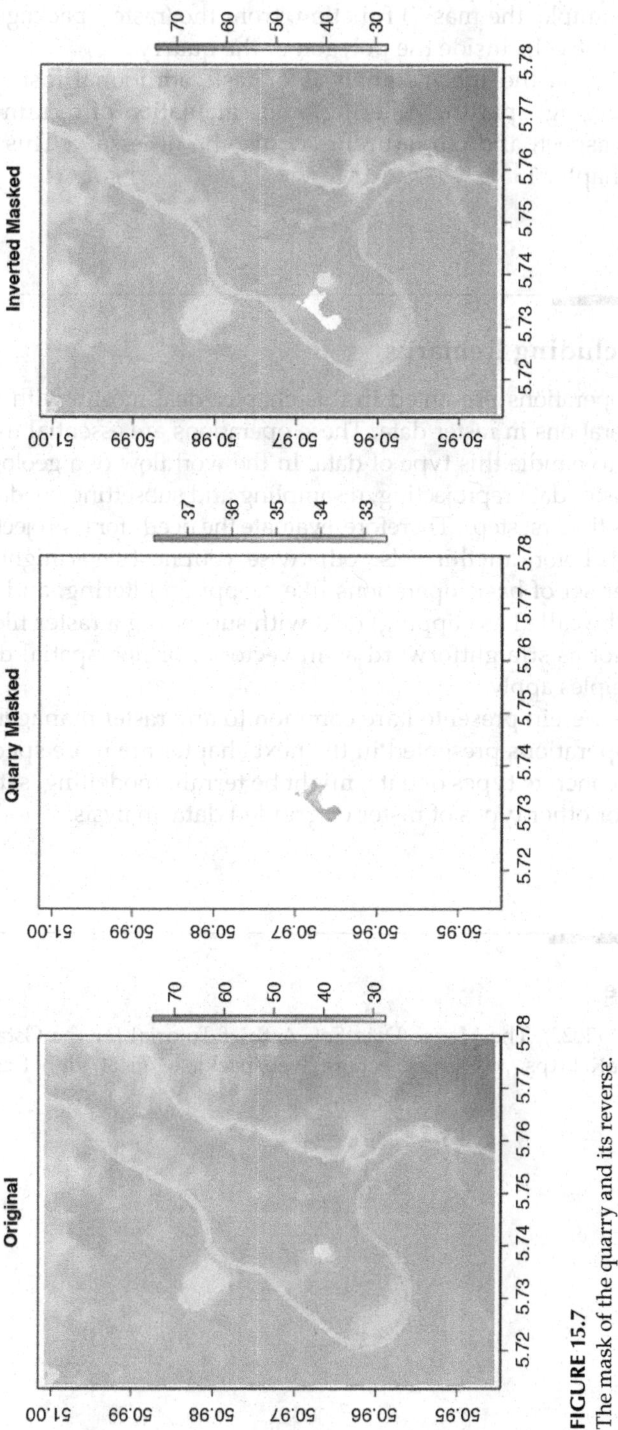

FIGURE 15.7
The mask of the quarry and its reverse.

In this example, the mask() function, from the 'raster' package, is used to retain only the cells inside the polygon of the quarry.

Depending on the specific analysis or task, additional raster operations such as merging, splitting, filtering, and calculation of terrain derivatives (e.g. slope, aspect, and curvature) may also be necessary. This will be the object of Chapter 16.

15.7 Concluding Remarks

The basic operations presented in this chapter deal mostly with the pre-processing operations in raster data. These operations are essential in any project that needs to handle this type of data. In the workflow of a geological project that uses raster data reprojecting, resampling and subsetting need always to be regarded as the first steps. Therefore, evaluate the needs for reproject and resample your data before anything else; otherwise, your next steps might be doomed.

The other set of basic operations like cropping, filtering, and masking (it could also be called as clipping) deal with subsetting a raster file or dataset. They are not as straightforward as in vectorial or non-spatial data, but the same principles apply.

The steps herein presented are common to any raster management operation. The operations presented in the next chapter are more specialised and applied to concrete types of data, might be terrain modelling, satellite image handling, or other types of raster or gridded data analysis.

Reference

Pebesma, E. (2022). The Meuse Data Set: A Brief Tutorial for the Gstat R Package. ViennaR. https://cran.r-project.org/web/packages/gstat/vignettes/gstat.pdf

16

Terrain Operations

Digital elevation models (DEMs), which are grid-based representations of the surface of the earth, are a primary data source for terrain analysis. DEMs are used to obtain information about elevation, slope, aspect, and curvature, which provide insights of the terrain morphology. Let us look at these different components, their meanings, and how to calculate and use them in a geomorphological or geological analysis.

Slope is a measure of the steepness of the terrain and is defined as the rate of change of elevation with respect to distance. Slope can be calculated using different methods, such as the simple difference method or the partial derivative method.[1] The simple difference method calculates the slope between adjacent grid cells, while the partial derivative method uses calculus to estimate the slope. Slope is used in a variety of applications, such as landslide analysis, hydrological modelling, land-use planning, and terrain classification.

Aspect is a measure of the direction in which a slope is facing and is typically measured in degrees from north. Aspect is calculated by determining the slope direction of each grid cell in the DEM.

Other terrain operations that are used to obtain better visualisations of the terrain include hillshade and viewshed images. Hillshade is the method used for shading the terrain to simulate the effects of light and shadows. Viewshed analysis is a method for determining what areas of the terrain are visible from a given location.

The 'raster' and 'terra' packages have various functions for calculating most of these terrain derivatives from elevation data.

Let's return to the example of Pico Island to reveal some of its terrain geomorphological features.

The following code is used to retrieve the Pico Island elevation 'pico_elev' variable and mask it to the island boundaries defined as the 'pico_sf_valid' variable. The final result is stored in the 'pico_mask' variable.

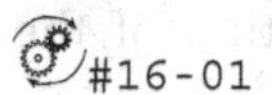

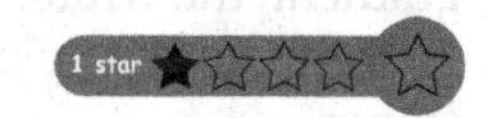


```
#16-01

# Load the libraries
library(sf)
library(terra)
```

[1] https://www.witpress.com/Secure/elibrary/papers/RM11/RM11013FU1.pdf

DOI: 10.1201/9781032651880-18

```
library(osmdata)
library(elevatr)

# Create the dataset
osm_query = opq(bbox = getbb("Ilha do Pico, Portugal")) %>%
  add_osm_feature(key="name", value = "Ilha do Pico")

# Retrieve the geometry
pico = osmdata_sf(osm_query)

# Use the outline of the island
pico_out = pico$osm_multipolygons$geometry

# Create an sf object
pico_out_sf = st_sf(geometry = pico_out)
st_crs(pico_out_sf) = 4326

# Retrieve elevation data for a larger region
pico_elev = get_elev_raster(pico_out_sf, z = 12)

# Sometimes part of the osmdata is not valid. Make it valid
pico_sf_valid = st_make_valid(pico_out_sf)

# Rasterize
pico_rast = rasterize(pico_sf_valid, pico_elev)

# Create a mask
pico_mask = mask(pico_elev, pico_rast)
```

16.1 Slope of Terrain

Slope is the rate of change of elevation over distance, typically expressed as a percentage or degree (Tang & Pilesjö, 2011). With the 'terra' package, one can calculate slope using the terrain() function, with the parameter 'v = "slope"' which returns a raster object containing slope values in degrees or percent. In the following example, the slope colour is graded in grey colour gradient, the whiter the higher the slope.

One colour types of gradients are good for the visualisation of the slopes.

#16-02

```
# Calculate slope in degrees
pico_slope = terrain(pico_mask, opt = "slope", unit = "degrees")
```

```
# View the results
plot(pico_slope, col = gray(seq(0, 1, length = 90)),
main="Pico Island slope (in degrees)")
plot(pico_out, add = TRUE)
```

In this example for creating the colour palette, the gray() function from the 'grDevices' package is used.

The resulting plot is shown in Figure 16.1.

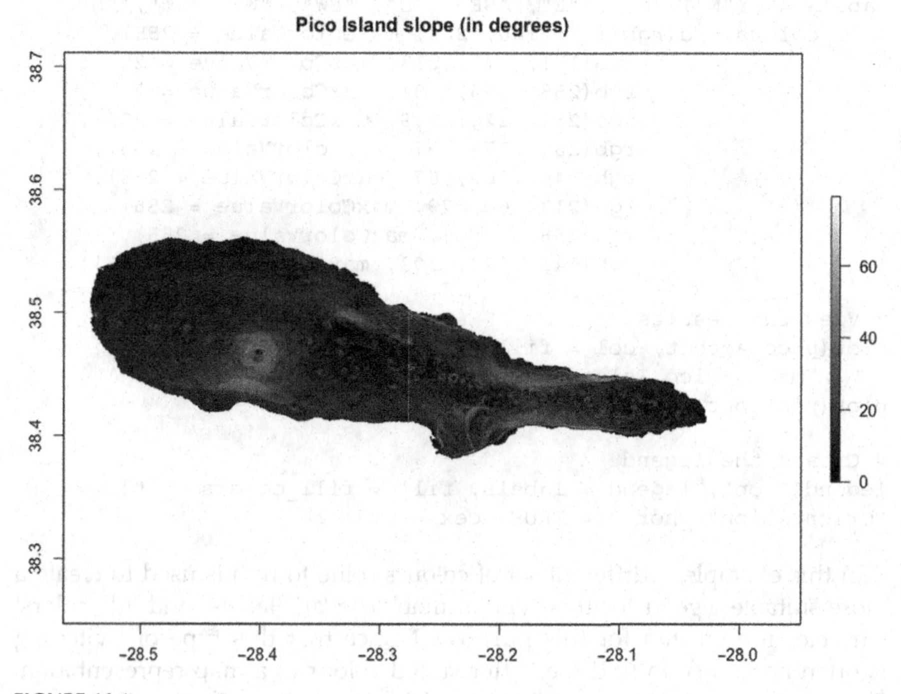

FIGURE 16.1
Pico Island slope (in degrees).

A first analysis in this plot reveals that the higher slopes (lighter colours) are located in the vicinity of the main volcano sides and that there are also strong slopes in the eastern part of the island (Nunes, 2020). Another observable feature is the small secondary cones of the main volcano that have high slopes and also appear very clearly in the image.

16.2 Aspect of Terrain

Aspect is the direction in which a slope faces (Ritter, 1987), typically expressed in degrees clockwise from north. To calculate the 'aspect' using the terrain() function, one must use the 'opt = "aspect"' parameter, which returns a 'raster' object containing aspect values in degrees.

```
#16-03

# Calculate aspect of the terrain
pico_aspect = terrain(pico_mask, opt = "aspect")

# Colours and breaks for the plot
at = c(0, 45, 90, 135, 180, 225, 270, 315, 360)
labels = c("N", "NE", "E", "SE", "S", "SW", "W", "NW", "N")
fill_colors = c(rgb(141, 160, 203, maxColorValue = 255),
                rgb(171, 217, 233, maxColorValue = 255),
                rgb(255, 255, 191, maxColorValue = 255),
                rgb(254, 224, 139, maxColorValue = 255),
                rgb(253, 174, 97, maxColorValue = 255),
                rgb(244, 109, 67, maxColorValue = 255),
                rgb(213, 62, 79, maxColorValue = 255),
                rgb(158, 1, 66, maxColorValue = 255),
                rgb(141, 160, 203, maxColorValue = 255))

# View the results
plot(pico_aspect, col = fill_colors, legend = FALSE,
      main="Pico Island aspect (Orientation of slopes)")
plot(pico_out, add=TRUE)

# Create the legend
legend("top", legend = labels, fill = fill_colors, title =
"Orientation", horiz = TRUE, cex = 0.6)
```

In this example, a different set of colours (blue to red) is used to create a more suitable legend for this type of map. The 'at', 'labels', and 'fill_colors' variables are created for this purpose. Notice that this type of twitching is often necessary to find the better suited colour to a map representation. Sometimes a predefined gradient is available and suits perfectly; other times, like in this example, we opted by creating a custom colours variable.

The result is presented in Figure 16.2.

The analysis of this map is straightforward revealing that in the north part of the island the slopes are majorly north facing, whereas in the southern part of the island the slopes are southerly facing. Around the main peak volcano, the slopes are facing in every direction.

16.3 Hillshade of Terrain

Hillshade is a measure of how much a surface is illuminated by the sun (Horn, 1981). It is often used in cartography to create three-dimensional-like visualisations of the terrain. Hillshade can be computed using the shade() function from the 'raster' package.

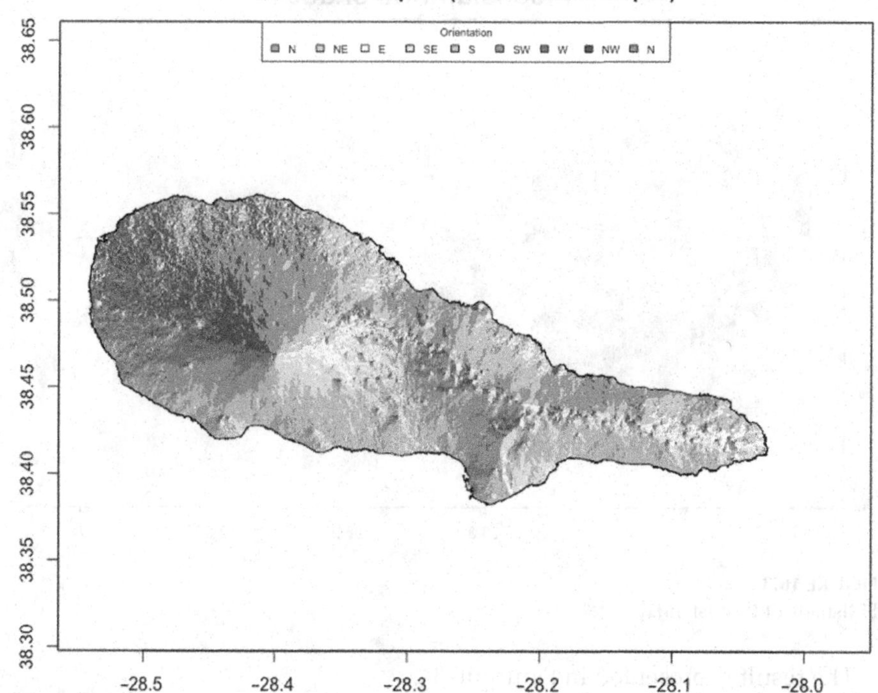

FIGURE 16.2
Aspect of terrain from Pico Island.

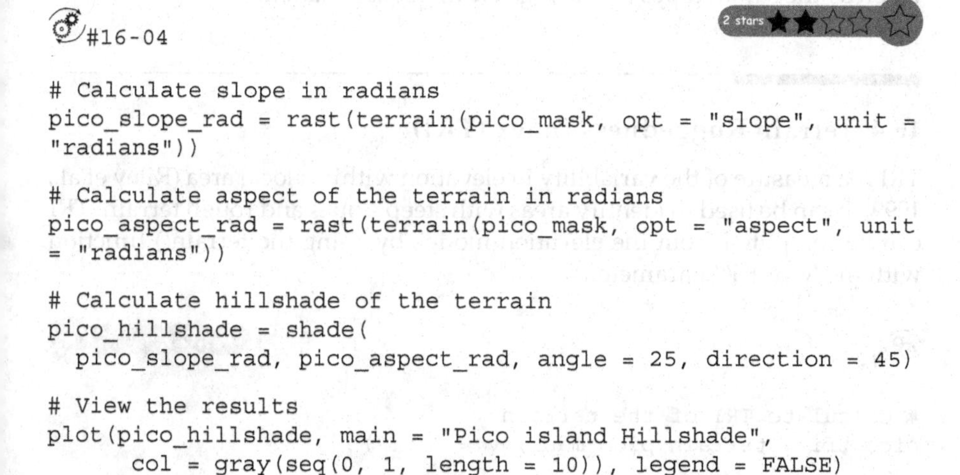

#16-04

```
# Calculate slope in radians
pico_slope_rad = rast(terrain(pico_mask, opt = "slope", unit =
"radians"))

# Calculate aspect of the terrain in radians
pico_aspect_rad = rast(terrain(pico_mask, opt = "aspect", unit
= "radians"))

# Calculate hillshade of the terrain
pico_hillshade = shade(
  pico_slope_rad, pico_aspect_rad, angle = 25, direction = 45)

# View the results
plot(pico_hillshade, main = "Pico island Hillshade",
      col = gray(seq(0, 1, length = 10)), legend = FALSE)
```

To compute the hillshade, one must define as input data the slope of the terrain and the aspect, in radians. Complementary one must define which is the elevation angle of the sun (0 means in the horizon) and its direction relative to the north in a clockwise sense. These values must be in degrees.

FIGURE 16.3
Hillshade of Pico Island.

The result is presented in Figure 16.3.

In this case, the direction of the sun selected is 45 degrees, i.e. in a north-eastern direction. The angle of the sun selected is 25 degrees. Therefore, the slopes in a northwest-southeast direction, facing northeast, are illuminated, whereas the same slopes but facing southwest are shaded.

16.4 Terrain Ruggedness Index (TRI)

TRI is a measure of the variability in elevation within a local area (Riley et al., 1999). It can be used to identify areas with steep slopes and rough terrain. TRI can be computed from the elevation model, by using the terrain() function with the 'v = "tri"' parameter.

#16-05 1 star

```
# Calculate TRI of the terrain
pico_tri = terrain(pico_mask, v = "tri")

# View the results
plot(pico_tri,
      col = topo.colors(n = 10),
      main = "Pico Island Ruggedness")
```

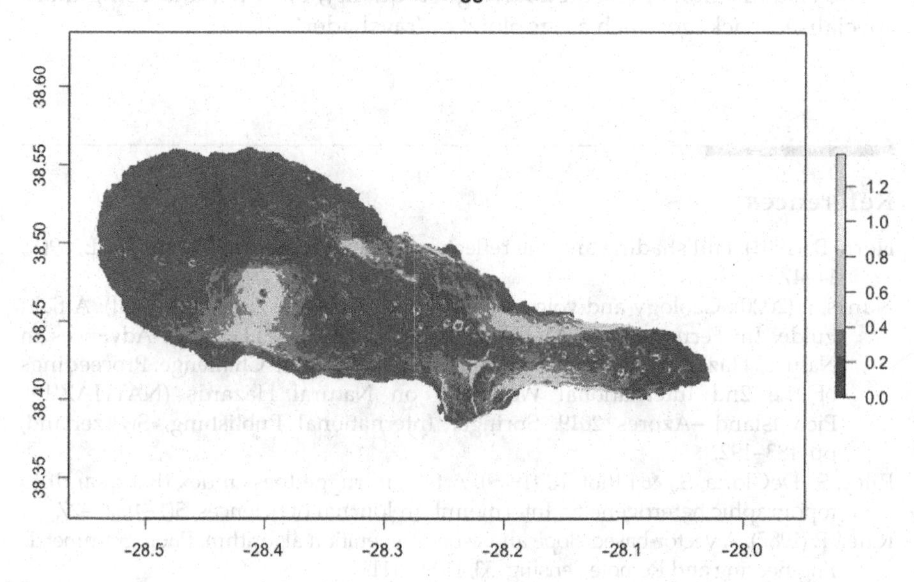

FIGURE 16.4
The Terrain Ruggedness Index (TRI) for Pico Island.

In this case, the topo.colors() function from the 'grDevices' package is used to create a range of 10 colours.

The resulting plot is presented in Figure 16.4.

In this example, one can notice that the TRI value is higher near the volcano top and in the eastern part on the island in both slopes facing north and south.

Other measures of properties of the terrain are available depending on the necessities. These are examples and you can always look for new packages and functions that retrieve other measures.

16.5 Concluding Remarks

I am happy whenever you are happy! These maps are beautiful, easy to accomplish, and very informative. The geomorphological analysis is the ground level of any geographic or geological analysis.

The terrain analysis is mostly done in raster data and always starts by harnessing the DEM for the study area. After some pre-processing that subsets the data from the study area, the 'terra' package has all the necessary tools to make terrain analysis. Slope, aspect, hillshade and ruggedness are calculated and explained as an appetiser. The reader is invited to conduct its own searches and calculate other terrain parameters.

The visualisations presented have good quality, even without using more specialised packages such as 'ggplot2' or 'rayshader'.

References

Horn, B. (1981). Hill shading and the reflectance map. Proceedings of the IEEE, 69(1), 14–47.

Nunes, J. (2020). Geology and volcanology of Pico Island (Azores, Portugal): A field guide. In: Fernandes, F., Malheiro, A., & Chaminé, H.I. (Eds.), Advances in Natural Hazards and Hydrological Risks: Meeting the Challenge: Proceedings of the 2nd International Workshop on Natural Hazards (NATHAZ'19), Pico Island—Azores 2019. Springer International Publishing, Switzerland, pp. 183–192.

Riley, S., DeGloria, S., & Elliot, R. (1999). A terrain ruggedness index that quantifies topographic heterogeneity. Intermountain Journal of Sciences, 5(1–4), 23–27.

Ritter, P. (1987). A vector-based slope and aspect generation algorithm. Photogrammetric Engineering and Remote Sensing, 53, 1109–1111.

Tang, J., & Pilesjö, P. (2011). Estimating slope from raster data: A test of eight different algorithms in flat, undulating and steep terrain. In: Brebbia, C. A. (Ed.), *River Basin Management VI* (Transactions on Ecology and the Environment, Vol. 146). WIT Press, pp. 143–154. https://doi.org/10.2495/RM110131

17

Working with Satellites

Satellite images are – literally – the zenith of earth observation programs and remote sensing applications. In the realm of open science policy, the governments and national agencies are investing in making satellite images available for everyone, which delivers an enormous potential for earth sciences, namely geology, marine sciences, or atmospheric studies.

A broad assortment of packages and functions for utilising satellite data is available in the R environment. These packages provide multiple applications in geology, permitting to perform tasks such as image preprocessing, feature extraction, image classification, or time-series analysis and change detection. The algorithms available provide a series of useful tools for a variety of applications, including land use and land cover identification, geological mapping, natural resource management, or environmental monitoring.

Among the many approaches that one can utilise for retrieving geological information from satellite images, machine learning and deep learning are the rising stars as they can provide valuable information on terrain characterisation (e.g. geomorphology) and classification (e.g. geology, soil type, land use, and land cover). There are also applications that use satellite data, for visualising and analysing spatiotemporal data, often related with risk analysis, management, and mitigation or with deeper understanding of climate evolutions and changes.

17.1 Satellite Packages

For working with the main satellite constellations, i.e. Landsat and Sentinel, the packages that are commonly used include the following:

1. 'openeo' package[1]: This package facilitates the interaction with Open Earth Observation (openEO)-compliant backends. It enables users to effortlessly access, process, and analyse vast earth observation datasets. It provides a unified API (Application Programming Interface)

[1] For more information consult: https://open-eo.github.io/openeo-r-client/index.html

DOI: 10.1201/9781032651880-19

that streamlines the execution of complex earth observation (EO) workflows, eliminating the need for direct data management and leveraging the power of openEO's distributed processing capabilities.

2. 'rgee' package[2]: This package provides an interface to Google Earth Engine (GEE), which is a cloud-based platform for processing and analysing remote sensing data. It allows users to easily access and process large volumes of satellite imagery and provides various functions for filtering, masking, and analysing data.
3. 'rasterVis' package: This package is an extension of the 'raster' package that provides additional functions for visualising raster data. It includes functions for creating interactive maps, adding legends, and customising colour palettes.

For this book, the examples for treating data will be based upon the Sentinel-2 and Landsat-8 satellite images. Nevertheless, there are other available data from several space agencies that can also be used and treated with the techniques presented. However, our experience demonstrates that these two are the most used and are the ones that provide the most reliable time and space information, usable in geology, and, yes, they are for free.

17.1.1 Sign In ... Into a Satellite Journey

Having free access to satellite images is a great resource for researchers, businesses, and government agencies. Among the most popular repositories of satellite imagery, one must consider the European Space Agency (ESA) Copernicus website and the United States Geological Survey (USGS) EarthExplorer services.

To gain access to these services, users must first register for an account. To register for an ESA account, it is necessary to go to the ESA website[3] and register. After providing the necessary individual information, users will be asked to accept the terms and conditions of the service. Once registered, there is access to Sentinel-2 data, as well as data from other ESA missions. Don't forget to retain in a safe place your username and password as you will need it.

Similarly, to register for a USGS EarthExplorer account,[4] one must first go to the EarthExplorer website and create a new account. Once registered, users can access a variety of satellite data, including data from LANDSAT[5], MODIS[6],

---

[2] For more information consult: https://rdocumentation.org/packages/rgee/versions/1.0.7

[3] https://scihub.copernicus.eu/

[4] https://earthexplorer.usgs.gov/

[5] The Landsat program represents a series of earth-observing satellite missions jointly managed by NASA and the USGS, providing continuous global coverage of the earth's land surface since 1972, crucial for environmental monitoring, agriculture, and land use planning.

[6] The MODIS satellite carries instruments that observe earth's atmosphere, oceans, and land in high detail across multiple spectral bands, supporting environmental and climate research.

or ASTER[7] missions. The registration process is simple and straightforward, and the data can be downloaded for free.

17.2 Working with Sentinel Images

The Copernicus Data Space Ecosystem[8] is a pivotal gateway to the repositories of open and freely accessible earth observation data. At its core, this ecosystem encompasses a wealth of information garnered from the Copernicus Sentinels Missions, Copernicus Contributing Missions, and federated datasets, collectively contributing to a comprehensive understanding of our planet's dynamic processes.

The Copernicus Browser serves as a user-friendly and intuitive interface, inviting the users to embark on an exploration of satellite imagery. With navigation capabilities, this browser interface offers an accessible means for geologists and other earth scientists to engage with and interpret earth observation data.

Furthermore, within the Copernicus Data Space Ecosystem, the openEO initiative plays a crucial role by providing standardised interfaces for simplified access and processing of earth observation data. Through its versatile tools, users can effortlessly create and customise workflows, enabling them to either construct new analytical processes or seamlessly integrate them into existing frameworks. The emphasis on minimal coding requirements ensures that users can unlock the full potential of earth observation data with efficiency.

Among the different ESA satellites, the Sentinel-2 constellation is particularly useful for geologists with its multispectral sensors. This constellation comprises two active satellites: Sentinel-2A and Sentinel-2B. Each satellite has a MultiSpectral Instrument (MSI[9]) that captures imagery across multiple spectral bands.

The Sentinel-2 MSI captures data that is provided to the end users with different levels of processing, commonly referred to as levels of images:

Level-0 (L0): Raw Data – this is the unprocessed data as acquired by the satellite's sensors. It includes raw telemetry and sensor data without any corrections.

[7] The ASTER (Advanced Spaceborne Thermal Emission and Reflection Radiometer) is a satellite instrument that captures high-resolution images of earth in multiple wavelengths to map and monitor the planet's surface temperature, emissivity, reflectance, and elevation.

[8] https://dataspace.copernicus.eu/

[9] The MultiSpectral Instrument (MSI) is an optical imaging sensor used in the Sentinel-2 satellite missions, which are part of the European Space Agency's (ESA) Copernicus program. The satellites equipped with MSI capture high-resolution optical imagery of the earth's surface across various spectral bands.

Level-1C (L1C): It is Top of Atmosphere (TOA) Reflectance that has undergone radiometric and geometric corrections to convert the raw digital numbers into top-of-atmosphere reflectance values. However, it does not include terrain correction.

Level-2A (L2A): It is Bottom of Atmosphere (BOA) Reflectance that includes additional processing to correct for atmospheric effects, providing surface reflectance values. L2A processing includes atmospheric corrections for gases, aerosols, and Rayleigh scattering as well as terrain correction. This means L2A data offers a more accurate representation of the earth's surface reflectance by removing the influence of the atmosphere and adjusting for the earth's topography, making it highly suitable for a wide range of applications including environmental monitoring, agriculture, and land cover mapping.

These levels represent different stages of processing, each providing progressively more refined and usable data for different applications. Level-2A data is often preferred for tasks such as land cover classification, vegetation monitoring, and other applications where atmospherically corrected, and terrain-corrected data are essential.

As stated, Sentinel-2 is particularly useful for geologists because of its multispectral sensors with 13 spectral bands, which can be used to identify different types of rocks and minerals. For example, the near-infrared (NIR) bands can be used to identify vegetation, while the shortwave infrared (SWIR) bands can be used to help in identifying water and iron-bearing minerals.

Sentinel-2 images are considered to be of high resolution, with a pixel size of 10 metres for the visible and NIR bands and 20 metres for the SWIR bands. This means that the images are suitable to map geological features with a fair amount of detail. For example, Sentinel-2 images can be used to identify faults, folds, and other geological structures. The Sentinel-2 images are also very frequent, with new images being acquired every 5 days. This means that the images can be used to monitor changes in the earth's surface over time, e.g., to track the movement of glaciers and landslides.

17.2.1 Register to Download

Although freely available, the Copernicus program asks its users to provide some basic information to be able to visualise and download the available data. To register and download Sentinel-2 images, follow these steps:

Register in the Copernicus dataspace website:

- Go to the website (https://dataspace.copernicus.eu/).
- Click on the 'Register' button.

- Enter your email address, choose a password, and enter the required information, accepting the Terms and Conditions.
- Click on the 'Register' button again to complete the registration process.

Verify your email address:

- After receiving an email with a verification link, click on the link to confirm your email address.

Login to dataspace:

- Once your email address is verified, you will be able to login to the website.
- On the top, there is a menu that allows you to Explore and Analyse the available data.
- Remember to verify your account's personal information, such as name, organisation, and country.

Start downloading Sentinel-2 images:

- Go to the Copernicus Browser window and verify the information that is available to your region of interest and the timespan for the study.

17.2.2 Verifying the Available Images

The Copernicus Browser (Figure 17.1) allows the user to define all the parameters necessary to retrieve the images. The bar on the left is used to define the parameters of the images to be searched, including single date or time range, cloud cover, sentinel type, and image data type. The main area

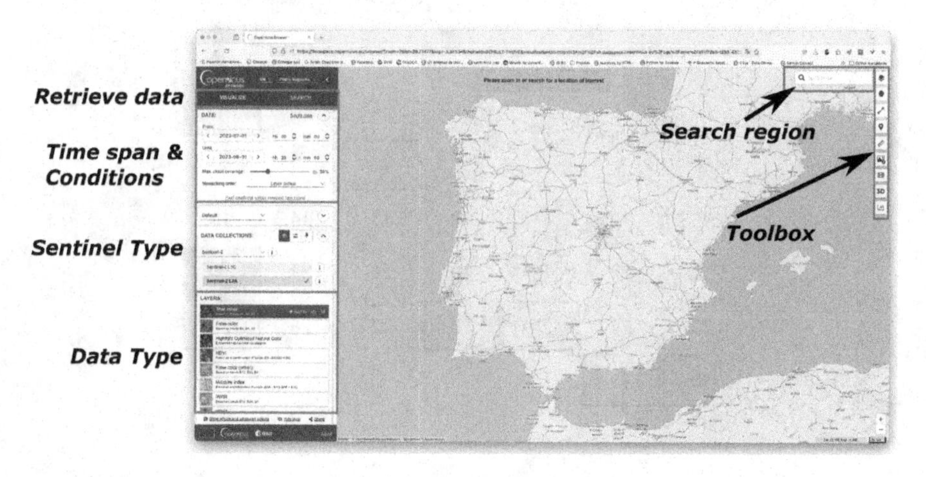

FIGURE 17.1
The Copernicus Browser window and its options.

of the window presents a map to visualise the area of interest, and there is a search box in top right to look for a specific location. There is also a toolbox with possibilities for further configuring the visualisation options or special downloads.

The Retrieve data area has two possibilities 'VISUALISE' and 'SEARCH' data. In the 'VISUALISE' the user can define the conditions indicating if it is searching for a single date or a Time range. In the case of single date, the calendar window shows the dates in which images are available and the user can indicate the threshold for the cloud coverage (Figure 17.2). When using a time range, the available images will be displayed in the 'SEARCH' tab.

After selecting the conditions in the left panel, and if the zoom is correct one can use the option 'Find products for current view' to verify the availability of images with the conditions selected (Figure 17.3).

From the available images, the user can click in a specific one to retrieve further details (Figure 17.4).

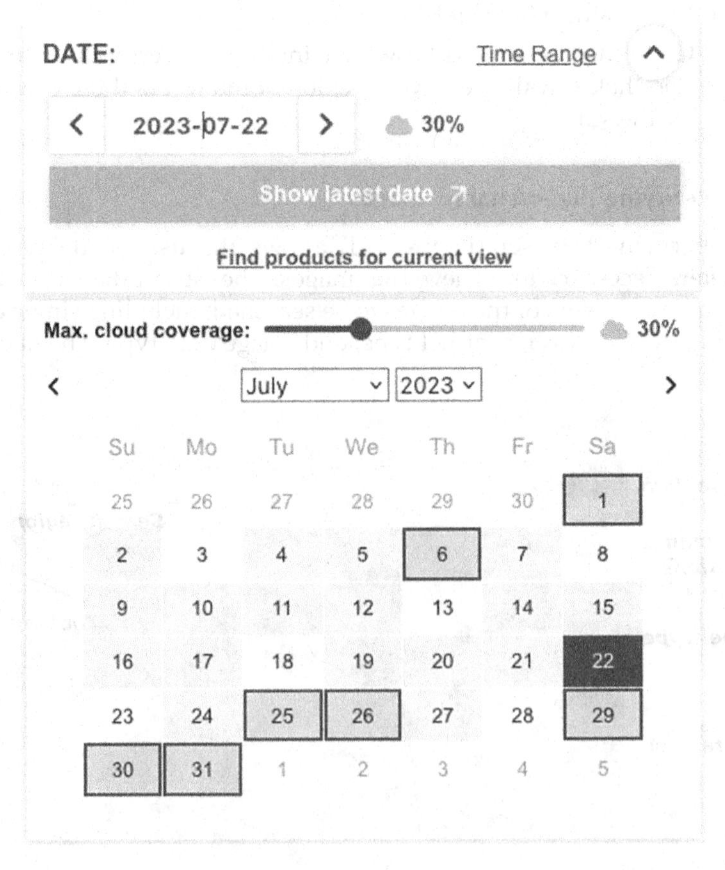

FIGURE 17.2
The Copernicus Browser date selector and maximum cloud coverage.

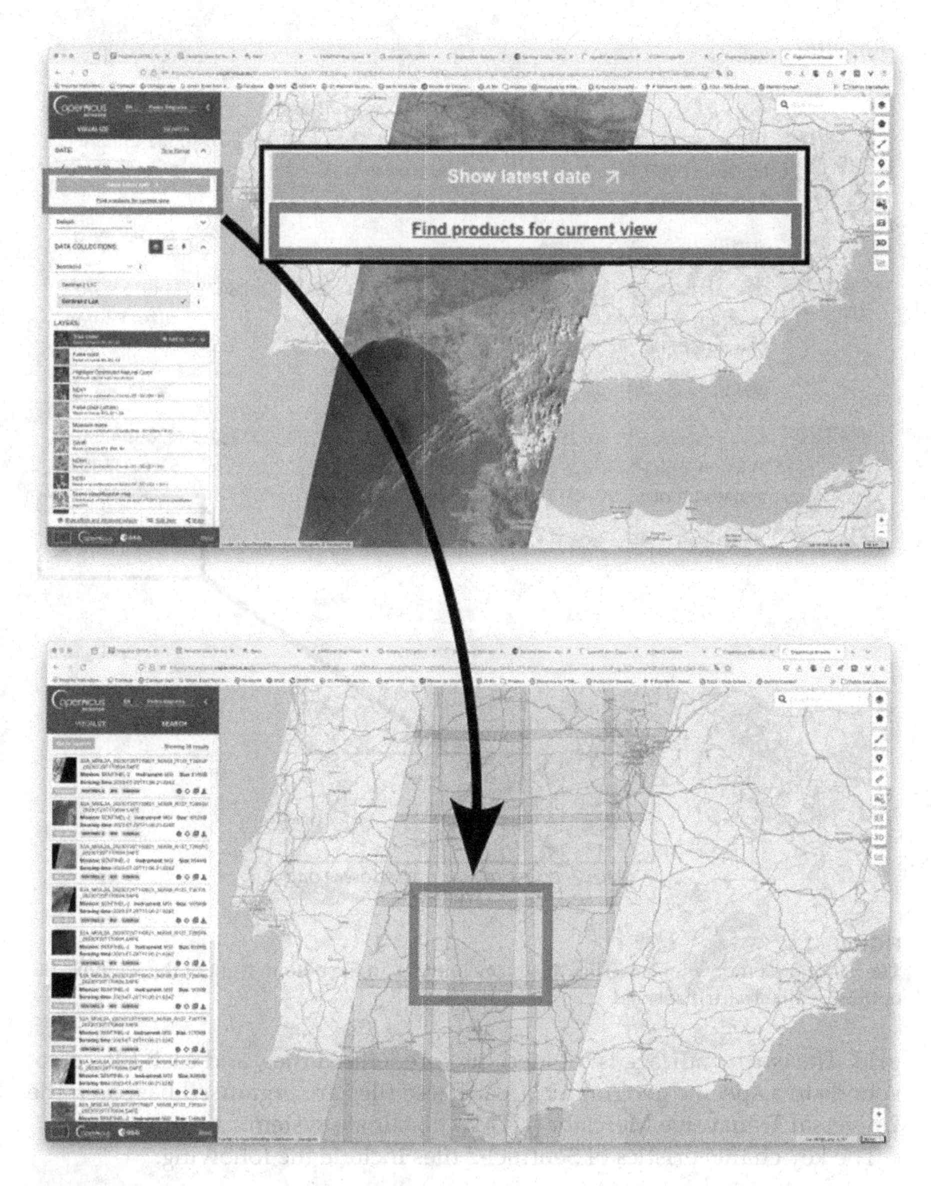

FIGURE 17.3
The Search tab displays the available images.

Selecting one Product/Tile in the bottom right appears the options to verify the metadata, zoom to the specific tile, add the tile to the workspace, or directly download the tile (Figure 17.4).

The structure of these files is somewhat complex and for beginners might look cryptic. The Sentinel-2A and Sentinel-2B satellites acquire imagery in a systematic and coordinated way, covering the earth's surface in a structured

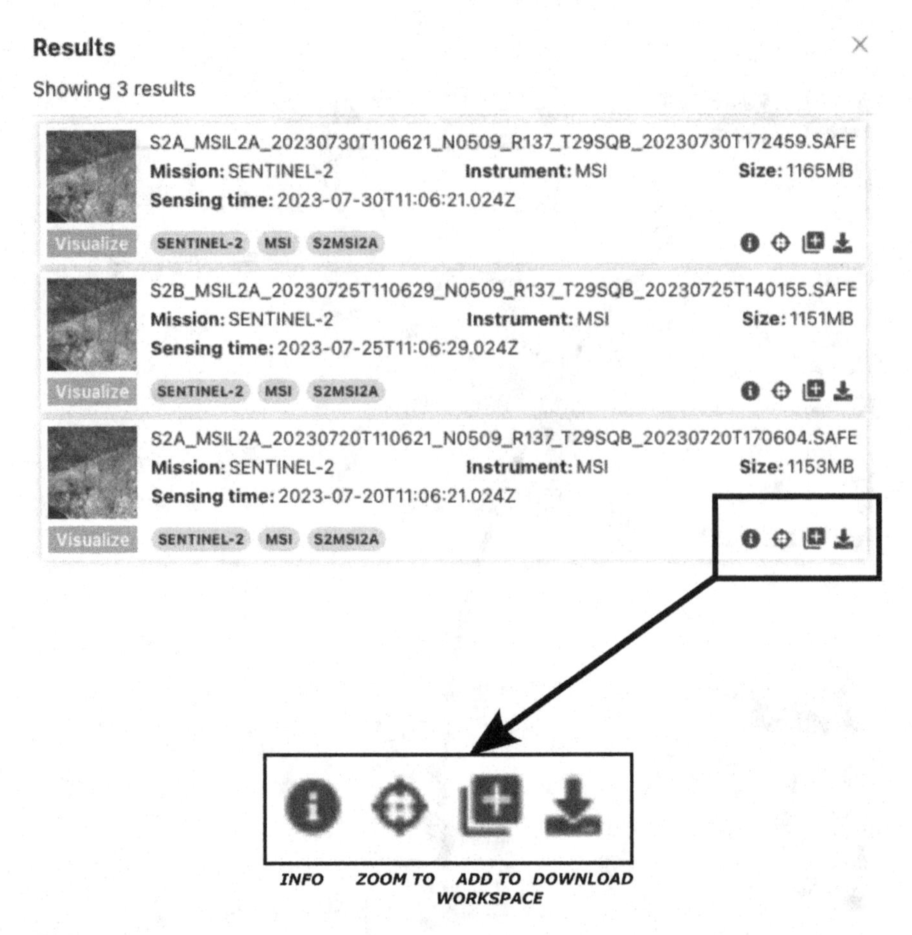

FIGURE 17.4
The Selection window for selecting the image options. In the bottom right of the window, the options available are displayed.

grid. The earth's surface is divided into predefined tiles, and each tile corresponds to a specific geographic area. These tiles are organised based on the Universal Transverse Mercator (UTM) coordinate system.

The key characteristics of Sentinel-2 tiles include the following:

Grid Structure: The earth is divided into a grid of tiles (see Figure 17.3), and each tile is identified by a unique tile identifier. The tiles are organised in a path-row system, where a path corresponds to a longitudinal strip, and a row corresponds to a latitudinal strip. Each tile is associated with a specific path and row.

Tile Size: Each Sentinel-2 tile covers a standard size on the ground, which is approximately 100 kilometres × 100 kilometres. The size of the tiles ensures systematic and consistent coverage across the earth's surface.

UTM Zones: The tiles are organised based on the UTM coordinate system, which divides the earth into zones for mapping and analysis. Each UTM zone is associated with specific path and row combinations, and tiles within a zone are identified by their path and row numbers.

Overlap: Adjacent tiles often have an overlap (see Figure 17.3) to ensure seamless and continuous coverage. This overlap facilitates data continuity and reduces artefacts in the imagery.

Tile Naming Convention: Tiles are commonly referred to by their UTM zone, path, and row. For example, a Sentinel-2 tile might be named like 'T29SQB' where 'T29' is the UTM zone, and 'SQB' corresponds to the specific path and row (cf. Figure 17.5).

These tiles and their systematic organisation allow for efficient and comprehensive coverage of the earth's surface. Users can specify the tiles they are interested in, making it easier to access and analyse specific geographic areas of interest.

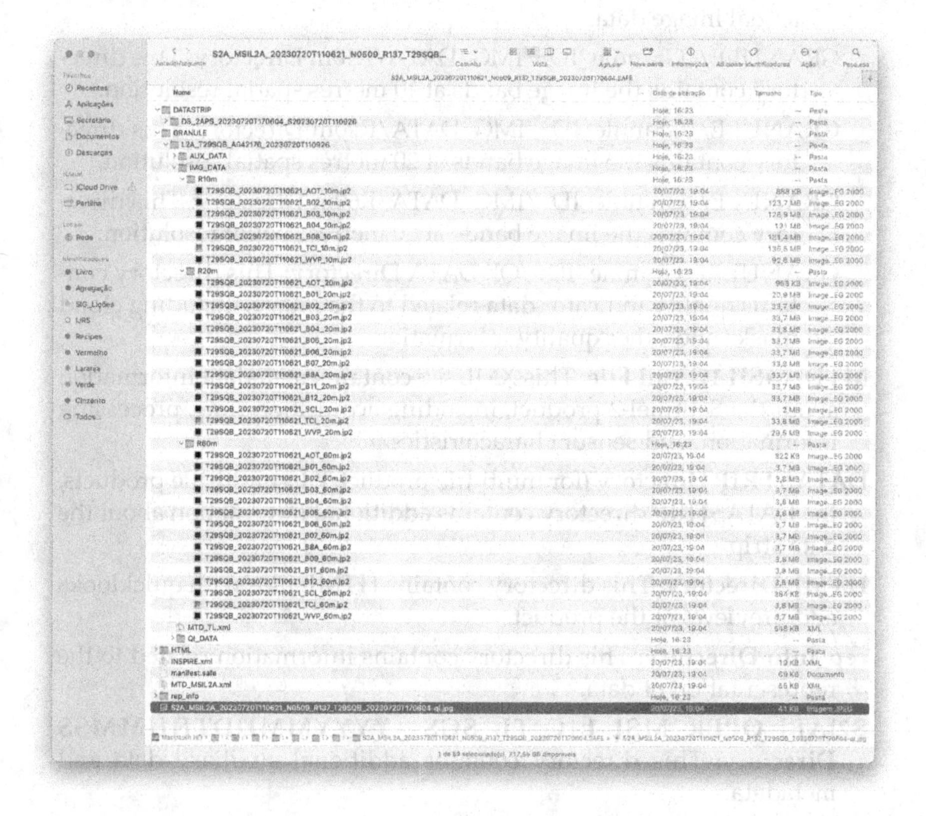

FIGURE 17.5
The file structure of the 'zip' file downloaded from the Copernicus website.

17.2.3 Downloading the Data

After selecting the data of interest, it is possible to download a 'zip' file with all the information of the corresponding image. The Sentinel-2 'zip' file is organised in a specific directory structure within the file downloaded from the Copernicus website.

This structure ensures that the data is organised in a logical and consistent manner, making it easier for users to access and process the data. By understanding the structure of the Sentinel-2 image data 'zip' file, users can easily access and process the data for their specific applications.

The Sentinel-2 image data 'zip' file typically contains the following directories and files (see Figure 17.5):

GRANULE Directory: It contains subdirectories named with the unique identifier of each granule (observation). Each granule corresponds to a specific geographic area.

GRANULE/<Granule_ID>/IMG_DATA Directory: Inside each granule's directory, there is an IMG_DATA directory containing the actual image data.

GRANULE/<Granule_ID>/IMG_DATA/R10m Directory: This directory contains the image bands at 10 metres spatial resolution.

GRANULE/<Granule_ID>/IMG_DATA/R20m Directory: This directory contains the image bands at 20 metres spatial resolution.

GRANULE/<Granule_ID>/IMG_DATA/R60m Directory: This directory contains the image bands at 60 metres spatial resolution.

GRANULE/<Granule_ID>/QI_DATA Directory: This directory contains quality indicator data related to the image, including cloud masks and other quality information.

MTD_MSIL1C.xml File: This XML file contains metadata information about the Sentinel-2 product, including acquisition time, processing information, and sensor characteristics.

DATASTRIP Directory (for multi-tile products): In multi-tile products, the DATASTRIP directory contains additional information about the data strip.

HTML Directory: This directory contains HTML files with quicklooks and previews of the product.

rep_info Directory: This directory contains information related to the representation of data.

S2[AB]_OPER_MSI_L1C_TL_SGS__YYYYMMDDTHHMMSS Directory: This directory contains additional auxiliary data and metadata.

Figure 17.5 shows an example of a window from the file explorer with the 'zip' file structure and corresponding files.

Using the Copernicus Browser, it is also possible to download only a partial part of the tile scene or area of interest. For this, in the Copernicus Browser zoom to the area of interest and in the toolbox (on the right of the Copernicus Browser) select the download icon. This opens a window with several options (Figure 17.6). In the top part there are three tabs that allow us to download the 'Basic' image, the image for 'Analytical' purposes, and the image for 'High resolution printing'.

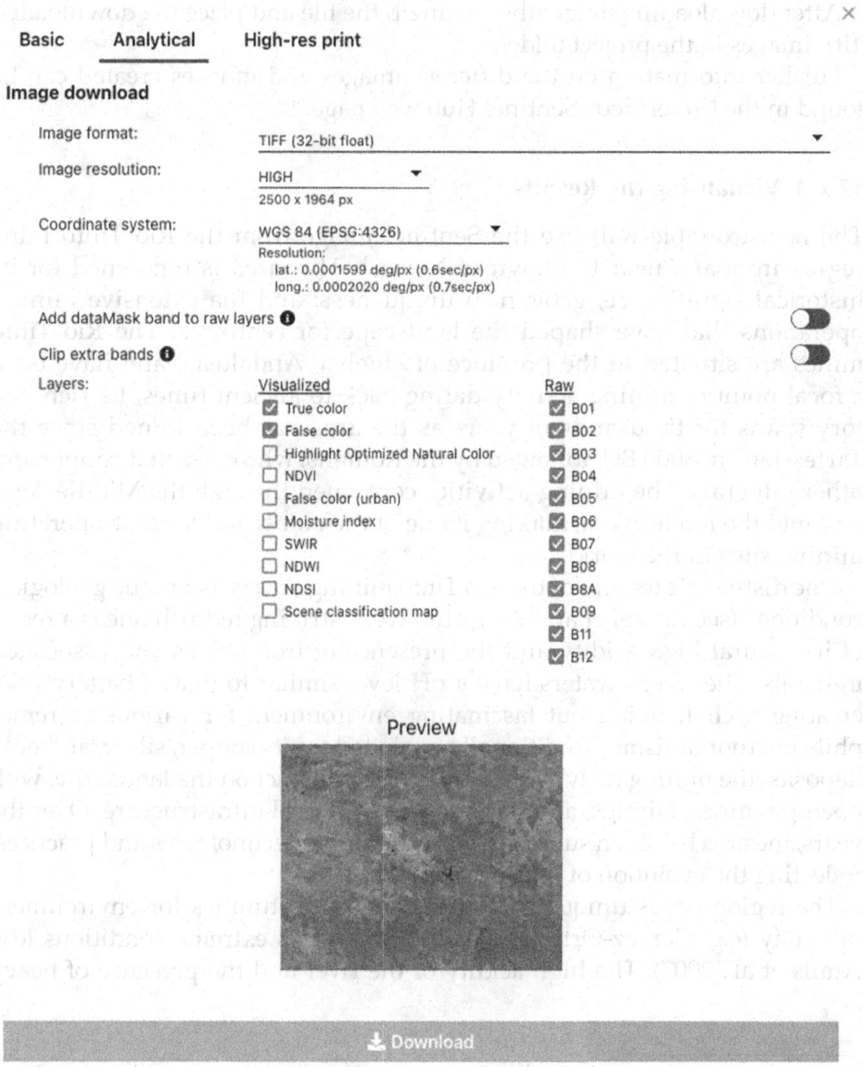

FIGURE 17.6
The download window for an area of interest.

The example in Figure 17.6 selects the 'Analytical' option that allows downloading all the bands and the image compositions 'True Color' (Bands 4, 3, and 2) and 'False Color' (Bands 8, 4, and 3). It also allows the download of individual bands. For better quality, use the High Resolution in the Tagged Image File Format (TIFF) (32-bit float) format. Also take note of the coordinate reference system (CRS) used (it is suggested to use the universal WGS84 –, i.e. EPSG = 4326). The options 'Add dataMask band to raw layers' and 'Clip extra bands' should be kept off to create clean 'tiff' images without extra bands or information.

This possibility to download the analytical image is faster and suitable for most of the case studies.

After downloading, remember to unzip the file and place the downloaded 'tiff' images in the project folder.

Further information on the different images and indexes created can be found in the Copernicus Sentinel Hub web page.[10]

17.2.4 Visualising the Results

The next example will use the Sentinel image from the Rio Tinto mine region in Spain, near the town of Nerva.[11] This area is renowned for its historical significance, geological uniqueness, and the extensive mining operations that have shaped the landscape for centuries. The Rio Tinto mines are situated in the province of Huelva, Andalusia, and have been a focal point of mining activity dating back to ancient times. Its rich history spans for thousands of years as the area has been mined since the Tartessians in 3000 BC, followed by the Romans, who extracted copper and other minerals. The mining activities continued through the Middle Ages and into the modern era, making it one of the oldest and longest-operating mining sites in the world.

One distinctive feature of the Rio Tinto mining area is its unique geological conditions (see Leistel et al., 1997). The river's striking reddish hue is a result of its natural high acidity and the presence of iron oxides and associated minerals. The river's waters have a pH level similar to that of battery acid, creating a challenging but fascinating environment for various extremophile microorganisms. Traditionally exploited for its copper, silver, and gold deposits, the mining activity has left a visible impact on the landscape, with open-pit mines, tailings, and remnants of historical infrastructure. Over the years, the area has been subject to various mining technologies and practices, reflecting the evolution of mining methods.

The region poses unique challenges and opportunities for environmental study (e.g. Gómez-Ortiz et al., 2014) due to its extreme conditions (e.g. Amils et al., 2007). The high acidity of the river and the presence of heavy

[10] https://custom-scripts.sentinel-hub.com/custom-scripts/sentinel/sentinel-2/

[11] The image for the Rio Tinto can be accessed at https://link.dataspace.copernicus.eu/wwk

metals have led to the development of research initiatives exploring extremophiles and their potential implications for astrobiology.

This Rio Tinto region example will be used in further examples for raster data analysis and treatment. The first step is to retrieve a list of files that were downloaded[12] and its full path.

For this, use the list.files() function. Don't forget to define the working directory to the path where the downloaded and zipped files are.

#17-01

```
# Load the Library
library(terra)

# Retrieve the tif files
files = list.files(pattern = ".tiff",  full.names = TRUE)

# View the files
files
```

The result is as follows:

```
 [1] "./2023-07-20-00:00_2023-07-20-23:59_Sentinel-2_L2A_B01_
(Raw).tiff"
 [2] "./2023-07-20-00:00_2023-07-20-23:59_Sentinel-2_L2A_B02_
(Raw).tiff"
 [3] "./2023-07-20-00:00_2023-07-20-23:59_Sentinel-2_L2A_B03_
(Raw).tiff"
 [4] "./2023-07-20-00:00_2023-07-20-23:59_Sentinel-2_L2A_B04_
(Raw).tiff"
 [5] "./2023-07-20-00:00_2023-07-20-23:59_Sentinel-2_L2A_B05_
(Raw).tiff"
 [6] "./2023-07-20-00:00_2023-07-20-23:59_Sentinel-2_L2A_B06_
(Raw).tiff"
 [7] "./2023-07-20-00:00_2023-07-20-23:59_Sentinel-2_L2A_B07_
(Raw).tiff"
 [8] "./2023-07-20-00:00_2023-07-20-23:59_Sentinel-2_L2A_B08_
(Raw).tiff"
 [9] "./2023-07-20-00:00_2023-07-20-23:59_Sentinel-2_L2A_B09_
(Raw).tiff"
[10] "./2023-07-20-00:00_2023-07-20-23:59_Sentinel-2_L2A_B11_
(Raw).tiff"
[11] "./2023-07-20-00:00_2023-07-20-23:59_Sentinel-2_L2A_B12_
(Raw).tiff"
[12] "./2023-07-20-00:00_2023-07-20-23:59_Sentinel-2_L2A_B8A_
(Raw).tiff"
```

[12] The image for the Rio Tinto can be downloaded at https://link.dataspace.copernicus.eu/wwk

```
[13] "./2023-07-20-00:00_2023-07-20-23:59_Sentinel-2_L2A_
False_color.tiff"
[14] "./2023-07-20-00:00_2023-07-20-23:59_Sentinel-2_L2A_True_
color.tiff"
```

For plotting the result, use the plot() function. This code snippet plots the red Band in a grey scale (file number 4).

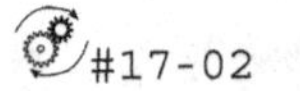#17-02

```
# Load individual raster
RT_red = rast(files[4])

# Plot the greyscale Red Band
plot(RT_red, col=grey.colors(255), main = "Rio Tinto Red Band")
```

The resulting plot is presented in Figure 17.7.
Note that the values are between 0 and 1.

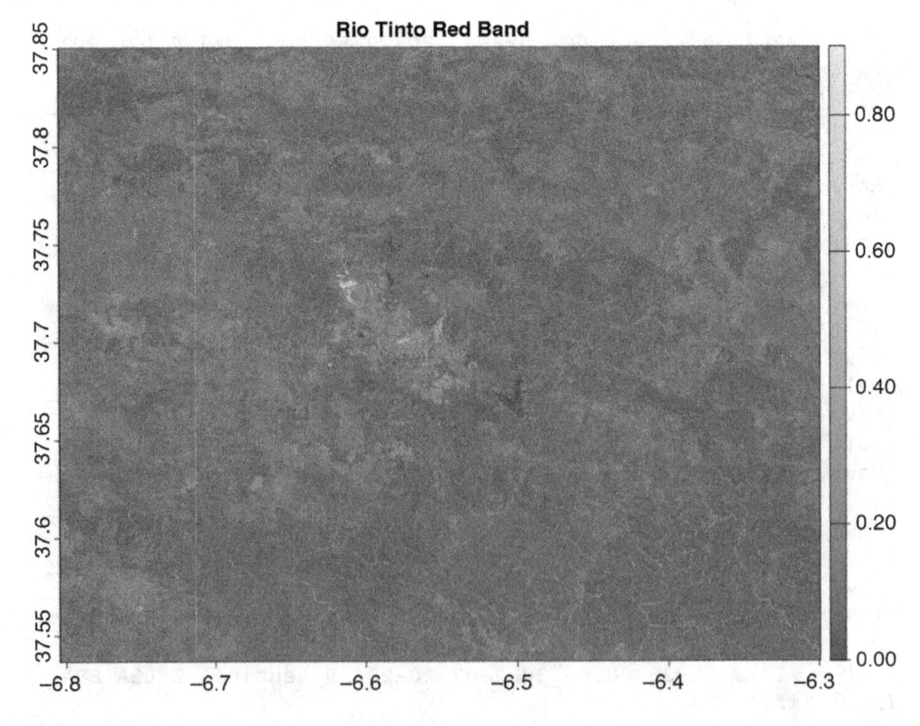

FIGURE 17.7
The Red Band (Band 4) from the Rio Tinto Mine area.

This 'RT-red' is a SpatRaster object:

```
class       : SpatRaster
dimensions  : 1964, 2500, 1  (nrow, ncol, nlyr)
resolution  : 0.0002020111, 0.000159902  (x, y)
extent      : -6.804314, -6.299286, 37.53668, 37.85073  (xmin,
xmax, ymin, ymax)
coord. ref. : lon/lat WGS 84 (EPSG:4326)
source      : 2023-07-20-00:00_2023-07-20-23:59_Sentinel-2_
L2A_B04_(Raw).tiff
name        : 2023-07-20-00:00_2023-07-20-23:59_Sentinel-2_
L2A_B04_(Raw)
```

In the download option (see Figure 17.6), we selected to download the "True Color" composite image, i.e. a 'tiff' with four bands where the first three are the bands 4, 3, and 2, and the fourth is 'alpha', i.e. transparency band. This file can be used to create a true colour image.

The composite of images created with plotRGB() function is an image created from three bands where the bands are as follows: Red = Band1, Green = Band2, and Blue = Band 3. An alternative way to plot red, blue, and green (RGB) images is to first use colorize() function to create a single layer SpatRaster with a color-table and then use the plot() function.

#17-03

```
# Load individual raster
RT_true = rast(files[14])

# Plot the raster composition using the plotRGB() function
plotRGB(RT_true, r=1, g=2, b=3, axes= T, mar=c(2,2,3,2), main
="Rio Tinto area (True Color)", scale = 1)
```

The result is also a 'SpatRaster' object.

```
class       : SpatRaster
dimensions  : 1964, 2500, 4  (nrow, ncol, nlyr)
resolution  : 0.0002020111, 0.000159902  (x, y)
extent      : -6.804314, -6.299286, 37.53668, 37.85073
(xmin, xmax, ymin, ymax)
coord. ref. : lon/lat WGS 84 (EPSG:4326)
source      : 2023-07-20-00:00_2023-07-20-23:59_Sentinel-2_
L2A_True_color.tiff
names       : 2023-07~color_1, 2023-07~color_2, 2023-07~
color_3, 2023-07~color_4
```

The plot is presented in Figure 17.8.

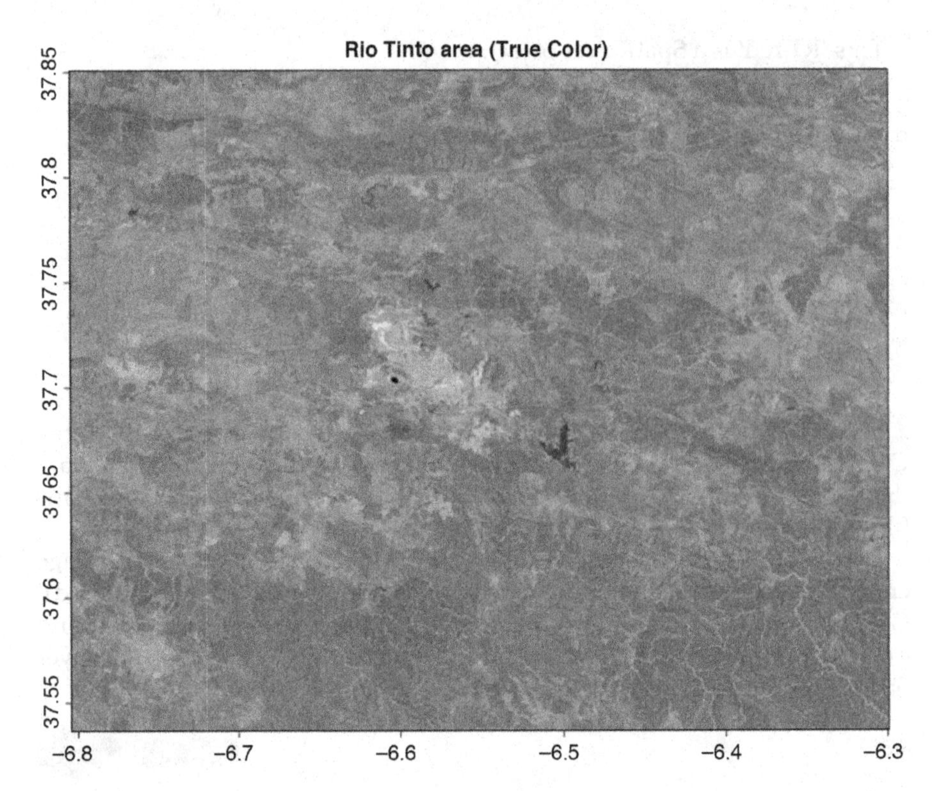

FIGURE 17.8
The Rio Tinto Mine area in true colour image from Sentinel-2 satellite.

17.3 Working with Landsat-8 Images

For the Landsat-8 images, we will use the EarthExplorer[13] site to directly download the images and work them in the R environment.

17.3.1 Download Landsat-8 Data

The same Rio Tinto Mine area is selected for this example. In the EarthExplorer site (Figure 17.9), the user can download each image as a 'tiff' file or create a bulk order that will be available in the user basket (see Figure 17.9) to download all the images together. This download is called a bulk download, and after submitted to the site, the user will receive an email with the link to download the files (unfortunately only in chrome or edge browsers).

[13] https://earthexplorer.usgs.gov/

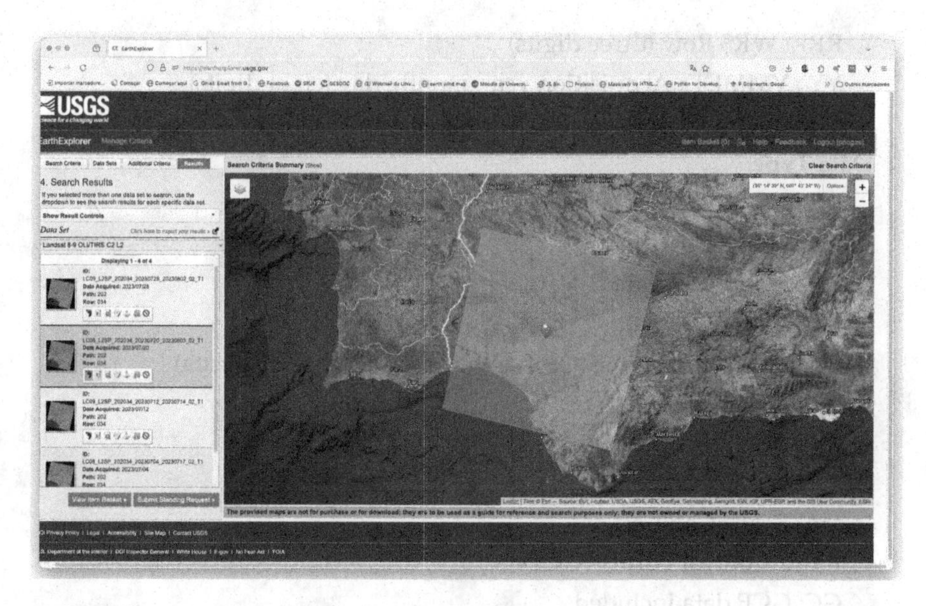

FIGURE 17.9
The download screen from the EarthExplorer site.

Save the images downloaded in a folder to use. The file naming convention for Landsat-8 files is a little complex, so dedicate a few moments to understand the file names and its structure.

17.3.2 Know Your Landsat-8 Data

When a complete set of files is downloaded from the EarthExplorer site, pay a little attention to understand the naming convention. The file names have a structure that provides essential information about the scene,[14] including its location, acquisition date, processing date, and sensor data. Here's a breakdown of the naming convention:

Scene Identifier:

L: Indicates Landsat satellite

C: Combined Operational Land Imager (OLI) and Thermal Infrared Sensor (TIRS) data

T: TIRS data only

O: OLI data only

08: Landsat-8 satellite

PPP: WRS (Worldwide Reference System) Path (three digits)

[14] See also https://www.usgs.gov/landsat-missions/landsat-provisional-surface-temperature

RRR: WRS Row (three digits)
YYYYMMDD: Acquisition year, month, day (eight digits)
yyymmdd: Processing year, month, day (six digits)

Data Type:

T1: Level-1 TIRS data
T2: Level-2 TIRS data
B1: Level-1 OLI data
B2: Level-2 OLI data
SP: Science Product (combined Level-1 OLI and TIRS data)

Scene Quality:

*SA: Scene Acquisition
*SG: Scene Geometry
*SC: Scene Calibration

Ground Control Points (GCPs):

GC: GCP data included

Product Quantity (PQ):

PQn: Product quantity (n)

Data Format:

TIFF: Tagged Image File Format
HDF: Hierarchical Data Format

For example, a Landsat-8 scene with the filename LC08_C202_034_20230923_20230924_T2_SP_SC_GCP_PQ01.TIF indicates the following:

Landsat Combined OLI and TIRS (LC08_C)
WRS Path: 202, WRS Row: 034
Acquisition date: 2023-09-23
Processing date: 2023-09-24
Level-2 TIRS data (T2)
Science Product (SP)
Scene Calibration (SC)
GCPs included
Product quantity: 1
Data format: Tagged Image File Format (TIFF)

The image downloaded for this exercise, also called a scene, is the one corresponding to the WRS Path = 202 and WRS ROW = 034 from the Landsat-8 collection (Figure 17.10).

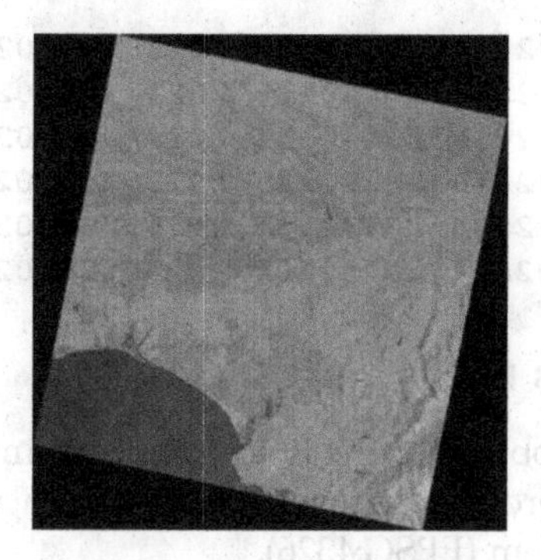

FIGURE 17.10
The Path=202 Row=034 image.

The file downloaded in the working folder for this exercise include the following:

File Name	Description
LC08_L2SP_202034_20230125_20230208_02_T1_SR_B1.TIF	coastal aerosol – ultraviolet
LC08_L2SP_202034_20230125_20230208_02_T1_SR_B2.TIF	Blue
LC08_L2SP_202034_20230125_20230208_02_T1_SR_B3.TIF	Green
LC08_L2SP_202034_20230125_20230208_02_T1_SR_B4.TIF	Red
LC08_L2SP_202034_20230125_20230208_02_T1_SR_B5.TIF	near infrared
LC08_L2SP_202034_20230125_20230208_02_T1_SR_B6.TIF	shortwave infrared 1
LC08_L2SP_202034_20230125_20230208_02_T1_SR_B7.TIF	shortwave infrared 2
LC08_L2SP_202034_20230125_20230208_02_T1_ST_B10.TIF	thermal infrared 1
Other data and metadata	

For this example, two images will be created i) a true colour composite and ii) the NDVI index image, for the Rio Tinto Mine study area.

As above, the variable 'files' is created with the list of files that correspond to the bands. This time the 'pattern = "_B"'is used to retrieve only files that contain the Bands, i.e. the ones whose names include the "_B" pattern.

#17-04

```
# Retrieve the Band files
files = list.files(pattern =  "_B", recursive = TRUE, full.
names = TRUE)

# Verify the result
files
```

The result is as follows:

```
[1] "./LC08_L2SP_202034_20230125_20230208_02_T1_SR_B1.TIF"
[2] "./LC08_L2SP_202034_20230125_20230208_02_T1_SR_B2.TIF"
```

```
[3] "./LC08_L2SP_202034_20230125_20230208_02_T1_SR_B3.TIF"
[4] "./LC08_L2SP_202034_20230125_20230208_02_T1_SR_B4.TIF"
[5] "./LC08_L2SP_202034_20230125_20230208_02_T1_SR_B5.TIF"
[6] "./LC08_L2SP_202034_20230125_20230208_02_T1_SR_B6.TIF"
[7] "./LC08_L2SP_202034_20230125_20230208_02_T1_SR_B7.TIF"
[8] "./LC08_L2SP_202034_20230125_20230208_02_T1_ST_B10.TIF"
```

17.3.3 Read Landsat-8 Data

To create a SpatRaster object for the RGB composite image, the rast() function is used. The function project() from the 'terra' package is used to convert the files to the WGS84 system (EPSG:4326).

#17-05

1 star

```
# Load the RGB bands to create the image
RT_L8_rgb = rast(c(files[4],files[3],files[2] ))

# View the result
plotRGB(RT_L8_rgb, r=1, g=2, b=3, axes= T, mar=c(2,2,3,2),
main ="L8- Rio Tinto area (True Color)")
```

The 'RT_L8_rgb' is a variable of type 'SpatRaster' with three bands. Notice that the download image in this case is in the UTM 29N (EPSG:32629) CRS.

```
class        : SpatRaster
dimensions   : 7711, 7581, 3  (nrow, ncol, nlyr)
resolution   : 30, 30  (x, y)
extent       : 628785, 856215, 4034685, 4266015  (xmin, xmax,
ymin, ymax)
coord. ref. : WGS 84 / UTM zone 29N (EPSG:32629)
sources      : LC08_L2SP_202034_20230125_20230208_02_T1_SR_
B4.TIF
               LC08_L2SP_202034_20230125_20230208_02_T1_SR_B3.TIF
               LC08_L2SP_202034_20230125_20230208_02_T1_SR_B2.TIF
names        : LC08_L2SP_~2_T1_SR_B4, LC08_L2SP_~2_T1_SR_B3,
LC08_L2SP_~2_T1_SR_B2
```

The resulting plot is the complete scene downloaded (Figure 17.11).

17.3.4 Plotting a Composite RGB Landsat-8 Image

Before continuing, the image will be cropped to the same extent of the Sentinel-2 example. For this, the Landsat-8 will be reprojected in the WGS84 (i.e. crs = 4326), after which the crop function is used. The plotRGB() function is used in the same way as above but with the difference that the RGB bands are in different positions in the 'RT_L8_crop' image. In the 'RT_L8_crop' variable, the red band is the number 4 index, the green is the index number 3, and the blue band is the index number 2.

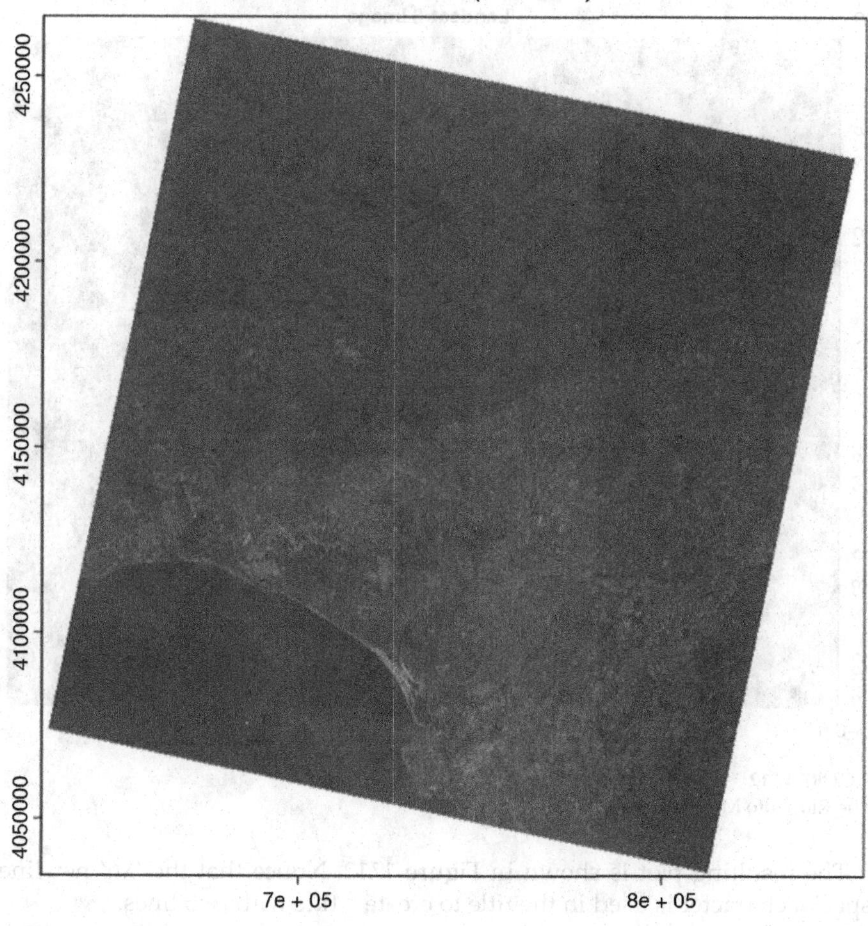

FIGURE 17.11
The complete scene from the Rio Tinto area from the Landsat-8 satellite.

#17-06

1 star

```
# Load the RGB bands to create the image
RT_L8_full = rast(files)

# Reproject Landsat-8 image
RT_L8_full_wgs84 = project(RT_L8_full, "EPSG:4326")

# Crop Landsat-8 image
RT_L8_crop = crop(RT_L8_full_wgs84, RT_red)

# View the result
plotRGB(RT_L8_crop, r=4, g=3, b=2, axes= T, mar=c(2,2,3,2),
      main ="L8- Rio Tinto area (True Color)\nLandsat-8 image")
```

FIGURE 17.12
The Rio Tinto Mine using the Landsat-8 image.

The resulting plot is shown in Figure 17.12. Notice that the '\n' newline special character is used in the title to create a title with two lines.

17.3.5 The NDVI

The NDVI is used to verify the healthiness of vegetation and its formula is as follows:

$$\text{NDVI} = (\text{NIR} - \text{RED}) / (\text{NIR} + \text{RED})$$

That is band 5 – band 4 divided by band 5 + band 4.

For this example, two variables 'red' and 'nir' are created from the cropped image. These variables contain the corresponding single bands. R is able to make calculations in rasters on a pixel-by-pixel basis; therefore, the operation is straightforward.

To plot the results, a colour ramp between blue, brown, and green is created.

```
#17-07

# Define the NDVI bands
red = RT_L8_crop$LC08_L2SP_202034_20230125_20230208_02_T1_SR_B4
nir = RT_L8_crop$LC08_L2SP_202034_20230125_20230208_02_T1_SR_B5

# Calculate the NDVI Index
ndvi = (nir - red)/(nir + red)

# Define the NDVI values
ndvi_breaks = c(-1,  0, 0.1, 0.2, 0.5, 0.8, 1)

# Define the corresponding colors
colors = c("lightblue","brown", "darkgreen", "green3",
"green2", "green")

# Create the color palette
ndviPalette = colorRampPalette(colors)

# View the result
plot(ndvi, col = ndviPalette(7), breaks= ndvi_breaks,
      main="Calculated NDVI for Rio Tinto Mine area")
```

The resulting plot is shown in Figure 17.13.

FIGURE 17.13
The NDVI image calculated for the Rio Tinto Mine using the Landsat-8 image.

Note that the stronger the green, the healthier the vegetation. Water bodies are below zero values; hence they are blue. Soil usually has values lower than 0.3. These values are used for reference, but, naturally, they need to be fine-tuned to a better vision of the terrain reality. The appropriate intervals for NDVI depend on the specific application and the type of vegetation being analysed. However, some general guidelines can be followed:

NDVI Interval	Description
–1 to –0.1	Bare soil, rocks, water
0.1 to 0.2	Sparse vegetation, barren land
0.2 to 0.3	Low vegetation cover
0.3 to 0.4	Moderate vegetation cover
0.4 to 0.5	High vegetation cover
0.5 to 0.6	Very high vegetation cover
0.6 to 1	Dense, continuous vegetation

17.4 Concluding Remarks

Satellite images provide geologists and earth scientists, in general, with tremendous tools for viewing earth from above. This overview, although limited to a predefined spatial resolution (10, 20, 30, 60, or 120 metres), depending on the sensor used, revisits the observation place with a usable timely visit, being of 16 days for Landsat-8 and 5 days for Sentinel-2.

Another advantage of the satellite images is that the observation is not limited to the visible part of the spectrum, going behind the red edge of the spectrum colours into the infrared, short-wave infrared, and even thermal part of the spectrum.

By combining specific bands (fragments of the spectrum), the geologist can view the earth with new eyes to highlight or unveil geological features. Another advantage of having the spectrum divided in small portions (the bands) is the possibility to create indexes of bands (ratios between bands) that can also be used as classifiers for specific features. The NDVI is the most common, allowing to highlight the vegetation healthier parts and defining usable thresholds for it. I am sure that this same approach after appropriate research will be paramount in the identification of geological features, mineral occurrences, or water presence in the earth surface.

References

Amils, R., González-Toril, E., Fernández-Remolar, D., Gómez, F., Aguilera, Á., Rodríguez, N., … & Sanz, J. L. (2007). Extreme environments as Mars terrestrial analogs: The Rio Tinto case. Planetary and Space Science, 55(3), 370–381.

Gómez-Ortiz, D., Fernández-Remolar, D. C., Granda, Á., Quesada, C., Granda, T., Prieto-Ballesteros, O., … & Amils, R. (2014). Identification of the subsurface sulfide bodies responsible for acidity in Río Tinto source water, Spain. Earth and Planetary Science Letters, 391, 36–41.
Leistel, J. M., Marcoux, E., Thiéblemont, D., Quesada, C., Sánchez, A., Almodóvar, G. R., … & Saez, R. J. M. D. (1997). The volcanic-hosted massive sulphide deposits of the Iberian Pyrite Belt Review and preface to the Thematic Issue: Review and preface to the Thematic Issue. Mineralium Deposita, 33, 2–30.

18

Putting It All to Work: Part II Rasters

In this wrap-up chapter, we will develop a case study where sentinel images from the Rio Tinto area are used to create a simple thematic map with some of the different features (i.e. land cover) observable in the terrain. For the different themes, some of the indices described in the satellite index database[1] will be used. For finalising, we will propose some considerations on the statistics of the different band sets.

18.1 Setting Up the Environment

The first step is to define the environment variables, load the study area with the st_read() function, and make all the transformations necessary to have the information in 'sp' and 'sf' formats for further processing.

#18-01

```
# Load the libraries
library(terra)
library(sf)
library(ggplot2)
library(rasterVis)

# Define the working folder
setwd("~/")
```

For this example, we will use the Sentinel-2 images downloaded as mentioned in Section 17.2.3 in Chapter 17. This time we will use the complete scene dataset downloaded (see Figure 17.5). Remember to set the working directory, using the setwd() function, to the folder where the full sentinel data is located.

For working with the several datasets, it is better to have the list of complete paths to the necessary files. In this step, the list.files() function is used to create these file lists. Specifically, the 20 metre band sets will be used. The

[1] https://www.indexdatabase.de/

 DOI: 10.1201/9781032651880-20

conversion to the 10 metre or 60 metre band sets is straightforward using the resample() function.

The 'files_20m' will contain all the files that are in the working pass that include the '20m.jp2' text string, whereas the files_bands will contain only the ones that have a pattern '_B' a digit ('\\d') a character ('\\w'), followed by the string '_20m.jp2'.

An inspection to the file names explains these patterns.

#18-02

```
# Define the lists of files
files_20 = list.files(pattern = "20m.jp2", full.names = TRUE,
recursive = T)

# Work with the 20m complete bandset
pattern_bands = "_B\\d\\w_20m.jp2"

files_bands = list.files(pattern = pattern_bands, full.names =
TRUE, recursive = T)

# View the results
files_20

files_bands
```

The output for the 'files_20' variable is as follows:

```
[1] "./GRANULE/L2A_T29SQB_A042176_20230720T110926/IMG_DATA/
R20m/T29SQB_20230720T110621_AOT_20m.jp2"
[2] "./GRANULE/L2A_T29SQB_A042176_20230720T110926/IMG_DATA/
R20m/T29SQB_20230720T110621_B01_20m.jp2"
[3] "./GRANULE/L2A_T29SQB_A042176_20230720T110926/IMG_DATA/
R20m/T29SQB_20230720T110621_B02_20m.jp2"
[4] "./GRANULE/L2A_T29SQB_A042176_20230720T110926/IMG_DATA/
R20m/T29SQB_20230720T110621_B03_20m.jp2"
[5] "./GRANULE/L2A_T29SQB_A042176_20230720T110926/IMG_DATA/
R20m/T29SQB_20230720T110621_B04_20m.jp2"
[6] "./GRANULE/L2A_T29SQB_A042176_20230720T110926/IMG_DATA/
R20m/T29SQB_20230720T110621_B05_20m.jp2"
[7] "./GRANULE/L2A_T29SQB_A042176_20230720T110926/IMG_DATA/
R20m/T29SQB_20230720T110621_B06_20m.jp2"
[8] "./GRANULE/L2A_T29SQB_A042176_20230720T110926/IMG_DATA/
R20m/T29SQB_20230720T110621_B07_20m.jp2"
[9] "./GRANULE/L2A_T29SQB_A042176_20230720T110926/IMG_DATA/
R20m/T29SQB_20230720T110621_B11_20m.jp2"
[10] "./GRANULE/L2A_T29SQB_A042176_20230720T110926/IMG_DATA/
R20m/T29SQB_20230720T110621_B12_20m.jp2"
[11] "./GRANULE/L2A_T29SQB_A042176_20230720T110926/IMG_DATA/
R20m/T29SQB_20230720T110621_B8A_20m.jp2"
[12] "./GRANULE/L2A_T29SQB_A042176_20230720T110926/IMG_DATA/
R20m/T29SQB_20230720T110621_SCL_20m.jp2"
```

```
[13] "./GRANULE/L2A_T29SQB_A042176_20230720T110926/IMG_DATA/
R20m/T29SQB_20230720T110621_TCI_20m.jp2"
[14] "./GRANULE/L2A_T29SQB_A042176_20230720T110926/IMG_DATA/
R20m/T29SQB_20230720T110621_WVP_20m.jp2"
[15] "./GRANULE/L2A_T29SQB_A042176_20230720T110926/QI_DATA/
MSK_CLDPRB_20m.jp2"
[16] "./GRANULE/L2A_T29SQB_A042176_20230720T110926/QI_DATA/
MSK_SNWPRB_20m.jp2"
```

Notice that some file names end with several types of information. The 'file_bands' variable is more straightforward.

```
[1] "./GRANULE/L2A_T29SQB_A042176_20230720T110926/IMG_DATA/
R20m/T29SQB_20230720T110621_B01_20m.jp2"
[2] "./GRANULE/L2A_T29SQB_A042176_20230720T110926/IMG_DATA/
R20m/T29SQB_20230720T110621_B02_20m.jp2"
[3] "./GRANULE/L2A_T29SQB_A042176_20230720T110926/IMG_DATA/
R20m/T29SQB_20230720T110621_B03_20m.jp2"
[4] "./GRANULE/L2A_T29SQB_A042176_20230720T110926/IMG_DATA/
R20m/T29SQB_20230720T110621_B04_20m.jp2"
[5] "./GRANULE/L2A_T29SQB_A042176_20230720T110926/IMG_DATA/
R20m/T29SQB_20230720T110621_B05_20m.jp2"
[6] "./GRANULE/L2A_T29SQB_A042176_20230720T110926/IMG_DATA/
R20m/T29SQB_20230720T110621_B06_20m.jp2"
[7] "./GRANULE/L2A_T29SQB_A042176_20230720T110926/IMG_DATA/
R20m/T29SQB_20230720T110621_B07_20m.jp2"
[8] "./GRANULE/L2A_T29SQB_A042176_20230720T110926/IMG_DATA/
R20m/T29SQB_20230720T110621_B11_20m.jp2"
[9] "./GRANULE/L2A_T29SQB_A042176_20230720T110926/IMG_DATA/
R20m/T29SQB_20230720T110621_B12_20m.jp2"
[10] "./GRANULE/L2A_T29SQB_A042176_20230720T110926/IMG_DATA/
R20m/T29SQB_20230720T110621_B8A_20m.jp2"
```

This last list contains only references to the working band set with 20 metre resolution information. We will use these images to run this example.

The next step is to create a 'SpatRaster' with all the bands, reproject it to the WGS84 coordinate reference system (CRS), and crop to the study area extent. We use a slightly modified version of the extent of the Rio Tinto study area from Chapter 17. In this case, the images are cropped to the extent (–6.70, –6.30, 37.55, and 37.85) using the ext() function followed by the crop() function.

```
#18-03

# Create a SpatRaster of 20m images
S2_20m = rast(files_bands)

# Convert to WGS84
S2_20m = project(S2_20m, "EPSG:4326")
```

```
# Define extent equal to Chapter 17 Rio Tinto cropped data
RT_extent = ext(-6.70, -6.30, 37.55, 37.85)

# Crop the image
RT_20m = crop(S2_20m, RT_extent)
```

In the next step, the name of the bands in the 'RT_20m' variable is changed to correspond to something humanly understandable, using the names() function.

It is advisable to verify that the image cropped corresponds to the intended study area. For this, a true colour combination is plotted.

#18-04

```
# Rename the SpatRaster bands
names(RT_20m) = c('CIR', 'blue', 'green', 'red', 'RE1', 'RE2',
'RE3', 'SWIR1', 'SWIR2', 'NIR')

# Plot the True Colour band composition
par(mfrow=c(1,1))
plotRGB(RT_20m, r="red", g="green", b="blue",
      axes=TRUE, mar=c(2,1,3,1), main ="Rio Tinto area (True
Color)")
```

Figure 18.1 displays the result obtained.

18.2 Combining Bands

Creating band combinations[2] is a common way to visualise remote sensing data in a single image, typically using different spectral bands or indices to highlight different features or characteristics of the study area. Four composite images that are commonly used are as follows:

IR false colour (B8, B4, and B3): The infrared (IR) false colour composite image uses the near-infrared (NIR), Red, and Green spectral bands to highlight vegetation, water, and urban areas. Vegetation appears bright red, water appears dark blue or black, and urban areas appear grey or white. This composite image is useful for mapping and monitoring vegetation cover, urban areas, and water bodies.

[2] See more at https://custom-scripts.sentinel-hub.com/custom-scripts/#sentinel-2 and https://gisgeography.com/sentinel-2-bands-combinations/

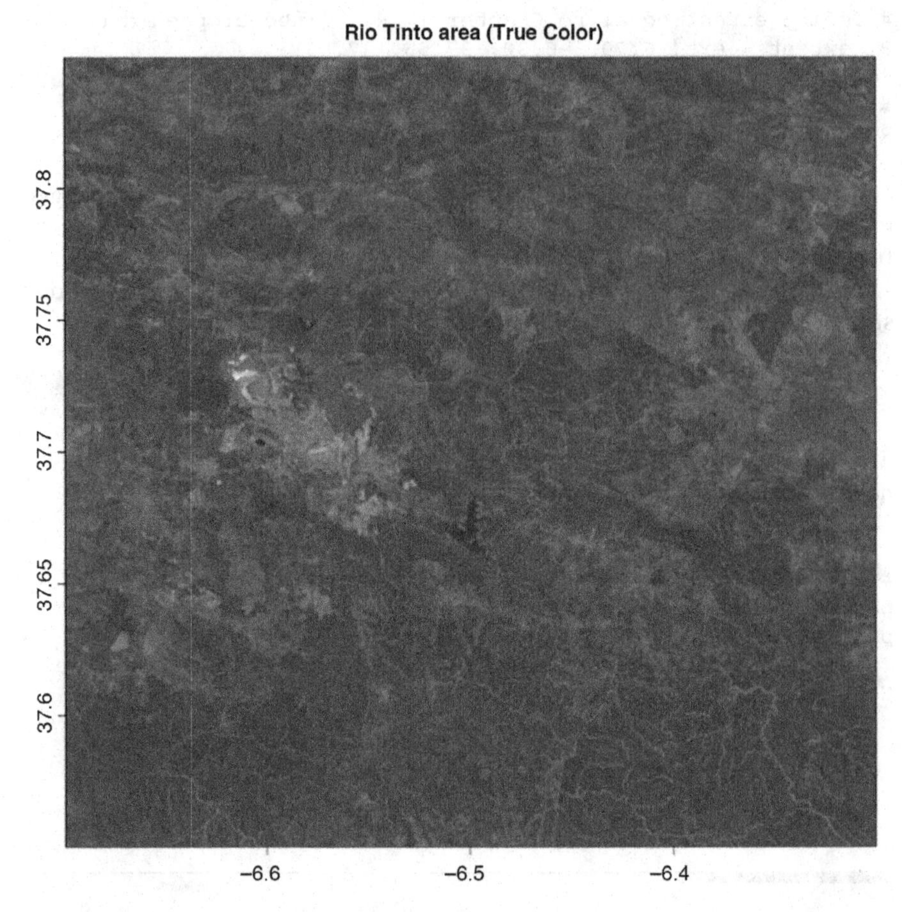

FIGURE 18.1
The Rio Tinto study area in a True Colour image.

SWIR (B12, B8, and B4): The short-wave infrared (SWIR) spectral bands are indeed useful for estimating soil moisture content and mapping vegetation health, as well as for distinguishing between different surface features such as snow and ice. Additionally, as mentioned, SWIR reflectance properties vary for different rock types, making it possible to use SWIR data to map geological features and mineral deposits. Moreover, SWIR data can be used to identify and map areas affected by wildfires and to monitor post-fire vegetation recovery.

Vegetation (B8, B5, and B2): The composite image is useful for mapping and monitoring vegetation cover and health, as well as for identifying and mapping water bodies and aquatic ecosystems. The composite is also useful for distinguishing between different land cover types and for identifying areas of land use change and urban areas.

Geology (B12, B11, and B2): The composite image uses the Blue, SWIR1, and SWIR2 spectral bands to highlight geological features and mineral deposits in the study area. The SWIR bands are sensitive to mineralogical and lithological properties of rocks and minerals and can be used to map mineral assemblages, alteration zones, and other geological features. The Blue band is sensitive to the presence of iron oxide minerals, which are common in many types of rocks and can be used to distinguish between different rock types. In the B12, B11, and B2 composite image, rocks and minerals with high SWIR reflectance appear bright while those with low reflectance appear dark. Iron oxide-rich rocks and minerals appear blue or blue-green while other minerals and rocks appear in shades of grey, brown, and green. This composite image is useful for mapping and characterising geological features and mineral deposits and for identifying areas of potential mineral resources.

In this example, an image comparing the four composites is created, using the proposed indexes.

```
#18-05

# Plot examples of combinations of bands
par(mfrow=c(2,2))
plotRGB(RT_20m, r="NIR", g="red", b="green",
      axes=TRUE, mar=c(2,1,3,1), stretch="lin",
      main ="IR False Color") # 8 4 3

plotRGB(RT_20m, r="SWIR2", g="NIR", b="red",
      axes=TRUE, mar=c(2,1,3,1),stretch="lin",
      main ="SWIR") # 12 8 4

plotRGB(RT_20m, r="NIR", g="RE3", b="blue",
      axes=TRUE, mar=c(2,1,3,1),stretch="lin",
      main ="Vegetation") # 8 5 2(4)

plotRGB(RT_20m, r="SWIR2", g="SWIR1", b="blue",
      axes=TRUE, mar=c(2,1,3,1),stretch="lin",
      main ="Geology") # 12 11 2
```

The resulting image is presented in Figure 18.2.

18.3 Calculating Indices

The range of spectral bands provided by Sentinel-2 satellite data can be used to extract different types of information about the study area. In addition to visual interpretation, satellite data can be analysed using

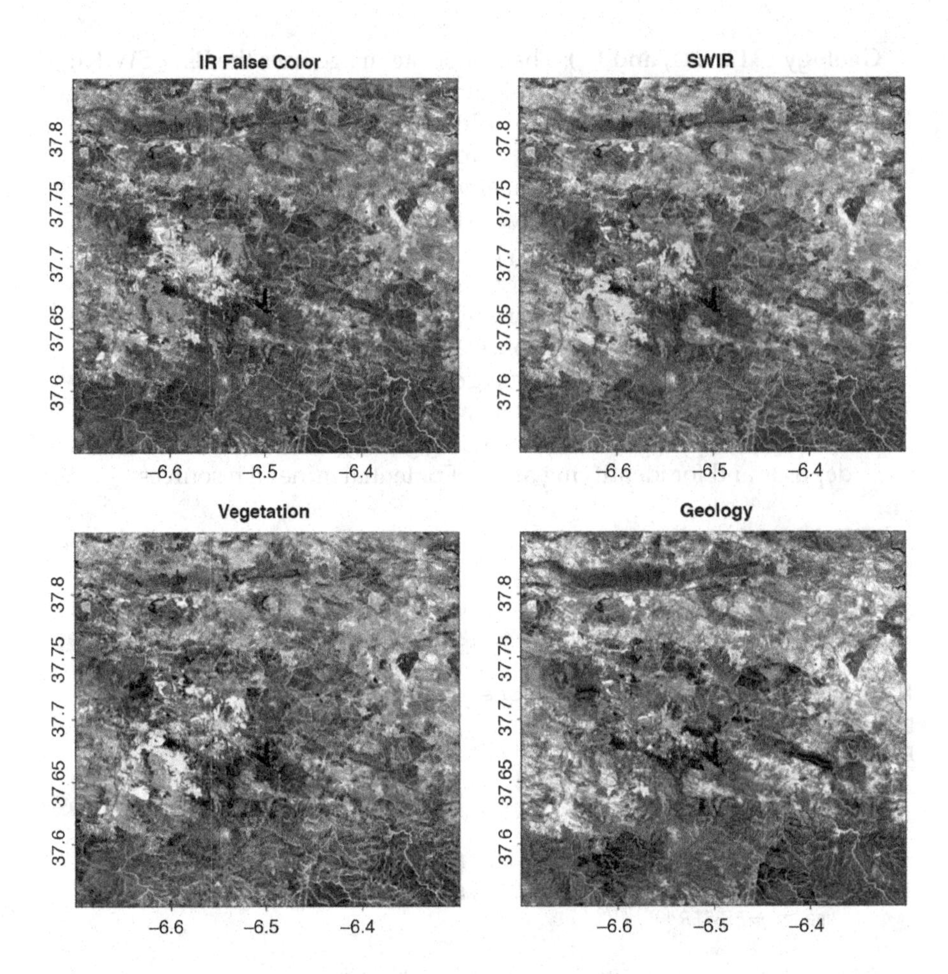

FIGURE 18.2
The different composite images of the Rio Tinto Study area.

indices, which are mathematical combinations of spectral bands or reflectance values.[3]

Here's a brief introduction to some commonly used indices in Sentinel-2 image analysis:

Normalised Difference Vegetation Index (NDVI): The NDVI is a widely used vegetation index that uses the Red and NIR spectral bands to estimate vegetation cover and health. Healthy vegetation absorbs more light in the Red band and reflects more light in the NIR band, resulting in a high NDVI value. This index can be used

[3] For more information, see https://www.indexdatabase.de/ and https://custom-scripts.sentinel-hub.com/custom-scripts/#sentinel-2

to monitor vegetation growth, identify areas of vegetation stress or damage, and estimate vegetation productivity.

Normalised Difference Water Index (NDWI): The NDWI is an index that uses the Green and NIR spectral bands to estimate water content and distribution. Water absorbs more light in the NIR band and reflects more light in the Green band, resulting in a high NDWI value. This index can be used to identify and map water bodies, monitor changes in water resources, and detect areas affected by floods or drought.

Moisture Stress Index (MSI): As water content in vegetation canopy leaves increases, the absorption at wavelengths around SWIR2 also increases, resulting in a high MSI value. The absorption at Red is used as a reference since it is nearly unaffected by changes in water content. The MSI is useful for canopy stress analysis, productivity prediction and modelling, fire hazard analysis, and studies of ecosystem physiology. The values of the MSI index range from 0 to more than 3, with the common range for green vegetation being 0.4–2.

Bare Soil Index (BSI): The BSI is an index that uses the Red and SWIR spectral bands to estimate the amount of bare soil in the study area. Soil reflects more light in the SWIR band and absorbs more light in the Red band, resulting in a high BSI value for bare soil. This index can be used to monitor soil erosion, identify areas of land degradation, and assess the impact of land use change on soil quality.

Chlorophyll Vegetation Index (CVI): The CVI is an index that uses the Red and NIR spectral bands to estimate the chlorophyll content and health of vegetation. Healthy vegetation absorbs more light in the Red band and reflects more light in the NIR, resulting in a high CVI value. This index can be used to monitor vegetation health and productivity and to identify areas of vegetation stress or damage.

It is a simple step to calculate the indices in R. The script below demonstrates the calculation of the referred indeed.

```
#18-06

# NDVI (Normalised Difference Vegetation Index)
NDVI = (RT_20m$NIR - RT_20m$red) / (RT_20m$NIR + RT_20m$red) #
(B05 - B04) / (B05 + B04)

# NDWI (Normalised Difference Water Index)
NDWI = (RT_20m$green - RT_20m$NIR) / (RT_20m$green +
RT_20m$NIR) # (B03 - B08) / (B03 + B08)
```

```
# Filter the NDWI to a threshold
WATER = NDWI > 0.05 # Water bodies definition
WATER[WATER == 0 ] = NA

# MSI (Moisture Soil Index)
MSI = RT_20m$SWIR2 / RT_20m$red # B11 / B08

# BSI (Bare Soil Index)
BSI = ( (RT_20m$SWIR1 + RT_20m$red) - (RT_20m$NIR +
RT_20m$blue)) / ( (RT_20m$SWIR1 + RT_20m$red) + (RT_20m$NIR +
RT_20m$blue)) # BSI = ((B11 + B04) - (B08 + B02)) / ((B11 +
B04) + (B08 + B02))

# CVI - Chlorophyll Vegetation Index
CVI = RT_20m$NIR * (RT_20m$red / (RT_20m$red^2)) # CVI
```

In this script, and to exemplify the use of threshold to define features in the area, the NDWI is used to limit water NDWI > 0.05 and not water NDWI ≤ 0.05.

Two figures are created from this data for a better visualisation of the results. The first plot in Figure 18.3 displays the true colour, NDVI, NDWI, and water threshold.

#18-07

```
# Create a plot
par(mfrow = c(2,2))
plotRGB(RT_20m, r="red", g="green", b="blue",
      axes=TRUE, mar=c(2,1,3,1), stretch ="lin",
      main ="True Color")
plot(NDVI, main="NDVI")
plot(NDWI, main="NDWI")
plot(WATER, main="WATER", col="blue3")
```

Figure 18.3 displays the resulting image.

For the other indexes, the code is similar (see code #18-06). To visualise the results, as above, the plot() function is used.

#18-08

```
# More plots
plotRGB(RT_20m, r="red", g="green", b="blue",
      axes=TRUE, mar=c(2,1,3,1), stretch ="lin",
      main ="True Color") # 4 3 2
plot(MSI, main="MSI")
plot(BSI, main="BSI")
plot(CVI, main="CVI")
```

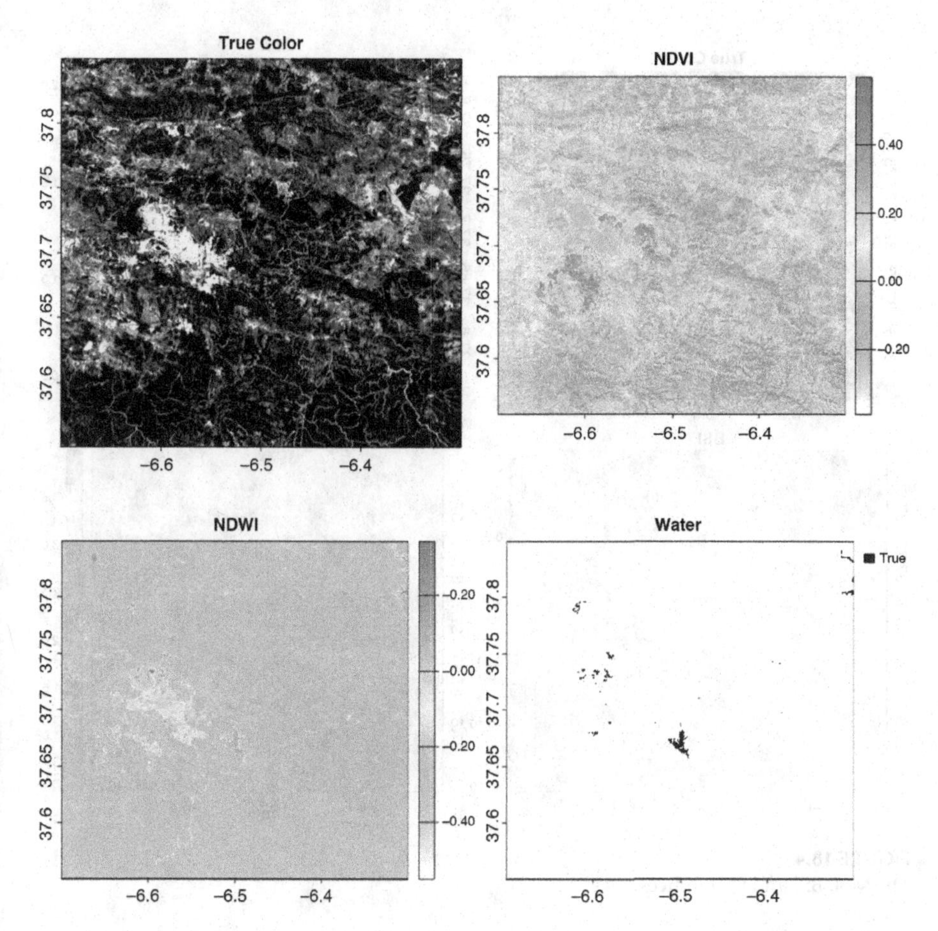

FIGURE 18.3
Water and vegetation indices. The water bodies are delineated using the NDWI with a threshold.

The resulting image is shown in Figure 18.4.

18.4 A Thematic Map

To create a thematic map effectively, it's essential to accurately identify features by applying appropriate thresholds to various indices and composite data. In the following example, we will demonstrate how to use selected indices to generate a thematic map. It's important to note that this example serves primarily for illustrative purposes. For the results to be robust and suitable for scientific research or academic publication, the chosen

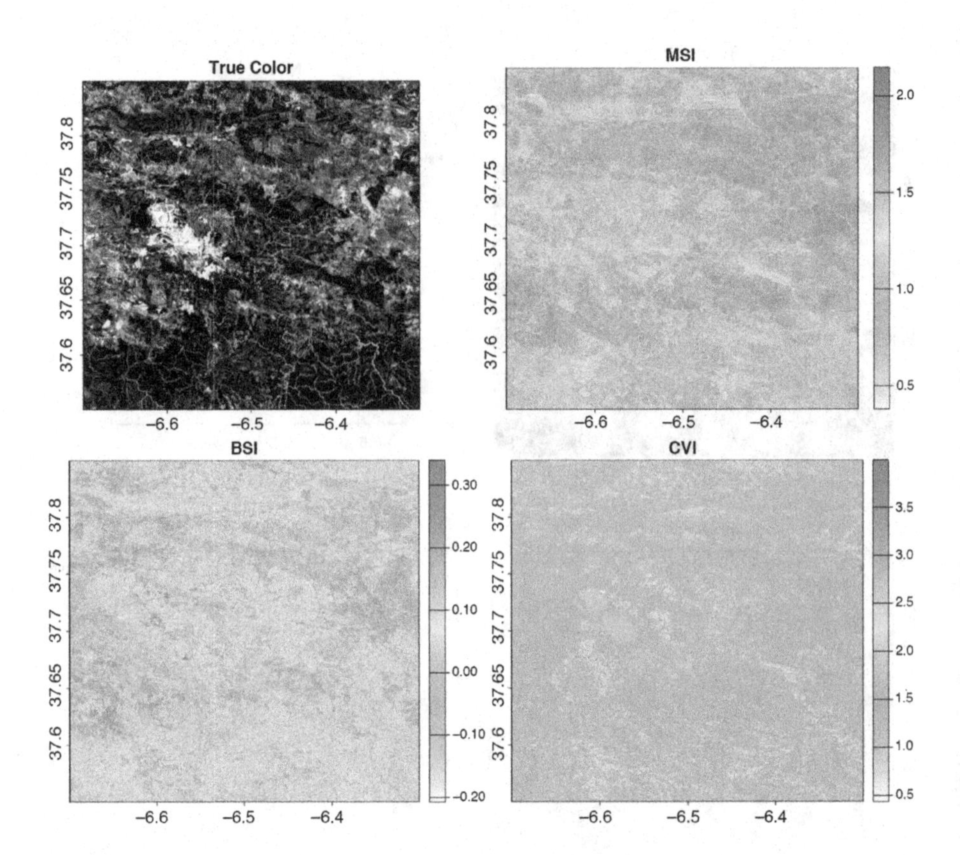

FIGURE 18.4
The MSI, BSI and CVI indices.

criteria and thresholds require careful adjustment and validation. This process ensures that the thematic map accurately represents the features of interest and meets the rigorous standards necessary for scholarly work.

In this example, we focus on three key land cover elements: water, vegetation, and soil. Each element is identified by specific indices that describe its presence within an area. Specifically, we use the following criteria based on index values to categorise each feature: (i) NDWI values greater than 0.05 indicate water areas, (ii) CVI (Crop Vegetation Index) values above 2.5 signify vegetation, and (iii) BSI values exceeding 0.05 are used to identify soil. The thematic map we aim to create will visually represent the distribution of these three land cover elements by integrating the areas defined by these index thresholds.

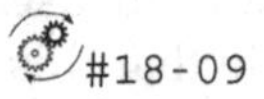

```
# THRESHOLDS
par(mfrow=c(2,2))
# The base image
```

```
plotRGB(RT_20m, r="red", g="green", b="blue",
      axes=TRUE, stretch ="lin", main ="True Color") # 4 3 2

# Water
water = NDWI > 0.05
plot(water, col=c("grey","blue"), main="Water")

# Vegetation
vegetation = CVI > 2.0
plot(vegetation, col=c("grey","green"), main="Vegetation")

# Soil
soil = BSI > .05
plot(soil, col=c("grey","brown"), main="Soil")
```

The resulting combined plot is shown in Figure 18.5.

To create the thematic map, we must merge the three separate maps, each representing a distinct land cover feature (water, vegetation, and soil) into a

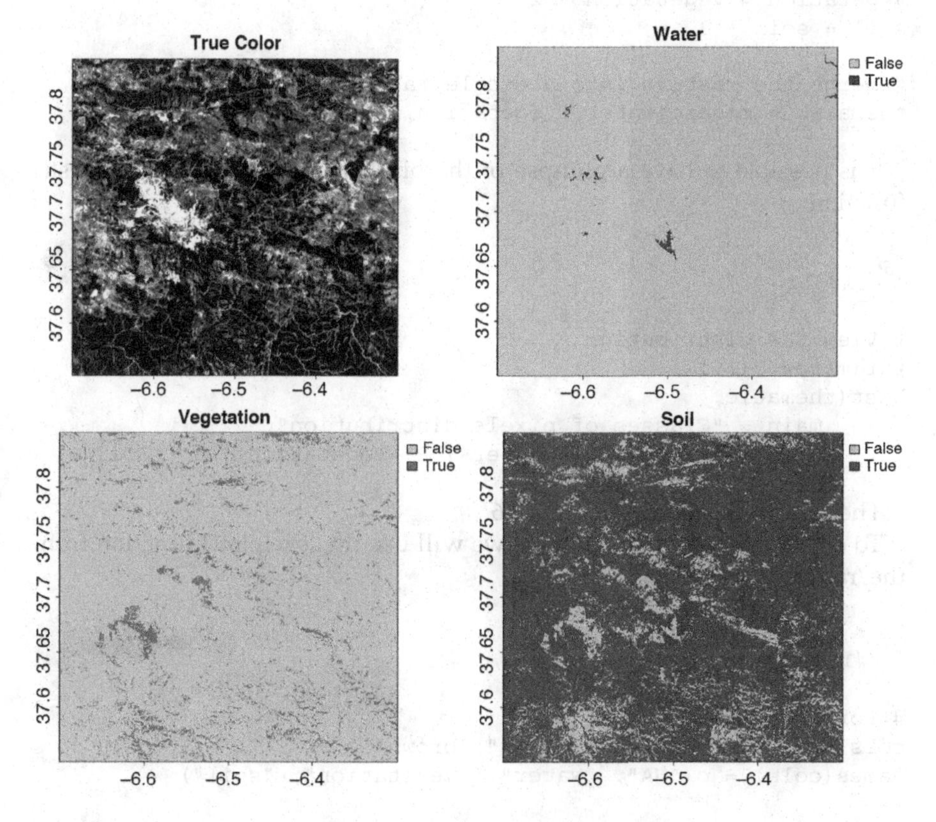

FIGURE 18.5
Combined plot of the identified features.

single map. The mosaic() function from the 'raster' package in R can facilitate this process.

It's important to note that some areas might have pixels classified under more than one feature category (for instance, a pixel could be identified both as soil and vegetation). In such cases, a decision rule needs to be applied to resolve the classification ambiguity. In this example, we prioritise classifications based on a predefined order where the highest value takes precedence: (1) for water, (2) for vegetation, and (3) for soil. Consequently, if a pixel qualifies as both vegetation and soil, it will be categorised as soil since soil has a higher assigned value (3) than vegetation (2). This approach ensures a consistent method for classifying pixels that might otherwise fall into multiple categories.

#18-10

```
# Prepare the data for creating a combined map
water = water * 1 # yes I know :)
vegetation = vegetation * 2
soil = soil * 3

# Merge the rasters into a single raster
thematic = mosaic(water, vegetation, soil, fun = "max")
```

It is possible to have a glimpse of the pixel distribution using the hist() function.

#18-11

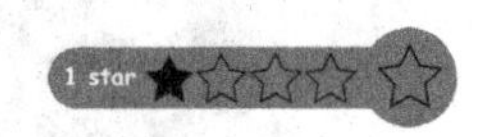


```
# View the distribution
par(mfrow=c(1,1))
hist(thematic,
      main = "Classes of pixels distribution",
      xlab = "0 - NA | 1- Water | 2- Vegetation | 3 - Soil")
```

The resulting plot is in Figure 18.6.

To create a more adequate map, we will use the levelplot() function from the 'rasterVis' package.

#18-12

```
# For the legend
cols = c("grey","blue","green","brown")
names(cols) = c("NA", "water", "vegetation", "soil")

# Plot raster
levelplot(thematic, main = "Thematic Map of Rio Tinto",
```

```
            margin = list(c(1,0,0,1)),
            col.regions = cols,
            colorkey = list(at = 0:4, space = "bottom",
                  labels = list(at=0.5:3.5, labels =
names(cols)),
                  title = "Landuses",
                  title.control = list(side = "bottom")))
```

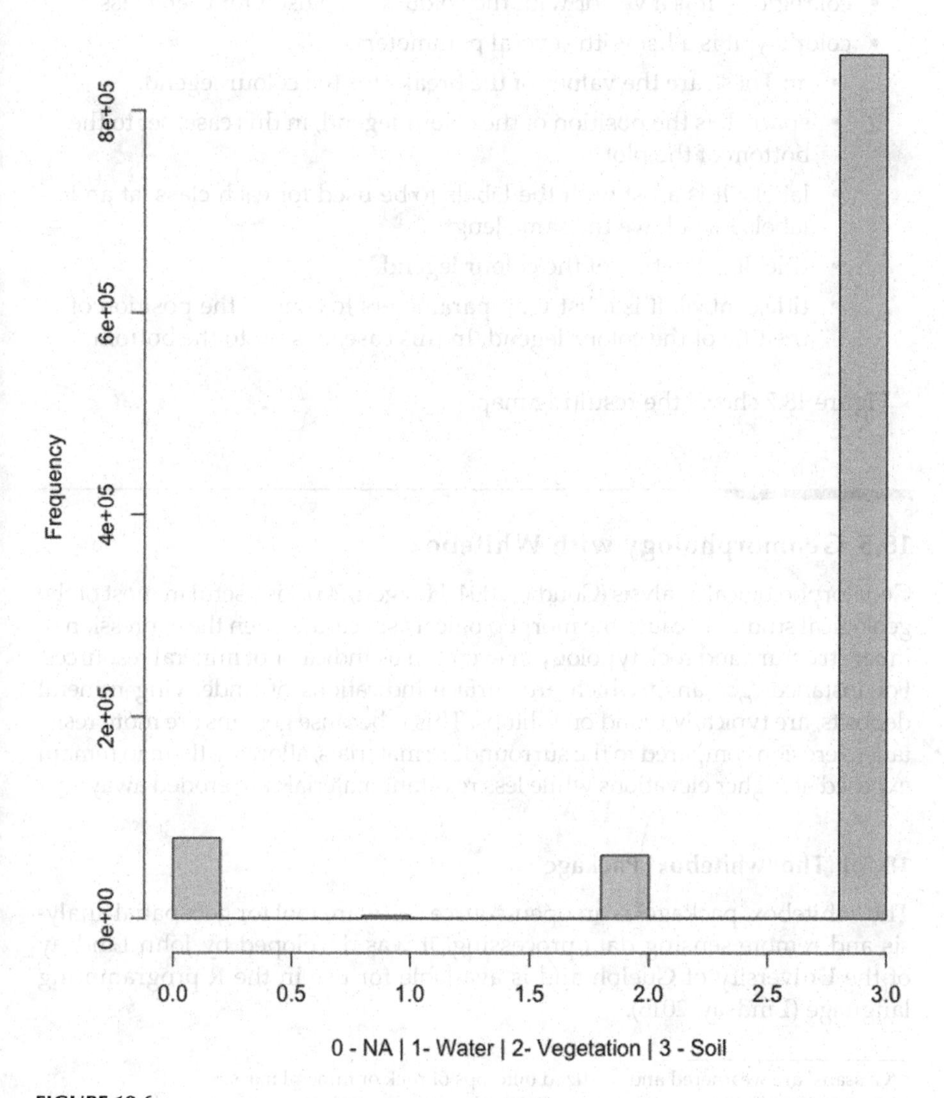

FIGURE 18.6
The distribution of the pixels in the thematic map.

The levelplot() function, from the 'rasterVis' package, uses several parameters to configure the result. First the colours are defined and named.

In this case, the levelplot() is called with the parameters:

- thematic: It is the raster data to be plotted.
- main: It is the title of the plot.
- margin: It is the size of the margins. It is a list with four elements: c(bottom, left, top, right).
- col.regions: It is a vector with the colours to be used for each class.
- colorkey: It is a list with several parameters:
  - at: These are the values of the breaks for the colour legend.
  - space: It is the position of the colour legend, in this case, set to the bottom of the plot.
  - labels: It is a list with the labels to be used for each class. at and labels must have the same length.
  - title: It is the title of the colour legend.
  - title.control: It is a list with parameters to control the position of the title of the colour legend. In this case, it is set to the bottom.

Figure 18.7 shows the resulting map.

18.5 Geomorphology with Whitebox

Geomorphological analysis (Goudie, 2004; Huggett, 2007) is useful in most of the geological studies because the morphological aspects are often the expression of inner structure and rock typology or even can be indicator of mineral resources. For instance, 'gossans',[4] which are surface indications of underlying mineral deposits, are typically found on hilltops. This is because gossans are more resistant to erosion compared to the surrounding materials, allowing them to remain exposed at higher elevations while less resistant materials are eroded away.

18.5.1 The 'whitebox' Package

The 'whitebox' package[5] is an open-source software tool for geospatial analysis and remote sensing data processing. It was developed by John Lindsay of the University of Guelph and is available for use in the R programming language (Lindsay, 2016).

[4] 'Gossans' are weathered and oxidized outcrops of rock or mineral masses.

[5] https://github.com/giswqs/whiteboxR; see also for a short introduction https://cran.rstudio.com/web/packages/whitebox/vignettes/demo.html

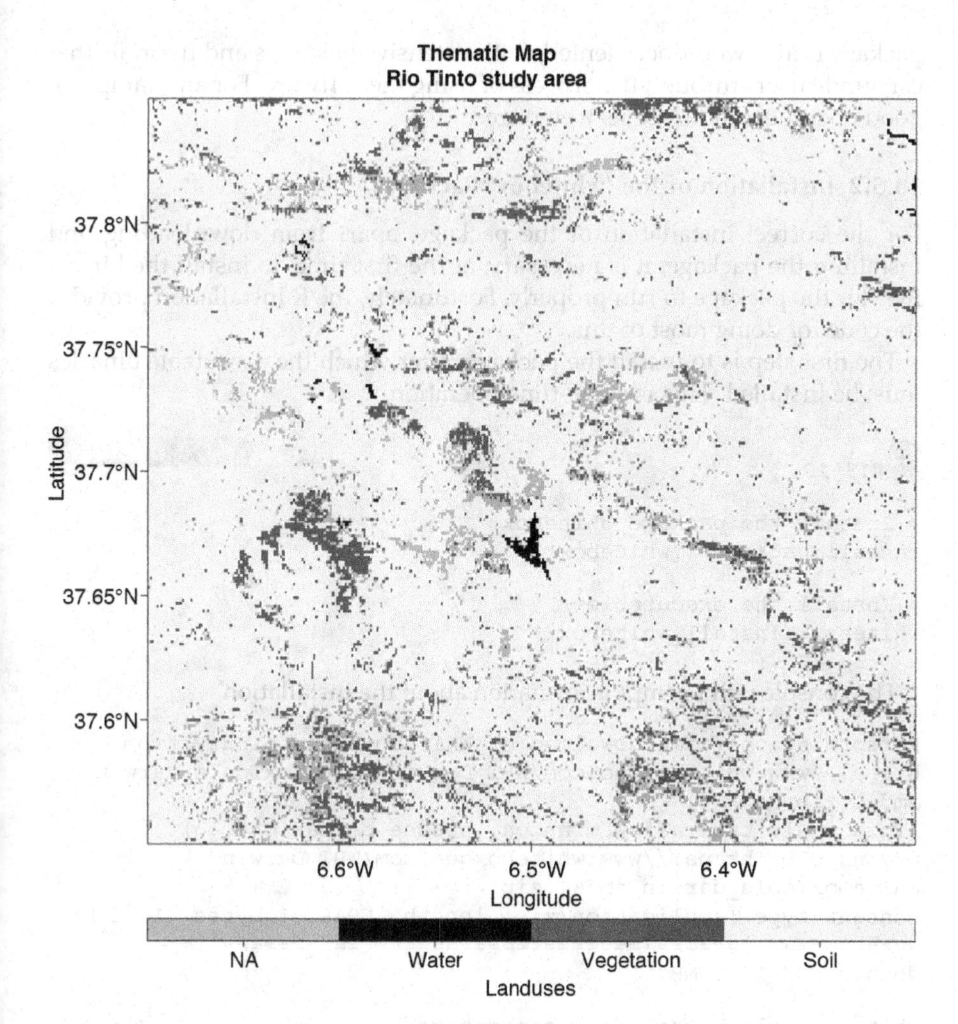

FIGURE 18.7
Simple thematic map from the Rio Tinto region.

The package is designed to provide a range of geospatial analysis and processing tools that are not available in other open-source geographic information system (GIS) software. It includes a variety of tools for terrain analysis, hydrological modelling, and spatial statistics. The package also provides tools for image processing and analysis, making it a useful tool for a range of applications, including geomorphological analysis, natural resource management, environmental monitoring, or land use planning.

One of the key features is its ability to handle large datasets. It uses efficient algorithms that are optimised for high-performance computing and can process large datasets with ease. This makes it a useful tool for processing high-resolution satellite imagery and other large geospatial datasets. The package is also well-documented, with extensive help files and tutorials that can guide users through the process of using the software. For an example of good documentation, see its web page.[6]

18.5.2 Installation of the 'whitebox' Package

For the correct installation of the package, apart from downloading and installing the package, it is necessary, at the first time, to install the binary files for the package to run properly. Fortunately, the R installation provides the code for doing most of this.

The first step is to install the package, after which the executable binaries must be installed. This is a one-time operation.

#18-13

```
# Install the package whitebox
install.packages("whitebox")

# Install the executables
whitebox::install_whitebox()
```

The console will prompt information about the installation.

```
Performing one-time download of WhiteboxTools binary from
https://www.whiteboxgeo.com/WBT_Darwin/WhiteboxTools_darwin_
amd64.zip
(This could take a few minutes, please be patient...)
trying URL 'https://www.whiteboxgeo.com/WBT_Darwin/
WhiteboxTools_darwin_amd64.zip'
Content type 'application/zip' length 16948571 bytes (16.2 MB)
==================================================
downloaded 16.2 MB

WhiteboxTools binary is located here:
/Library/Frameworks/R.framework/Versions/4.2/Resources/
library/whitebox/WBT/whitebox_tools
You can now start using whitebox
      library(whitebox)
      wbt_version()
```

18.5.3 A First on Geomorphological Analysis

For this exercise, a geomorphological analysis is proposed based on some of the functions from the 'whitebox' package combined with what was considered in the previous sections.

[6] https://www.whiteboxgeo.com/manual/wbt_book/available_tools/geomorphometric_analysis.html

The first step is to download the digital elevation model (DEM) for the study area using the 'elevatr' package. The data is then converted to a SpatialDataFrame to be plotted using the 'ggplot2' package. The conversion is made using the rasterToPoints() function from the 'raster' package. This function is used to extract the cell values and corresponding spatial coordinates from a raster object and convert them into a data frame. It is particularly useful for converting raster data into a format that can be more easily manipulated. For a better control on the names of the new data frame, new names are assigned to the created 'r_elev_df' variable.

After loading the libraries, the get_elev_raster() function is used to fetch the elevation for the study area and to plot it. This time, the ggplot() function will be used for creating publishing quality maps.

The 'ggnewscale' package is also introduced; it is designed to extend the functionality of the 'ggplot2' package and allows the creation of new scales for additional aesthetic mappings in a 'ggplot2' without overwriting the existing ones. Before running the code, remember to install it as well.

In the ggplot() function, notice that the scales::number_format() function is used to define the decimal places in the axis labels.

```
#18-14

# Load the libraries
library(sf)
library(terra)
library(raster)
library(whitebox)
library(elevatr)
library(ggplot2)
library(ggnewscale)

# Retrieve elevation data for a larger region
RT_elev = get_elev_raster(RT_20m, z=10)

# Convert to a SpatialPointsDataFrame
r_elev_pts = rasterToPoints(RT_elev, spatial  = TRUE)

# Then to a 'conventional' dataframe
r_elev_df  = data.frame(r_elev_pts)
names(r_elev_df) = c("Elevation", "x", "y","optional")

# The colour palette
col_elev = terrain.colors(10)

# The plot
ggplot() +
  geom_raster(data = r_elev_df, aes(x = x, y = y, fill =
Elevation)) +
```

```
  scale_fill_gradientn(name = "Elevation (m)", colors = col_
elev) +
  ggtitle("Rio Tinto area\nElevation") +
  labs(x = "Longitude", y = "Latitude") +
  coord_fixed() +
  scale_x_continuous(labels = scales::number_format(accuracy =
0.01)) +    # DECIMAL PLACES in X and Y
  scale_y_continuous(labels = scales::number_format(accuracy =
0.01))
```

The resulting plot is shown in Figure 18.8.

There are some settings necessary to work with the 'whitebox' package. After loading the library, the system must be initialised with the wbt_init() function and it is convenient to define the working folder with the wbt_wd() function. Remember to adjust the working directory path to the one in your computer.

Whitebox works with calls to externally existing applications. Therefore, all the functions need to have an input file that is read and an output file that is written. Hence, one must save the elevation data as a '.tif' file in the working folder.

FIGURE 18.8
Elevation of the Rio Tinto Area.

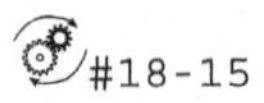

```
#18-15
```

```
# Whitebox initialization
wbt_init()
```

```
# Working folder for whitebox
wbt_wd("wb")
setwd("wb")

# Save the elevation to the working folder
writeRaster(RT_elev, filename="RT_elev.tif",format="GTiff",
overwrite = TRUE)
```

A way of better visualising the terrain is to create a composition mixing the elevation colours with the hillshade values to create a third-dimension effect. The wbt_multidirectional_hillshade() function creates a raster with the hillshade effect.

In this example, the hillshade is combined with a raster with a transparency of 30%, i.e. alpha = 0.7. For this in the second geom_raster() function, the 'alpha' opacity parameter is set to 0.7, i.e. a transparency of 30%.

As the functions are executed in order, the first raster with the hill shade is fully opaque (i.e. 'alpha = 1'), whereas the elevation is drawn on top of it, with the selected transparency. This can be tuned in a trial-and-error mode.

#18-16 3 stars

```
# Create the hillshade
wbt_multidirectional_hillshade("RT_elev.tif", "hillshade.tif")

# Read the result
r = raster("hillshade.tif")

# Convert to a df for plotting
r_hill_pts = rasterToPoints(r, spatial = TRUE)
r_hill_df  = data.frame(r_hill_pts)

# Color scale for the hillshade
col_hill = grey(0:10/10)

# Combined elevation and hillshade
ggplot() +
  geom_raster(data = r_hill_df,      # HILLSHADE
          aes(x = x, y = y, fill = hillshade),
          show.legend = FALSE) +
  scale_fill_gradientn(name = "Hillshade", colors = col_hill,) +
# GREYS
  new_scale_fill() +
  geom_raster(data = r_elev_df,                # ELEVATION
          aes(x = x, y = y, fill = Elevation),
          alpha = 0.7) +
  scale_fill_gradientn(name = "Elevation", colors = col_elev) +
# Terrain Colors
  coord_fixed() +
```

```
  labs( x = "Longitude", y = "Latitude",
      title = "Rio Tinto area", subtitle="Elevation with
Hillshade") +
  scale_x_continuous(labels = scales::number_format(accuracy =
0.01)) +    # DECIMAL PLACES in X
  scale_y_continuous(labels = scales::number_format(accuracy =
0.01))
```

The resulting plot is shown in Figure 18.9.

FIGURE 18.9
Combined elevation and hillshade map.

Another useful feature that can be extracted from a DEM is the contour lines. They can be created as shapefiles using the wbt_contours_from_raster() function. In this case, the contour intervals are of 100 metre and the lines are smoothed with a smooth filter of 9. An odd integer value is required to the smooth filter applied to the x-y position of vertices in each contour. For instance, the recommended values are 3, 5, 7, 9, 11, and so on. Using larger values will result in smoother contour lines.

#18-17 3 stars

```
# Contours
wbt_contours_from_raster(
            input="RT_elev.tif",
            output="contours.shp",
            interval=100,
            smooth=9)
# Read the shapefile
RT_contours = st_read("contours.shp")
```

```
# Combined elevation and hillshade
ggplot() +
  geom_raster(data = r_hill_df,     # HILLSHADE
          aes(x = x, y = y, fill = hillshade),
          show.legend = FALSE) +
  scale_fill_gradientn(name = "Hillshade", colors = col_hill,)
+ # GREYS
  new_scale_fill() +
  geom_raster(data = r_elev_df,                # ELEVATION
          aes(x = x, y = y, fill = Elevation),
          alpha =.7) +
  scale_fill_gradientn(name = "Elevation", colors = col_elev)
+ # Terrain Colours
  geom_sf(data=RT_contours, colour="brown")+
  coord_sf() +
  labs( x = "Longitude", y = "Latitude", title ="Rio Tinto
area", subtitle="Elevation/Hillshade/Contours") +
  scale_x_continuous(labels = scales::number_format(accuracy =
0.01)) +    # DECIMAL PLACES in X
  scale_y_continuous(labels = scales::number_format(accuracy =
0.01))
```

The resulting map is shown in Figure 18.10.

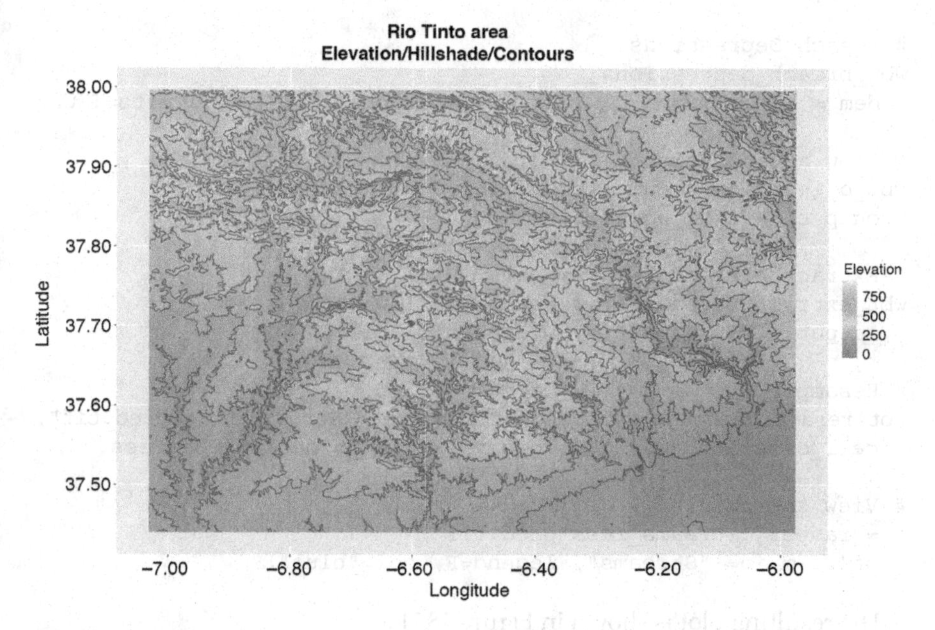

FIGURE 18.10
Elevation, Hillshade, and contours for the Rio Tinto area.

18.5.4 Streams and Waterflow

A DEM is a great tool not only to understand the terrain but also to extract the features of a landscape, might them be rivers, basins, or other naturally occurring landforms. The 'whitebox' package provides the tools for creating these features. For the moment, the purpose is to identify streams and rivers and the corresponding basins.

The workflow for this task includes smoothing the DEM with the wbt_feature_preserve_smoothing() function to remove any noise on the data, followed by the filling of small depressions with the wbt_breach_depressions() function. These depressions might prevent the identification of flow directions.

The last step is to calculate the flow accumulation with the wbt_d_inf_flow_accumulation() function, followed by the stream extraction from the flow accumulated, with the wbt_extract_streams() function.

For visualisation purposes, the wbt_resample() function is used to retrieve a 'raster' object with a pixel size equivalent to the resolution of the DEM input raster.

#18-18 3 stars

```
# Smooth
wbt_feature_preserving_smoothing(
  dem = "RT_elev.tif", output = "smoothed.tif")

# Breach Depressions
wbt_breach_depressions(
  dem = "smoothed.tif", output = "breached.tif", fill_pits=TRUE)

# Flow accumulation
wbt_d_inf_flow_accumulation( input = "breached.tif",
  output = "flow_accum.tif")

# Extract streams
wbt_extract_streams(flow_accum = "flow_accum.tif",
  output = "Streams.tif", threshold = 1)

# Resample
wbt_resample(input="Streams.tif", output="Streams_resampled.tif",
  cell_size = 0.00002) # This value is in decimal degrees

# View the result
r = raster("Streams_resampled.tif")
plot(r, main= "Streams", legend=F, col="blue")
```

The resulting plot is shown in Figure 18.11.

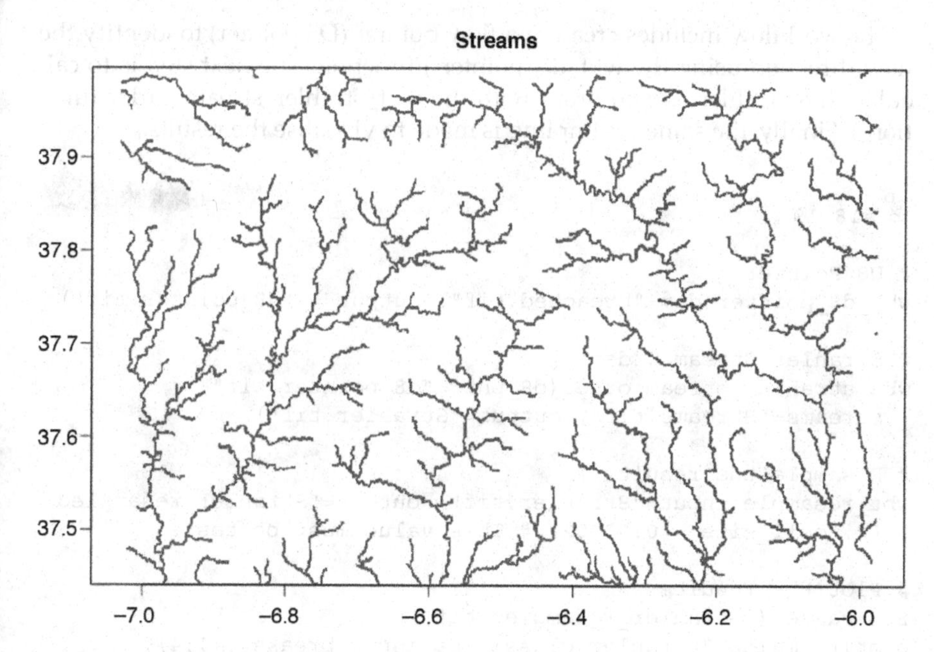

FIGURE 18.11
The rivers and streams from the RT area.

18.5.4.1 Strahler Classification

The analysis of streams is often based on the Strahler classification that is a system used to categorise streams and rivers based on their size and the number of tributaries they have. It was first proposed by Arthur Strahler in 1952 and is widely used in hydrology and geomorphology (Strahler, 2013).

The system is based on a hierarchical structure, where the smallest streams are designated as first-order streams, and larger streams that are formed by the confluence of two or more first-order streams are designated as second-order streams. Similarly, third-order streams are formed by the confluence of two or more second-order streams and so on. The largest streams are typically designated as tenth-order or higher.

The Strahler classification system also considers the number of tributaries that each stream or river has. However, if a lower-order stream joins a higher-order stream, the resulting stream is assigned the higher order. This ensures that the stream hierarchy reflects the true size and complexity of the river network.

Strahler classification is often used to analyse the characteristics of rivers and streams, such as their drainage area, channel slope, and flow rate. It can also be used to identify potential sources of pollution or to assess the impacts of land use changes on the river network.

The workflow includes creating a flow pointer (D8 pointer) to identify the flow direction using the wbt_d8_pointer() function. The next step is to calculate the Strahler stream order with the wbt_strahler_stream_order function(). Finally, the same resampling is made to visualise the results.

#18-19 3 stars

```
# D8 Pointer
wbt_d8_pointer(dem="breached.tif", output = "d8_pointer.tif")

# Strahler Stream Order
wbt_strahler_stream_order(d8_pntr="d8_pointer.tif",
  streams="Streams.tif", output="Strahler.tif")

# Resample the result
wbt_resample(input="Strahler.tif", output="Strahler_resampled.
tif", cell_size = 0.00002) # This value must be tested

# Plot the results
r = raster("Strahler_resampled.tif")
plot(r, main= "Strahler classification", breaks=c(1:4),
      col=c("blue","green","red"))
```

The resulting plot is shown in Figure 18.12.

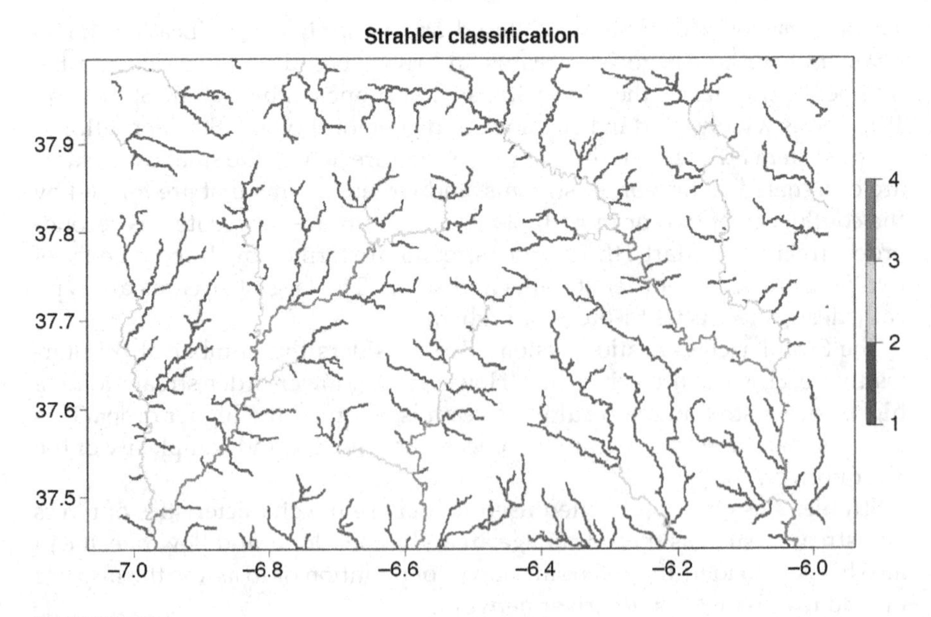

FIGURE 18.12
The Strahler classification of the streams in the Rio Tinto area.

18.5.5 Basins and Rivers

The last step in this hydrological analysis is to outline the basins. Again, the 'whitebox' package has a simple way of outlining the basins with the wbf_basins() function. For this example, we will combine the result of the basins with the rivers divided by the Strahler classification.

#18-20 2 stars

```
# Basins
wbt_basins(d8_pntr = "d8_pointer.tif", output = "basin.tif")

# View the results
r = raster("basin.tif")
plot(r, main = "RT Basin and Streams", legend = F)
r1 = raster("Strahler_resampled.tif")
plot(r1, add=T, breaks=c(1:4), col=c("blue","green","red"))
```

The resulting plot is shown in Figure 18.13.

18.5.6 Geomorphs, a Collection of Geographic Features

The geomorphs tool in the 'whitebox' package provides a powerful way to analyse and classify the different types of landforms in a DEM, which can be useful in a range of applications.

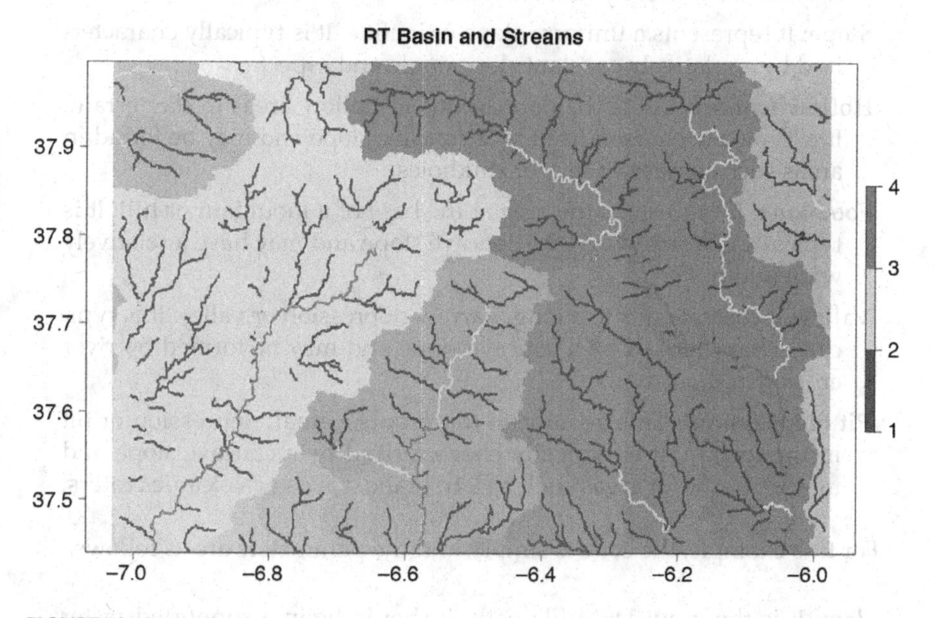

FIGURE 18.13
The basins and the rivers from the Rio Tinto area.

The wbt_geomorphons() function creates a raster image in which each pixel is classified into a specific type of geomorphic feature based on the topographic attributes of that location. The classification is based on the topographic features of each pixel in the DEM, including slope, aspect, and curvature.

The following is a brief explanation of each of the geomorphs and the landform types they represent:

Flat: It represents flat or nearly flat terrain. It has a relatively uniform slope and aspect and may be found in areas such as floodplains or alluvial fans.

Peak (summit): It corresponds to the highest point or summit of a mountain or hill. It is typically characterised by a steep slope and a small area at the top with relatively flat terrain.

Ridge: It outlines the linear features that run along the crest of a hill or mountain. It is typically characterised by a convex slope and may have a relatively narrow width.

Shoulder: It is the slightly lower area adjacent to a ridge. It is typically characterised by a convex slope and may have a slightly wider width than a ridge.

Spur (convex): It is when there is a protrusion or spur on the side of a mountain or hill. It is typically characterised by a convex slope and may have a relatively narrow width.

Slope: It represents a uniform slope or incline. It is typically characterised by a relatively constant slope angle and aspect.

Hollow (concave): It is the depressions or hollow areas in the terrain. It is typically characterised by a concave slope and may be found in areas such as river valleys or sinkholes.

Footslope: It represents the area at the base of a mountain or hill. It is typically characterised by a concave slope and may have a relatively wide width.

Valley: It corresponds to a long, narrow depression or valley. It is typically characterised by a concave slope and may be formed by river erosion or glaciation.

Pit (depression): This geomorph represents a small depression or pit in the terrain. It is typically characterised by a concave slope and may be found in areas such as karst landscapes or volcanic craters.

For this example, the code is simple, and the parameters are as follows:

dem: It is the input DEM file path. It should be in a supported raster format.

output: It is the output file path and name for the geomorphs raster grid.

search: This parameter specifies the number of cells in each direction that should be searched for determining the local terrain shape. The default value is 100, and larger values will result in longer processing times. In the plot, the 100 cells in the border of the image are not part of the whole classification.

threshold: It sets a threshold value for the difference in elevation between neighbouring cells. If the difference in elevation is greater than this value, the cell is considered part of a peak, while if it is less than this value, the cell is considered part of a pit. The default value is 0.0.

fdist: It sets the maximum distance that a cell can be from a flow path before it is no longer considered part of the same landform. The default value is 0, which means that all cells in the same landform will be grouped together.

skip: It is the number of cells that should be skipped when processing the input DEM. This can be used to speed up processing times by reducing the resolution of the input DEM. The default value is 0, which means that all cells will be processed.

forms: It is a Boolean value that determines whether the output raster grid should contain the landform types (Geomorphs) or the frequency of occurrence of each landform type.

residuals: It is a Boolean value that determines whether the output raster grid should contain the residuals (difference between the original DEM and the reconstructed DEM using the Geomorphs) or not.

```
#18-21

# GEOMORPHS
wbt_geomorphons(
  dem = "breached.tif",
  output = "Geomorphs.tif",
  search=100,
  threshold=0.0,
  fdist=0,
  skip=0,
  forms=T,
  residuals=F
)

# Colours for plot
col_geomorph = c("grey", "red", "orange", "green", "darkgreen",
"purple", "darkgreen",  "cyan", "lightblue","blue")
```

```
names(col_geomorph) = c("Flat", "Peak", "Ridge", "Shoulder",
"Convex", "Slope", "Concave", "Footslope", "Valley", "Pit")

# View the results
r = raster("Geomorphs.tif")
plot(r, col = col_geomorph, legend = F, main = "RT
Geomorphology")
legend("topright", horiz=F, legend = names(col_geomorph), fill
= col_geomorph, bg = "white", cex=0.8, title = "Geomorphons")
```

The resulting image is shown in Figure 18.14.

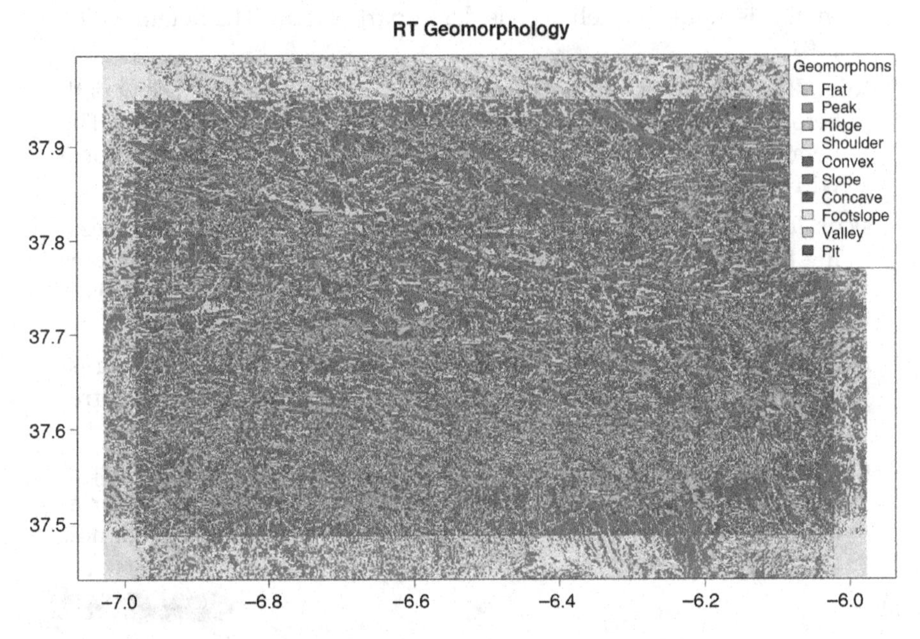

FIGURE 18.14
All the geomorphs from the RT area.

This image can become more intelligible if non-relevant information is removed. In this case for analysing the effects of the existing geomorphs, we suggest keeping only the ridges and valleys.

For this example, we use the 'ggplot' to have a publication quality result. Also, the data is filtered to include only the cells that are the ridges and the valleys.

```
#18-22

# Load the library
library(dplyr)
```

```
# Convert to a df for plotting using ggplot
r_pts = rasterToPoints(r, spatial = TRUE)
r_df  = data.frame(r_pts)

# Verify the groups
group =  group_by(r_df, Geomorphs)
c = summarize(group, counts = n())

# Filter for Ridge and Valley
r_df_filtered = filter(r_df, Geomorphs %in% c(3, 9)) # 3 = Ridge
and 9 = Valley

# New colours
col_geomorph_f = c("red","blue")
names(col_geomorph_f) = c("Ridge","Valley")

# Create the plot
ggplot() +
  geom_raster(data = r_df_filtered,  aes(x = x, y = y, fill =
factor(Geomorphs))) +
  ggtitle("Ridge and Valley of RT area") +
  scale_fill_manual(values = setNames(col_geomorph_f, nm =
levels(factor(r_df_filtered$Geomorphs))),
                    labels = names(col_geomorph_f[1:10]),
                    name = "Geomorphons") +
  coord_sf() +
  labs( x = "Longitude", y = "Latitude",
        title = "RT area", subtitle="Ridge and Valley") +
  scale_x_continuous(labels = scales::number_format(accuracy =
0.01)) + # DECIMAL PLACES in X
  scale_y_continuous(labels = scales::number_format(accuracy =
0.01))
```

The resulting plot is shown in Figure 18.15.

This image provides one interesting insight to understanding the geomorphology of the area. Although the main pattern is dominated by the rivers that flow from North to South, it is visible that the ridge and valley pattern is more in a E-W to WNW-ESE direction that is the direction of the main Variscan orogeny structuring in this region.

18.6 Concluding Remarks

Raster images constitute a great asset for data analysis constituting the ideal framework for many spatial operations. Its equally distributed pattern in a matrix-like structure facilitates many of the operations. As so, many types of spatial data are in the raster or grid format.

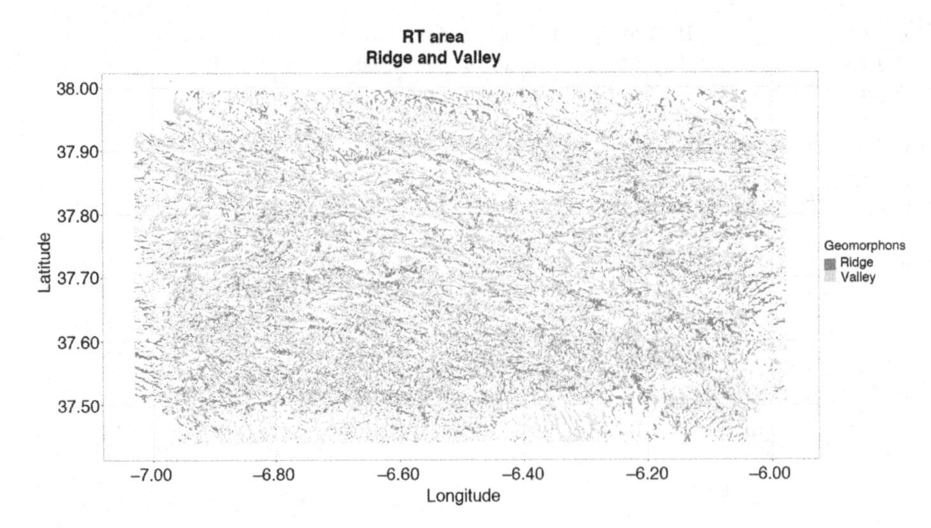

FIGURE 18.15
The ridge and valley pattern from the RT area.

Satellite images are one example of raster type of images, and with the open data movement is the support for numerous analyses that provide valuable new insight to understand the earth's features and its transformations, including many physical (e.g. temperature, surface reflectance, and atmospheric parameters). In geology, the applications are tremendous and only the surface of the iceberg has been revealed.

Another example of invaluable uses of raster images is the interpretation of geomorphological characteristics of a region or terrane. Starting from a simple DEM from a region, one can harness information on water flow, rivers and basins, and even extract information on the subtract of the region.

References

Goudie, A. (Ed.). (2004). Encyclopaedia of Geomorphology. Psychology Press, London, UK, p. 1156.

Huggett, R. (2007). Fundamentals of Geomorphology. Taylor & Francis, London, UK, p. 488.

Lindsay, J. (2016). Whitebox GAT: A case study in geomorphometric analysis. Computers & Geosciences, 95, 75–84. http://dx.doi.org/10.1016/j.cageo.2016.07.003

Strahler, A. (2013). Introducing Physical Geography (6th ed). John Wiley & Sons, New York, p. 672.

Part III

Spatial Statistics and Modelling

19

Introduction to Spatial Statistics and Modelling Applied to Geology

The spatial arrangement of geological features is usually not random. To uncover the underlying patterns of these features, it's essential to use spatial statistics and modelling. These techniques are applied in areas such as geological mapping, analysing the distribution of mineral resources, studying earthquake patterns, and offering insights for risk management. Many examples of these applications in earth sciences and beyond are documented in specialised literature, including works by Davis (1986), Gaetan & Guyon (2010), and Kalkhan (2011).

The extensive toolboxes for spatial statistics analysis that are at our disposal in R can be used in an everyday base for a geologist's work, nurturing proficiency to analyse data or create analytical and thematic maps.

An example of the use of spatial statistics to address societal challenges, namely in tackling the challenge of finding new sources for raw materials, is the work done by exploration geologists who apply this methodologies to map the distribution of mineral deposits. They pinpoint areas likely to have high mineralisation potential and forecast possible locations of new mineral discoveries (e.g. Kumar & Dimitrakopoulos, 2022; Shi et al., 2023). The spatial analysis problems are relevant in the case of mineral deposits whose distribution is not random and might be related with hydrothermal fluid circulation, fault pattern of lithological constraints (e.g. Ford & Blenkinsop, 2008; Fouedjio et al., 2018).

Geologists can also team up with policy makers in defining potentially hazardous areas for the occurrence of earthquakes with a certain magnitude (e.g. Todes et al., 2021; Yang et al., 2023; Wei et al., 2023). The understanding of the earthquake patterns in a region is another public concern issue, as one might recall that the occurrence of the earthquakes obeys a natural pattern that depends on the geological history and geodynamic conditions of a region (e.g. Zhuang, et al., 2005; Vijay & Nanda, 2023).

In environmental geochemical studies where the samples might be collected in different patterns, from gridded to cluster or random, the spatial statistics allows not only to interpolate the data but also to preview the chemical values in new locations. The kriging methods are often used to provide very good results in these cases (e.g. Stojdl et al., 2017; Leenaers et al., 2020).

DOI: 10.1201/9781032651880-22

In a broader sense, spatial statistics can be used to answer a wide variety of questions about spatial data, including:

- How is the data distributed in space?
- Are there any spatial patterns in the data?
- What are the relationships between different spatial variables?
- How can we predict the value of a spatial variable at a new location?

There are many packages available for spatial statistics, and the combination of some of these packages provide solutions for data management, spatial data analysis, and spatial modelling. The previously presented 'sp' and 'sf' packages (see Sections 10.4.1 and 10.4.2 in Chapter 10) provide functions for creating and manipulating the spatial objects; the 'spatstat' package (see Section 19.1 in Chapter 19) provides functions for spatial point pattern analysis, whereas the 'gstat' package (see Chapters 21, 22, and 23) provides functions for spatial prediction and kriging.

19.1 The 'spatstat' Package

The 'spatstat' package[1] is a key asset for anyone investigating spatial statistics within the R framework. It offers the essential toolbox for exploring and making sense of various spatial patterns, making it an essential companion for those engaged in spatial analysis research. It includes a set of functions for analysing spatial point patterns, and it provides a wide range of tools for descriptive statistics, exploratory data analysis, model-fitting, and hypothesis testing, among other tasks (Baddeley et al., 2015).

Some of the key features include the following:

Point pattern analysis: It functions for characterising and comparing point patterns, such as the pair correlation function (pcf), the K-function (Kest), and the Ripley's L function (Lest). These functions allow us to quantify the spatial dependence, clustering, and dispersion of points in a point pattern.

Point process models: It functions for fitting and testing models for point processes, such as Poisson processes, Gibbs processes, and marked point processes. The 'spatstat' package also provides functions for simulation and prediction, such as the rPoissonCluster() function for generating realisations of a Poisson point process.

[1] https://spatstat.org/

Exploratory data analysis: It functions for visualising and summarising point pattern data and provides an extension of the plot() function for creating scatterplots of points, the summary function for computing summary statistics, and the quadratcount() function for computing the number of points in each quadrat.

Window functions: It functions for defining and manipulating the spatial window in which a point pattern is defined. For example, the owin() function creates an object of class 'owin' that defines a rectangular, circular, or polygonal window.

Spatial covariates: It functions for incorporating spatial covariates into point pattern analysis, such as the Kest() function for computing the K-function with covariates and the spatcov() function for fitting models for point processes with covariates.

A standout feature of 'spatstat' is its versatility in managing both uniform and non-uniform point patterns. This capability is essential for analysing point distributions that may be clustered, evenly spread, or display intricate patterns of interaction. Additionally, 'spatstat' offers a suite of tools for creating synthetic point patterns, visualising them through different methods, and evaluating their properties. The package also features advanced statistical techniques for conducting hypothesis tests, summarising patterns, and fitting models to data.

Furthermore, 'spatstat' seamlessly incorporates point process theory, facilitating comprehensive statistical evaluations of point pattern data. It offers methods for calculating intensity functions, applying Ripley's K-function (referenced in Ripley, 1998), and employing various metrics based on distances. These tools are vital for dissecting the spatial processes that give rise to the point patterns observed in many point pattern study cases.

Moreover, 'spatstat' extends beyond simple analyses and offers advanced functionalities such as the fitting of point process models to data. This allows us to explore spatial interaction, clustering, and inhibition effects within point patterns. By utilising point process models, one can make predictions, simulate new point patterns, and gain insights into the spatial processes at play.

References

Baddeley, A., Rubak, E., & Turner, R. (2015). Spatial Point Patterns: Methodology and Applications with R. Chapman and Hall/CRC Press, London. ISBN 9781482210200.

Davis, J. (1986). Statistics and Data Analysis in Geology. Wiley, New York.

Ford, A., & Blenkinsop, T. (2008). Combining fractal analysis of mineral deposit clustering with weights of evidence to evaluate patterns of mineralization: application to copper deposits of the Mount Isa Inlier, NW Queensland, Australia. Ore Geology Reviews, 33(3–4), 435–450.

Fouedjio, F., Hill, E., & Laukamp, C. (2018). Geostatistical clustering as an aid for ore body domaining: case study at the Rocklea Dome channel iron ore deposit, Western Australia. Applied Earth Science, 127(1), 15–29.

Gaetan, C., & Guyon, X. (2010). Spatial statistics and Modeling. Springer, New York, p. 297. https://doi.org/10.1007/978-0-387-92257-7

Kalkhan, M. (2011). Spatial Statistics: Geospatial Information Modeling and Thematic Mapping. CRC press, Taylor & Francis Group, Boca Raton, FL, p. 161.

Kumar, A., & Dimitrakopoulos, R. (2022). Updating geostatistically simulated models of mineral deposits in real-time with incoming new information using actor-critic reinforcement learning. Computers & Geosciences, 158, 104962. https://doi.org/10.1016/j.cageo.2021.104962

Leenaers, H., Burrough, P. A., & Okx, J. P. (2020). Efficient mapping of heavy metal pollution on floodplains by co-kriging from elevation data. In: Raper, J., (Ed.), Three Dimensional Applications in GIS. CRC Press, London, UK, pp. 37–50.

Ripley, B. (1998). Statistical Inference for Spatial Processes. Cambridge University Press, Cambridge, UK.

Shi, G., Wang, X., Wang, W., Liu, D., Liu, Q., Zhou, J., Chi, Q., & Liu, H. (2023). Nationwide concentration and spatial distribution of manganese with links to manganese mineralization in China. Journal of Geochemical Exploration, 244, 107130. https://doi.org/10.1016/j.gexplo.2022.107130

Stojdl, J., Matys Grygar, T., Elznicova, J., Popelka, J., Vachova, T., & Hosek, M. (2017). Kriging-a challenge in geochemical mapping. In: EGU General Assembly Conference Abstracts, Vienna, Austria, p. 3615.

Todes, J. P., Okal, E. A., & Kirby, S. H. (2021). Frequency-size distributions of Wadati-Benioff zone and near-boundary intraplate earthquakes: Implications for intermediate and deep seismicity. Physics of the Earth and Planetary Interiors, 321, 106707.

Vijay, R. K., & Nanda, S. J. (2023). Earthquake pattern analysis using subsequence time series clustering. Pattern Analysis and Applications, 26(1), 19–37.

Wei, X., Luan, X., Meng, F., Lu, Y., He, H., Qiao, J., … & Xue, Y. (2023). Deformation feature and tectonic model of the Timor Trough: New interpretation of the evolution and mechanism of Banda arc-continent collision. Tectonophysics, 862, 229958.

Yang, Y., Tang, C., Tang, C., Chen, M., Cai, Y., Bu, X., & Liu, C. (2023). Spatial and temporal evolution of long-term debris flow activity and the dynamic influence of condition factors in the Wenchuan earthquake-affected area, Sichuan, China. Geomorphology, 435, 108755.

Zhuang, J., Chang, C.-P., Ogata, Y., & Chen, Y.-I. (2005). A study on the background and clustering seismicity in the Taiwan region by using point process models. *Journal of Geophysical Research*, 110, B05S18, https://doi.org/10.1029/2004JB003157

20

Point Pattern Analysis

Point pattern analysis (PPA) represents the statistical methodology employed for examining the spatial arrangement of point data across various fields. This approach is instrumental in understanding the patterns and underlying processes that influence the distribution of discrete events or objects in space, ranging from natural phenomena to human-made structures. Examples of such applications include the mapping and analysis of mineral deposit locations, which are pivotal for mining and geological studies; the study of earthquake occurrences, aiding in the assessment of seismic risks and the understanding of tectonic processes; and the investigation of volcanic vent occurrences, which can provide insights into volcanic behaviour and potential hazards.

The work done in this area is showcased in works such as Diggle (1983), Boots & Getis (1988), and Illian et al. (2008), that exemplify the groundwork for the development of robust statistical techniques and models. These contributions have advanced the capability to not only describe the spatial patterns observed in point data but also to infer the spatial processes driving these distributions. Through the application of methods such as intensity estimation, cluster detection, and spatial autocorrelation analysis, researchers can identify patterns of aggregation or dispersion, test hypotheses regarding spatial randomness, and model the interaction between points.

Moreover, PPA facilitates the exploration of spatial relationships and interactions among points, enabling the identification of significant clusters or hotspots of activity. This is particularly useful in environmental science, urban planning, and seismic hazard studies, where understanding the spatial dynamics of contaminated soils, urban susceptible features, or earthquake locations is crucial.

20.1 Distribution Patterns

To understand the distribution of points in space – its patterns – a wide assortment of tools is present including functions for analysing and visualising the spatial point pattern data. These tools encompass techniques for

DOI: 10.1201/9781032651880-23

assessing the randomness or uniformity of point patterns, including the Ripley's K-function, pair correlation function (pcf), and the transformation of the Ripley's function (L-function), or even the Fry method.

The K-function provides a measure of the expected number of points within a given distance of a point in the pattern (Dixon, 2014). If the K-function for a point pattern deviates from the expected value for a random pattern, it suggests that the pattern may exhibit clustering or regularity. The 'pcf' provides a measure of the spatial dependence between pairs of points in a pattern and can be used to assess the presence of clustering or repulsion.

The generalisation of the K-function provides information about the spatial distribution of points around each point in the pattern. The Fry method (Fry, 1979) provides a cumulative distribution of the nearest neighbour (nn) distances for a point pattern and can also be used to assess the presence of clustering or regularity.

Parametric models, such as Poisson processes, log-Gaussian Cox processes, and determinantal point processes, can also be fitted to point pattern data using functions such as ppm(), kppm(), slrm(), and dppm(). These models can be used to explore the relationship between the point pattern and underlying covariates, such as environmental variables or geological features (Illian et al., 2008).

20.1.1 The K-Function

As stated, the K-function is central in spatial statistics due to its ability to report the spatial distribution of points in a point pattern dataset. Its use and applications in earth sciences, and more specifically in geology, can be varied, including:

Mineral deposit patterns: The patterns of clustering or regularity in the spatial distribution of mineral deposits provide insights into the geological processes that produced the deposits and can help guide exploration and mining activities (e.g. Carranza, 2009; Lisitsin, 2015).

Fault patterns: The spatial distribution of faults in a rock formation, its patterns of clustering or regularity, allows geologists to understand the structural properties of the rock and the processes that produced the faults (e.g. Shakiba et al., 2022).

Earthquake patterns: The K-function can be used to analyse the spatial distribution of earthquakes in a region. The underlying tectonic processes that produce earthquakes can be studied by understanding its distribution and patterns in space and time (Vijay & Nanda, 2023).

Geothermal system patterns: The spatial distribution of geothermal systems, such as hot springs and geysers, its patterns of clustering, or regularity, allows the understanding of the geological processes that produce these systems and the potential for geothermal energy production (e.g. Rodriguez-Gomez et al., 2023).

As so, the kest() function from the 'spatstat' package emerges as the quintessential tool for unravelling the complexities inherent in spatial point patterns. The function calculates an estimate of the so-called K-function, which quantifies the spatial dependence of points in a point pattern.

Spatial literacy implies the skill of observing point distribution and accurately describing its spatial arrangement. This entails using precise concepts and language to convey how objects are distributed in space, as often the concepts are diffusely used. There are three primary types of points distribution that describe the spatial pattern of objects:

Clustered: When objects are located near each other, it is referred to as a clustered pattern.

Dispersed: When objects are widely spaced, it is known as a dispersed pattern.

Random: When objects do not exhibit either a clustered or dispersed pattern, it is called a random pattern. This pattern is also known as a 'hypothetical' or 'normative' pattern.

The K-function is defined as the expected number of points, in a point pattern, that are within a specified distance 'r' of a randomly chosen point in the pattern. In other words, the K-function measures the spatial clustering of points in a point pattern, as a function of the distance 'r'.

If points are randomly distributed, the K-function will be equal to $\pi \times r^2$, which is the area of a circle with radius 'r'. If points are clustered, the K-function will be larger than $\pi \times r^2$ for small values of 'r', and if points are dispersed, the K-function will be smaller than $\pi \times r^2$ for small values of 'r'.

The kest() function calculates an estimate of the K-function for a given point pattern by dividing the point pattern into a grid of cells and counting the number of points in each cell. Its estimate is obtained by summing the number of points in each cell, that are within a distance 'r' of a randomly chosen point, in the pattern.

To interpret the results obtained, it is necessary to examine the plot of the K-function estimate.

A plot of the K-function shows how the expected number of points within a distance 'r' changes as it increases. If the plot is close to $\pi \times r^2$, then the points in the pattern are randomly distributed, and if the plot is above $\pi \times r^2$, then the points are clustered. If the plot is below $\pi \times r^2$, then the points are dispersed.

The shape of the plot can also provide information about the nature of the clustering or dispersion of the points in the pattern.

The following examples demonstrate how to create these types of spatial distributions and how the K-function behaves in each case. This provides insights to interpret the information provided by the K-function.

20.1.2 K-Function Interpretation

The kest() function from the 'spatstat' package creates a set of variables in response to the desired analysis. These variables plotted as lines are used to help interpret the results. It's important to keep in mind that the interpretation of the K-function plot depends on the context of the point pattern and may require additional information, analysis, and modelling to fully understand the spatial dependence of the points.

The 'kiso', 'ktrans', 'kbord', and 'kpois' lines in a plot of the K-function in 'spatstat' are reference lines that represent the expected values of the K-function for different models of spatial point patterns.

'kiso': It is a line that represents the expected value of the K-function for an isotropic point pattern, where points are randomly distributed in all directions. The 'kiso' line is a straight line with slope π, which represents the expected number of points in a circle of radius r.

'ktrans': It is a line that represents the expected value of the K-function for a translational point pattern, where points are randomly distributed along a line or a curve.

'kbord': It is a line that represents the expected value of the K-function for a point pattern that is confined to a region with a boundary. The 'kbord' line is a curve that depends on the shape and size of the region, and the distribution of points along the boundary.

'kpois': It is a line that represents the expected value of the K-function for a Poisson point process, where points are randomly distributed in space, with a constant intensity. For illustration purposes in the plots in Figures 20.1–20.4, this line is represented with a heavier stroke.

By comparing the plot of the K-function estimate for a given point pattern with these reference lines, one can gain insight into the spatial distribution of points in the pattern and make informed decisions about further analysis or modelling.

However, one must note that the K-function is dependent on the density of points in a spatial dataset. The dependency on point density arises from the fact that, in areas with higher point densities, the expected number of points within a given distance will also be higher. Therefore, the K-function values will generally be larger in regions with denser point patterns. This is important to consider because what may appear to be clustering in a region could simply be a result of higher point density.

Normalisation can be used to account for density in the K function analysis, and a common method is to divide the K-function values by the intensity or density of points in the region. This yields the L-function, which is a standardised version of the K function that is less affected by variations in point density. The L-function can help to identify true clustering or dispersion patterns without being confused by differences in density.

20.1.3 Clustered Points

If the points in a point pattern are clustered, the K-function estimate will be above $\pi \times r^2$ for small values of 'r'.

The plot of the K-function will show a curve that is above the line $\pi \times r^2$, indicating that the expected number of points within a distance r of a randomly chosen point in the pattern is larger than $\pi \times r^2$ for small values of 'r'.

The following code will exemplify the use of the kest() function.

Firstly, the code defines a rectangular window using the owin() function. This window is the observation window in a two-dimensional plane. It is used to define and describe the spatial extent within which the pattern of points is being analysed. It essentially specifies the bounds and shape of the area under investigation.

Secondly, a clustered pattern is generated and stored in the 'clust_ppp' variable. For creating a clustered pattern, the code starts by generating a random point pattern with 20 points inside the rectangular window using the runifpoint() function. These 20 points will be the seed for slightly moved points around them to create the clustered pattern around them.

Inside the while() loop, the code first selects one random point from the original point pattern using the sample() function. It then generates a jittered copy, i.e. a slightly randomly moved version, of the selected point using the rjitter() function with an amount of 5 units. Finally, it adds the jittered point onto the original set of points using the superimpose() function. The loop continues until the point pattern contains 100 points.

The code then calculates the K-function estimate for the pattern using the Kest() function.

#20-01

```
# Load the required packages
library(spatstat)

# Generate a point pattern with clustered points
set.seed(123)

# Define a rectangular window
win = owin(c(0,10), c(0,10))

# Generate a random uniform point pattern with 20 points
clust_ppp = runifpoint(20, win=win)
```

```
# Create 100 clustered points
npoint = 20
while(npoint < 100) {
  npoint = npoint + 1
  random_point = sample(1:20, 1)
  clust2_ppp = rjitter(clust_ppp[random_point], amount = 5)
  clust_ppp = superimpose(clust_ppp, clust2_ppp)
}

# Calculate the K-function
clust_k = Kest(clust_ppp)

# View the results
par(mfrow=c(1,2))
plot(clust_ppp, cex = 0.4, main="Clustered Distributed Points
(CDP)")
plot(clust_k, main = "K-Function for CDP")
lines(clust_k$r, pi * clust_k$r^2, col = "purple", lty = 1,
lwd = 3)
```

For a clustered point pattern, the K-function estimate should deviate from the reference line (heavy stroke), typically by being above the reference line (heavy stroke line in Figure 20.1). This indicates that the points in the pattern are more clustered than expected for randomly distributed points, with a higher number of close neighbours than would be expected by chance.

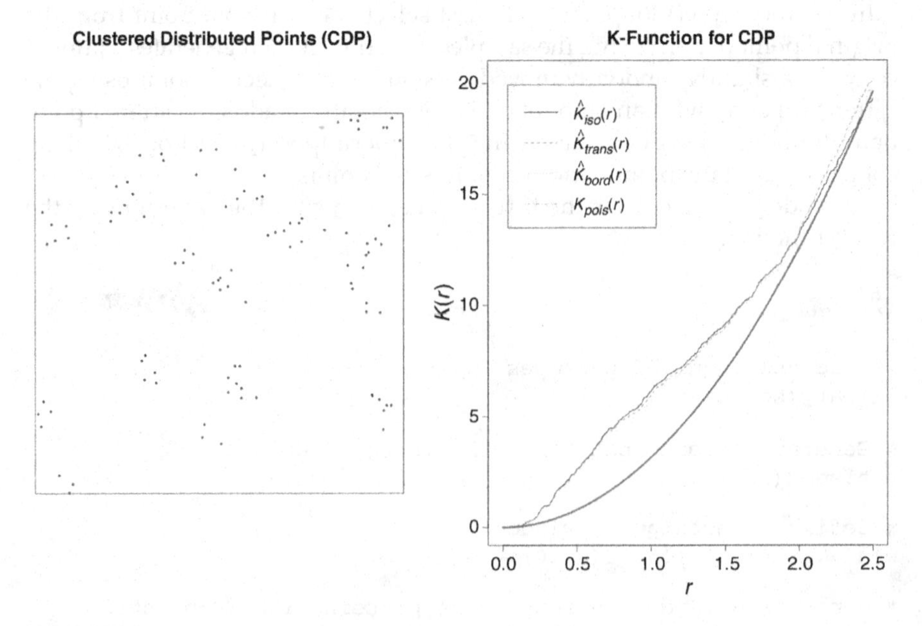

FIGURE 20.1
Clustered distribution of points and the Kest() function obtained.

20.1.4 Regular Points

To exemplify the behaviour of the K-function in a regular pattern of points, we used the same approach. This example shows how to generate a point pattern with regularly distributed points starting by creating a gridded distribution of points with a 10 × 10 pattern, by using the expand.grid() function from 'base' R.

This grid is 'randomised' using the rjitter() function. The K-function is then calculated, and the results are plotted.

#20-02

```
# Generate a point pattern with dispersed points
set.seed(123)

# Define a rectangular window
win = owin(c(0, 10), c(0, 10))

# Define the number of points in each direction
n = 10

# Generate a regular grid of points inside the window
x = seq(from = 0.5, to = 9.5, length.out=n)
y = seq(from = 0.5, to = 9.5, length.out=n)
xy = expand.grid(x=x, y=y)

# Create the ppp object and randomize it in a radius of 0.5
disp_ppp = as.ppp(xy, win)
disp_ppp = rjitter(disp_ppp, radius = 0.5)

# Calculate the K-function estimate
disp_k = Kest(disp_ppp)

# View the results
par(mfrow=c(1,2))
plot(disp_ppp, cex = 0.4, main="Dispersed Distributed Points
(DDP)")
plot(disp_k, main = "K-Function for DDP")
lines(disp_k$r, pi * disp_k$r^2, col = "purple", lty = 1, lwd
= 3)
```

For a regular point pattern, the K-function estimate should be below the reference line for randomly distributed points, which is given by $\pi \times r^2$. This indicates that the points in the pattern are apart ones from the others, with no significant clustering or randomness (see Figure 20.2).

20.1.5 Randomly Distributed Points

If the points in a point pattern are randomly distributed, the K-function estimate will be close to $\pi \times r^2$.

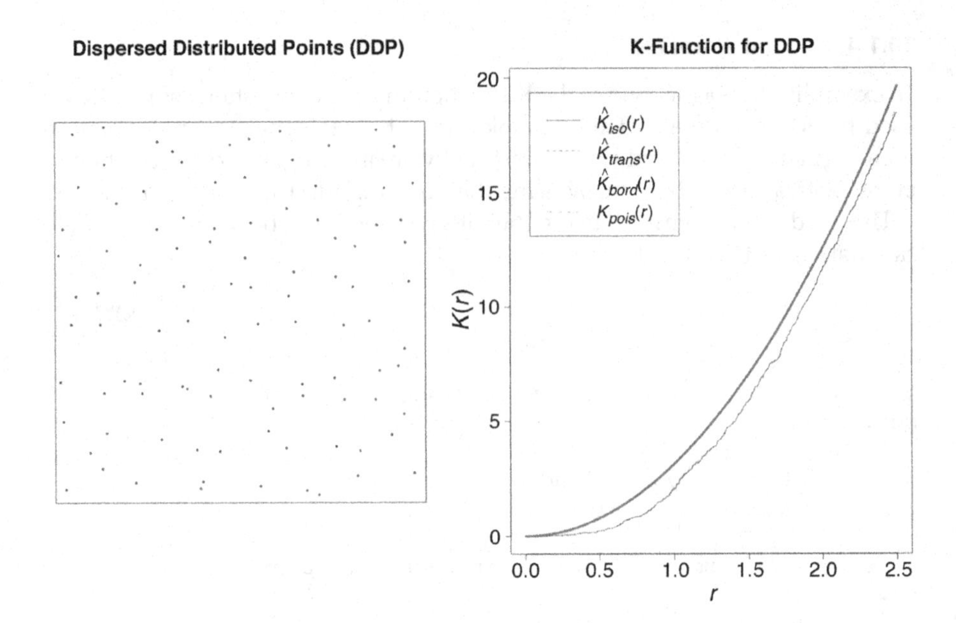

FIGURE 20.2
Dispersed distribution of points and the Kest() function obtained.

The plot of the K-function shows a nearly straight line, indicating that the expected number of points within a distance 'r' of a randomly chosen point, in the pattern, is close to πr^2.

This code example follows the previous ones, but for creating the randomly distributed points, it uses the rpoint() function from the 'spatstat' package.

#20-03

3 stars

```
# Generate a point pattern with randomly distributed points
set.seed(123)

# Define a rectangular window
win = owin(c(0,10), c(0,10))

# Generate a random point pattern with 100 points
rand_ppp = rpoint(100, win = win)

# Calculate the K-function estimate
rand_k = Kest(rand_ppp)

# View the results
par(mfrow=c(1,2))
plot(rand_ppp, cex = 0.4, main="Randomly Distributed Points
(RDP)")
```

```
plot(rand_k, main = "K-Function for RDP")
lines(rand_k$r, pi * rand_k$r^2, col = "purple", lty = 1, lwd
= 3)
```

The obtained results are plotted in Figure 20.3.

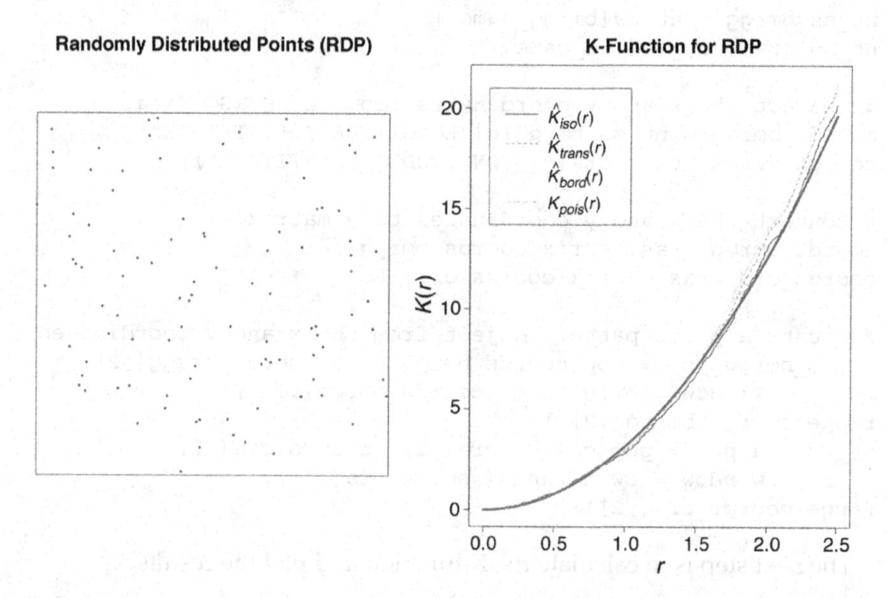

FIGURE 20.3
Randomly distribution of points and the Kest() function obtained. The thick dashed line represents the reference line.

Again, the Kest() function is used to calculate the K-function estimate, and the results are plotted in a graph.

The K-function estimate does not deviate from the reference line (in heavy stroke); it indicates that the points in the pattern are randomly distributed.

20.2 PPA for the Northern Territory Dataset

For a real example, let us return to the Northern Territory Stream Sediments dataset. We will look at the sample pattern distribution for the 'Bureau of Mineral Resources' (BMRGG) and the 'CRA Exploration' companies.

The first steps include subsetting the data and creating the necessary 'ppp' objects based on this data.

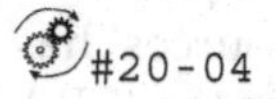#20-04 

```
# Read the File
nt_ss = read.csv("NT_SS/GEOCHEM_STREAM_SEDIMENTS.csv")
```

```
# Filter the dataset
bmrgg_name = grepl("^Bureau of Mineral", nt_ss$COMPANY)
cra_name = nt_ss$COMPANY =='CRA Exploration Pty Ltd.'

# Create the subset
nt_ss_bmrgg = nt_ss[bmrgg_name,]
nt_ss_cra = nt_ss[cra_name,]

# Extract the x and y coordinates from the BMRGG data
coords_bmrgg = nt_ss_bmrgg[c("LONGITUDE", "LATITUDE")]
coords_cra = nt_ss_cra[c("LONGITUDE", "LATITUDE")]

# Convert the x and y coordinates to a matrix
coords_bmrgg = as.matrix(coords_bmrgg)
coords_cra = as.matrix(coords_cra)

# Create a point pattern object from the x and y coordinates
nt_ss_bmrgg_ppp = ppp(coords_bmrgg[,1], coords_bmrgg[,2],
        window = owin(range(coords_bmrgg[,1]),
range(coords_bmrgg[,2])))
nt_ss_cra_ppp = ppp(coords_cra[,1], coords_cra[,2],
        window = owin(range(coords_cra[,1]),
range(coords_cra[,2])))
```

The next step is to calculate the K-function and plot the results.

#20-05

```
# Calculate the K-function estimate
nt_ss_bmrgg_k = Kest(nt_ss_bmrgg_ppp)
nt_ss_cra_k = Kest(nt_ss_cra_ppp)

# View the results
par(mfrow=c(2,2))
plot(nt_ss_bmrgg_ppp, cex = 0.4, main="BMRGG data points")
plot(nt_ss_bmrgg_k, main = "K-Function for BMRGG")
lines(nt_ss_bmrgg_k$r, pi * nt_ss_bmrgg_k$r^2, col = "purple",
lty = 1, lwd = 2)

plot(nt_ss_cra_ppp, cex = 0.4, main="CRA data points")
plot(nt_ss_cra_k, main = "K-Function for CRA")
lines(nt_ss_cra_k$r, pi * nt_ss_cra_k$r^2, col = "purple",
lty = 1, lwd = 2)
```

Figure 20.4 displays the results for the data collected by both companies.

It is readily visible from the plots that 'Bureau of Mineral Resources' has a fairly random distribution of stream sediment analysis, whereas the 'CRA exploration' has a notoriously clustered nature.

The K-functions obtained are a proof of this conclusion.

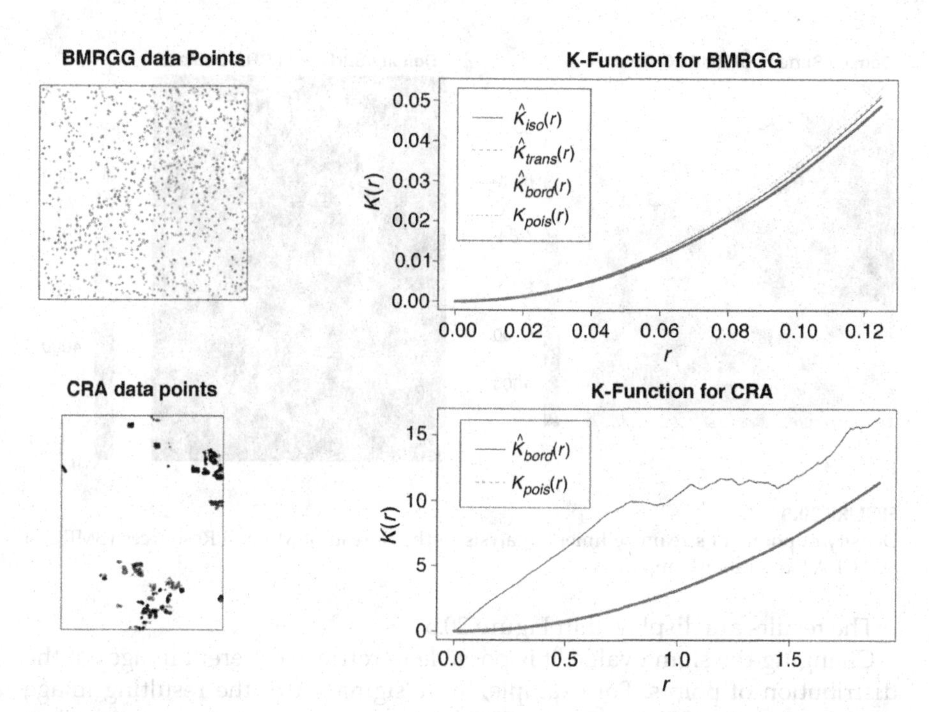

FIGURE 20.4
Data from Stream Sediments of Bureau of Mineral Resources (BMRGG) and CRA Exploration companies.

20.2.1 Point Density Analysis

Another approach to PPA analysis is to estimate the intensity (i.e., density) of points in a point pattern dataset across a spatial domain. It calculates the density of points in a given area by dividing the number of points by the area of the domain. The resulting density estimate can be visualised as a heatmap or contour plot, allowing for the identification of areas of high or low point density.

For this, the density.ppp() function can be used to create an image with the areas of high and low density of data.

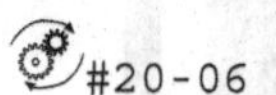

```
# Estimate the density of the dataset using a Gaussian kernel
bmrgg_dens = density.ppp(nt_ss_bmrgg_ppp, sigma = 0.05)
cra_dens = density.ppp(nt_ss_cra_ppp, sigma = 0.05)


# View the results
par(mfrow=c(1,2))
plot(bmrgg_dens, main = "Density Surface of BMRGG Dataset")
plot(cra_dens, main = "Density Surface of CRA Dataset")
```

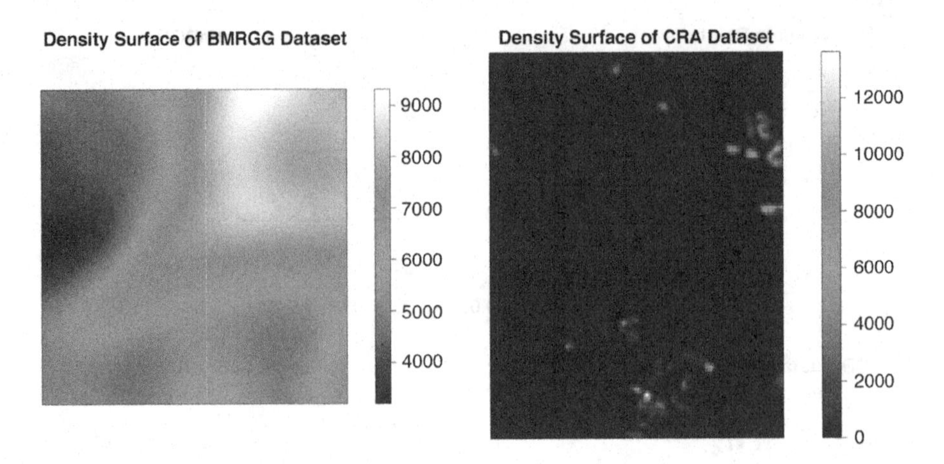

FIGURE 20.5
Density of points of stream sediment analysis in the Bureau of Mineral Resources (BMRGG) and CRA Exploration companies.

The results are displayed in Figure 20.5.

Changing the sigma value, it is possible to retrieve different images of the distribution of points. For example, for a 'sigma = 0.5', the resulting image changes significantly (see Figure 20.6).

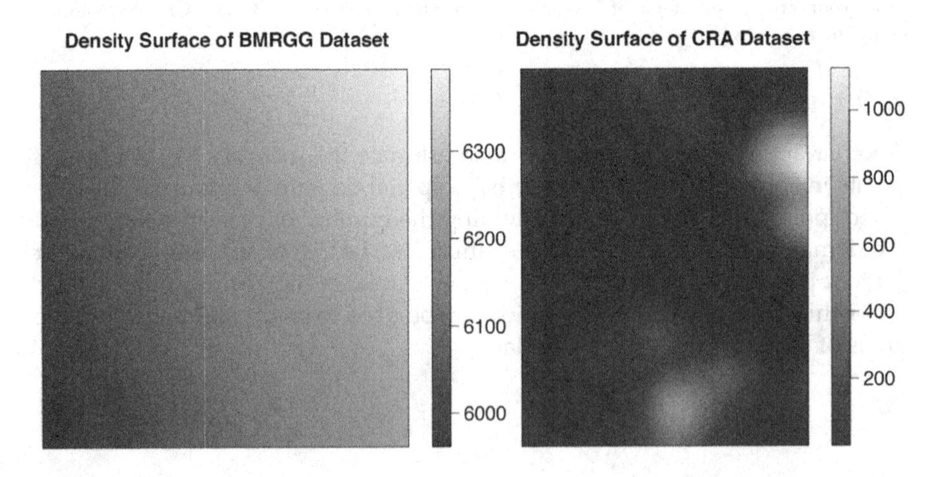

FIGURE 20.6
The resulting density image with a sigma = 0.5.

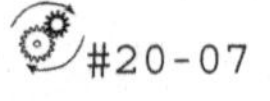

```
# Estimate the density of the dataset using a Gaussian kernel
bmrgg_dens = density.ppp(nt_ss_bmrgg_ppp, sigma = 0.5)
cra_dens = density.ppp(nt_ss_cra_ppp, sigma = 0.5)
```

```
# View the results
par(mfrow=c(1,2))
plot(bmrgg_dens, main = "Density Surface of BMRGG Dataset")
plot(cra_dens, main = "Density Surface of CRA Dataset")
```

It becomes clear that these functions can be used to identify not only point distribution but also patterns and trends in the data.

20.3 Earthquakes Data

Another example of freely available data is the earthquakes from all around the globe. This earthquake data can be extracted from the United States Geological Survey (USGS) earthquake catalogue (https://earthquake.usgs.gov/). This data is retrievable in the Comma-Separated Values (CSV) format, directly by querying the USGS server.

In this example, we will extract earthquakes with magnitude greater than 3 that occurred in the Mediterranean region from 2015 to 2020.

#20-08

```
# Load required packages
library(readr)
library(leaflet)

# Read in earthquake data
base_query = "https://earthquake.usgs.gov/fdsnws/event/1/
query?format=csv&"
time_interval = "starttime=2015-01-01&endtime=2020-12-31&"
mag_interval = "minmagnitude=3&"
lat_interval = "maxlatitude=45&minlatitude=30&"
lon_interval = "maxlongitude=45&minlongitude=-10"
query = paste0(base_query,time_interval, mag_interval,
lat_interval, lon_interval)

# Read in earthquake data
earthquakes = read_csv(query)

# Plot the earthquakes locations on an interactive map using
leaflet
leaflet(earthquakes) %>%
  addTiles() %>%
  addCircleMarkers(lng = ~longitude, lat = ~latitude,
                   radius = ~exp(mag) * .02, color = "red",
                   fillOpacity = 0.3)
```

In this example, we use the read_csv() function from the 'readr' package to obtain the earthquake data from the USGS earthquake catalogue and filter the data to include only earthquakes in the Mediterranean region.

The map with the retrieved results is presented in Figure 20.7.

20.3.1 PPA of Earthquakes

The visual analysis of the earthquake distribution indicates a clustered nature. Let us apply the previous tools to corroborate this.

```
#20-09

# Convert the longitude and latitude coordinates to a matrix
coords = earthquakes[c("longitude", "latitude")]
coords = as.matrix(coords)

# Create a point pattern object
med_eq_ppp = ppp(coords[,1], coords[,2], window =
owin(range(coords[, 1]), range(coords[,2])))

# Calculate the K-function
med_eq_k = Kest(med_eq_ppp)

# View the results
par(mfrow=c(1,2))
plot(med_eq_ppp, cex=.4, main="Earthquakes Med. Region")
plot(med_eq_k, main = "K-Function for Earthquakes")
lines(med_eq_k$r, pi * med_eq_k$r^2, col = "purple", lty = 1,
lwd = 2)
```

In this example, we use the ppp() function, from the 'spatstat' package, to create a point pattern object from the coordinates. The point pattern is plotted using the plot() function adapted to 'spatstat' objects, and the K-function is calculated using the Kest() function and plotted as well.

The output of this example is a scatterplot of the earthquake point pattern in the Mediterranean region and a plot of the K-function for this point pattern. These plots provide a graphical representation of the spatial distribution of earthquakes in the region and can be used to assess the spatial dependence, clustering, and dispersion of earthquakes in the Mediterranean region over the period from 2015 to 2020.

The result of the plot is in Figure 20.8.

The K-function plots clearly below the representative data indicating the clustered nature of the data.

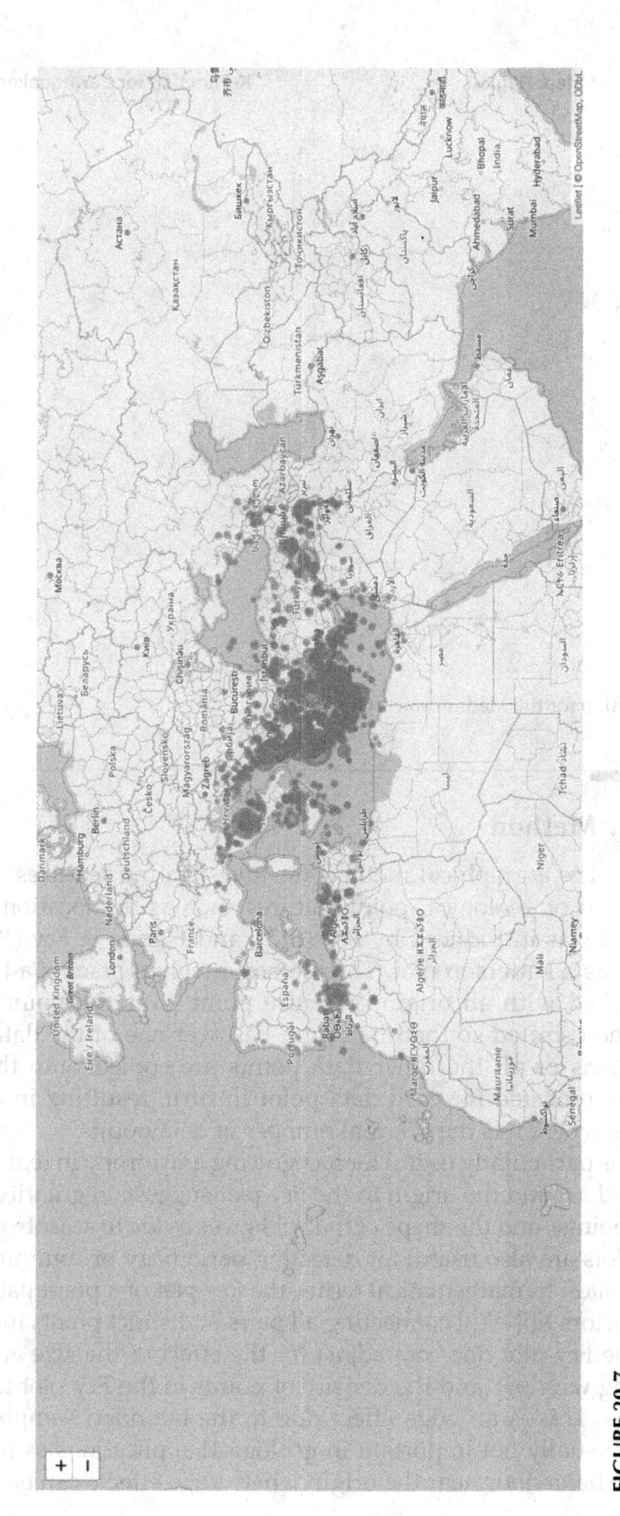

FIGURE 20.7
Earthquakes from the Mediterranean region between 2015 and 2020.

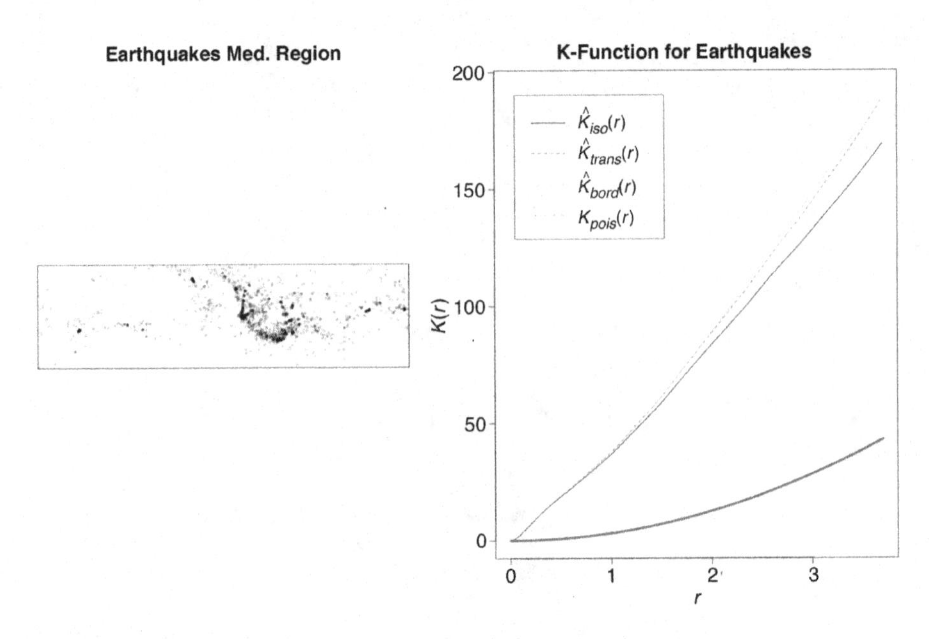

FIGURE 20.8
Plot of the K-function for the Mediterranean Region.

20.4 The Fry Method

The Fry method is a graphical method for investigating features of a spatial point pattern of geological point features such as the location of mineral deposits. It was introduced by Fry (1979) and Hanna & Fry (1979) and is also known as a Patterson plot. The method involves placing a transparent sheet, marked with an origin or centre point, over the point pattern. The sheet is then shifted so that the origin lies over one of the data points, and the positions of all the other data points are copied onto the sheet. This process is repeated for each data point in turn, resulting in a plot of n(n – 1) points, where n is the original number of data points.

The Fry plot is particularly useful for recognising anisotropy in regular point patterns. A void around the origin in the Fry plot suggests regularity (inhibition between points), and the shape of the void gives a clue to anisotropy in the pattern. Fry plots are also useful for detecting periodicity or rounding of the spatial coordinates. In mathematical terms, the Fry plot of a point pattern X is a plot of the vectors X[i] - X[j] connecting all pairs of distinct points in X.

Note that the Fry plot does not adjust for the effect of the size and shape of the sampling window, and the density of points in the Fry plot tapers off near the edges. This is an edge effect due to the bounded sampling window, which is usually not important in geological applications as interest is focused on the behaviour near the origin where edge effects can be ignored.

To correct the edge effect, other methods such as the Kmeasure() or Kest() functions can be used.

The Fry method when applied to cluster data can give misleading results.

In the example below, it is applied to the earthquakes with magnitude higher than 2 from an area of southwest (SW) Iberia.

#20-10

```
# Read in earthquake data
base_query = "https://earthquake.usgs.gov/fdsnws/event/1/
query?format=csv&"
time_interval = "starttime=1900-01-01&endtime=2019-12-31&"
mag_interval = "minmagnitude=2&"
lat_interval = "maxlatitude=39.0&minlatitude=36.5&"
lon_interval = "maxlongitude=-8.0&minlongitude=-10.0"
query = paste0(base_query, time_interval, mag_interval,
lat_interval, lon_interval)
SW_Iberia_eq = read_csv(query)

# Plot the earthquakes locations using leaflet
leaflet(SW_Iberia_eq) %>%
  addTiles() %>%
  addCircleMarkers(lng = ~longitude, lat = ~latitude, popup =
paste0("date: ",SW_Iberia_eq$time,"<br> mag: ",
SW_Iberia_eq$mag),
      radius = ~exp(mag) * 0.02,  color = "red", fillOpacity =
0.3)
```

The retrieved data is visualised in Figure 20.9.

After retrieving the data, the 'ppp' object is created and the fryplot() function is called, considering the full data and a window around the centre of the data.

#20-11

```
# Convert the longitude and latitude coordinates to a matrix
coords = SW_Iberia_eq[c("longitude", "latitude")]
coords = as.matrix(coords)

# Create a point pattern object from longitude and latitude
SW_Iberia_eq_ppp = ppp(coords[,1], coords[,2], window =
owin(range( coords[,1]), range(coords[,2])))

# View the results
par(mfrow = c(1, 2))
fryplot(SW_Iberia_eq_ppp, cex=.1, main = "Fry Plot of SW
Iberia Earthquakes")
fryplot(SW_Iberia_eq_ppp, cex=.2, width=2, main = "Window of
Fry Plot")
```

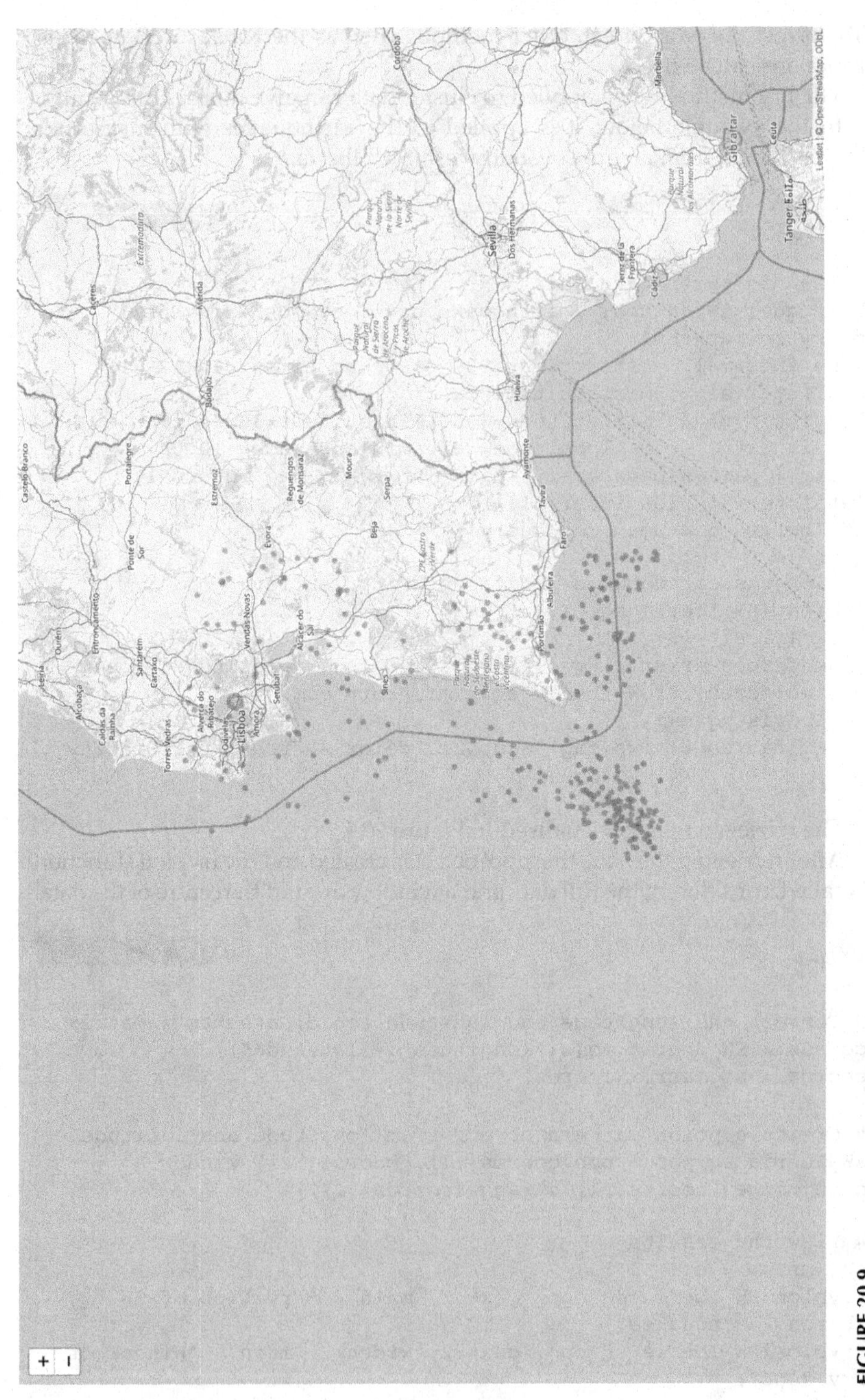

FIGURE 20.9
The earthquakes in southwest (SW) Iberia.

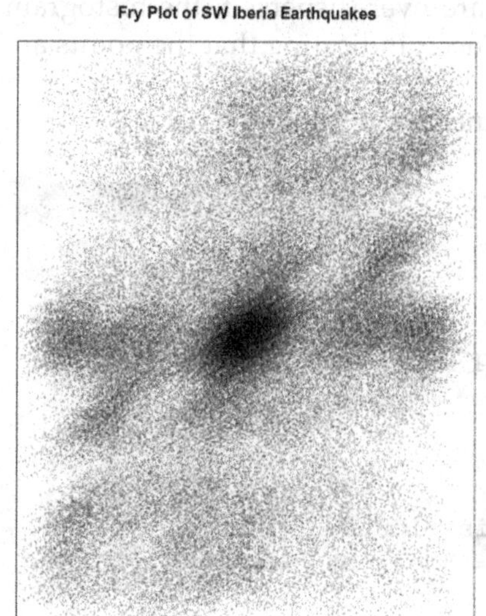

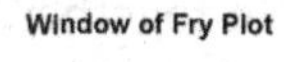

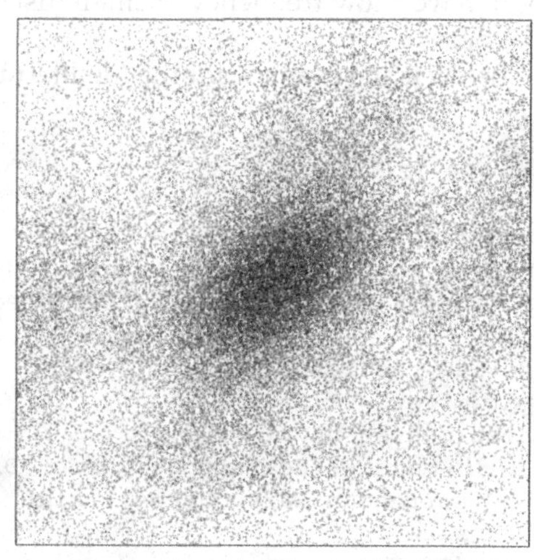

FIGURE 20.10
Fry plot of the earthquakes from southwest (SW) Iberia.

The resulting plots are represented in Figure 20.10.

Although there is some clusteredness in the data, it is also visible the existence of an ellipse with a northeast-southwest orientation of the main axis. This is necessarily related with the Africa-Europe collision direction that is the dominant geotectonic effect in this region (Fernández et al., 2019).

20.5 Nearest Neighbour and PPA

Another example of the process of analysing the distribution of data is the nearest neighbour (nn) approach. This is a measure of the distribution of distances between a point and its closest neighbour. This analysis can provide information about the spatial distribution of the points and can be used to assess the degree of clustering or over-dispersion in the pattern.

A histogram of the distances provides information about the spatial distribution of the points and can be used to assess the degree of clustering or over-dispersion in the pattern.

If the points are randomly distributed, the histogram should have a uniform distribution. If the points are clustered, the histogram of the nn distances will have a high frequency of small distances, indicating that the points are close together. If the points are over-dispersed, the histogram will have a low frequency of small distances, indicating that the points are far apart.

Below is the example for the Mediterranean and SW Iberia data.

```
#20-12

# Nearest Neighbors
med_eq_nn = nndist(med_eq_ppp)
SW_Iberia_nn = nndist(SW_Iberia_eq_ppp)

# View the results
par(mfrow = c(1, 2))
hist(med_eq_nn,
      main = "Nearest Neighbour Distribution \nfor
Mediterranean Earthquakes", xlab = "Nearest Neighbour
Distance")
hist(SW_Iberia_nn, main = "Nearest Neighbour Distribution \
nfor SW Iberia Earthquakes", xlab = "Nearest Neighbour
Distance")
```

The result for both datasets is shown in Figure 20.11.

In the case of the Mediterranean earthquakes, it has already been demonstrated that they have a clustered pattern. The nn approach has a histogram with a positively asymmetric distribution, with most of the points in the vicinity of each other. The SW Iberia example has, comparably, a more uniform distribution, therefore a better approximation to a random distribution pattern.

20.6 Concluding Remarks

PPA is fun. Although some concepts might be cryptic at the beginning, it is another day in the office for most of the geologists. Clustered, random, or gridded data are concepts that any structural geologist or exploration geologist are familiar with. The introductory tools presented in this chapter provide the spatial analyst with a way of measuring the type of distribution that is present in most of the geological situations.

The applications of PPA in geology are vast. The proposed statistical approach not only expands the geologist's perspective beyond mere visual assessments, transforming it from a subjective 'it looks like' perspective to a more robust and quantitative stance. Geologists can now engage in discussions backed by statistical evidence, confidently stating that 'the distribution follows the Poisson distribution'.

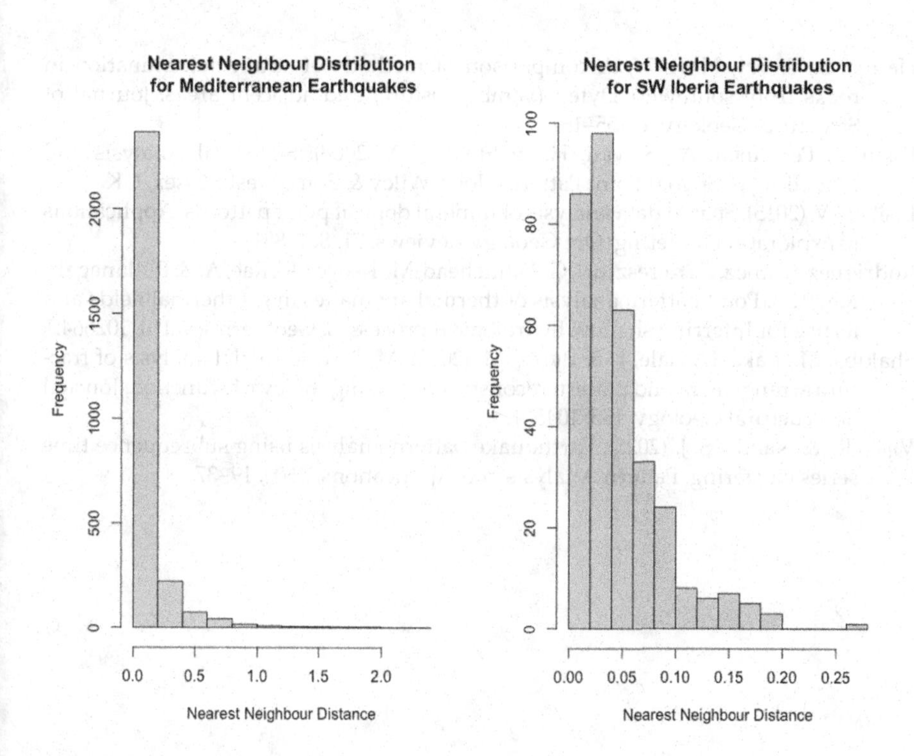

FIGURE 20.11
Histogram for the nearest neighbour distribution for the Mediterranean and southwest (SW) Iberia earthquakes datasets.

References

Boots, B., & Getis, A. (1988). Point Pattern Analysis. Reprint. Edited by Grant Ian Thrall. WVU Research Repository, 2020. Consulted in: https://researchrepository.wvu.edu/rri-web-book/13/

Carranza, E. (2009). Controls on mineral deposit occurrence inferred from analysis of their spatial pattern and spatial association with geological features. Ore Geology Reviews, 35(3–4), 383–400.

Diggle, P. (1983). Statistical Analysis of Spatial Point Patterns. Academic Press, London, UK.

Dixon, P. (2014). Ripley's K Function. Wiley StatsRef: Statistics Reference Online. https://doi.org/10.1002/9781118445112.stat07751

Fernández, M., Torne, M., Vergés, J., Casciello, E., & Macchiavelli, C. (2019). Evidence of segmentation in the Iberia–Africa plate boundary: A Jurassic heritage? Geosciences, 9(8), 343. https://doi.org/10.3390/geosciences9080343

Fry, N. (1979). Random point distributions and strain measurement in rocks. Tectonophysics, 60, 89–105.

Hanna, S., & Fry, N. (1979) A comparison of methods of strain determination in rocks from southwest Dyfed (Pembrokeshire) and adjacent areas. Journal of Structural Geology, 1, 155–162.

Illian, J., Penttinen, A., Stoyan, H., & Stoyan, D. (2008). Statistical Analysis and Modelling of Spatial Point Patterns. John Wiley & Sons, West Sussex, UK.

Lisitsin, V. (2015). Spatial data analysis of mineral deposit point patterns: Applications to exploration targeting. Ore Geology Reviews, 71, 861–881.

Rodriguez-Gomez, C., Kereszturi, G., Whitehead, M., Reeves, R., Rae, A., & Pullanagari, R. (2023). Point pattern analysis of thermal anomalies in geothermal fields and its use for inferring shallow hydrological processes. Geothermics, 110, 102664.

Shakiba, M., Lake, L., Gale, J., & Pyrcz, M. (2022). Multiscale spatial analysis of fracture arrangement and pattern reconstruction using Ripley's K-function. Journal of Structural Geology, 155, 104531.

Vijay, R., & Nanda, S. J. (2023). Earthquake pattern analysis using subsequence time series clustering. Pattern Analysis and Applications, 26(1), 19–37.

21

Interpolating Data

Interpolation of point data is used in geology to estimate values at unsampled locations based on known measurements at sampled locations. The 'gstat' package[1] (Gräler et al., 2016) is designed to, among others, provide functions for interpolating point data using various methods, such as inverse distance weighting (IDW), kriging, or bivariate spline interpolation.

IDW is an interpolation method that assigns weights to nearby points based on their distances to the unsampled location (Shepard, 1968; Hengl, 2007). The closer a point is to the unsampled location, the more weight it is given in the interpolation. This method assumes that the values of nearby points are more representative of the unsampled location than those farther away.

Kriging is a method that estimates values at unsampled locations by fitting a mathematical model to the known measurements and their spatial relationships (Goovaerts, 1997; Hengl, 2007; Bivand et al., 2013). The kriging model considers the spatial autocorrelation (see Figure 21.1) of the data and produces estimates with minimum error variance. It considers not only the distance between the known points and the unsampled location but also the spatial correlation between the values of the known points. This method is based on geostatistics and is known for its ability to capture the spatial variability of the data.

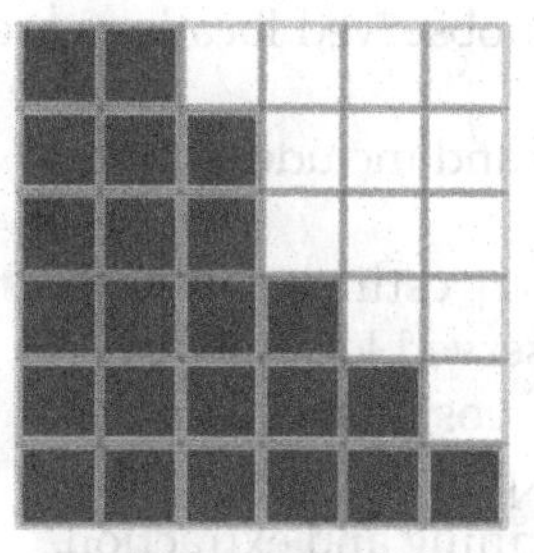

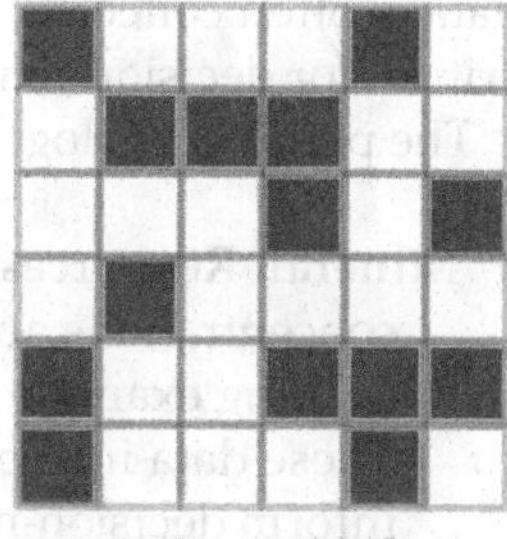

FIGURE 21.1
Types of spatial autocorrelation.

[1] https://github.com/r-spatial/gstat

DOI: 10.1201/9781032651880-24

Bivariate spline is employed as a piecewise-defined polynomial function. This surface-fitting technique involves dividing the data space into smaller regions and approximating the surface within each region using polynomial segments. A prevalent choice for interpolation is the bicubic spline, wherein cubic polynomials are utilised to represent the surface characteristics within each subdivision. This approach ensures continuity and smoothness, facilitating the creation of a coherent and flexible surface representation through a collection of interconnected polynomial segments. It can be used for modelling geomorphological landscapes or interpreting geochemical variations.

Interpolating point data has a wide range of geologic applications, such as estimating the concentration of mineral deposits across a mining site, predicting the distribution of geological features across a landscape, and modelling the spatial variability of soil properties for land management purposes (Atkinson, 2005; Lado et al., 2008; Lu & Wong, 2008; Stojdl et al., 2017).

It is important to keep in mind that the accuracy of the interpolated values depends on the spatial distribution and density of the sampled points as well as the chosen interpolation method and its parameters. Therefore, it is recommended to perform a thorough analysis and evaluation of the interpolated results before making any geologic inferences or decisions.

21.1 The 'gstat' Package

The 'gstat' package is the geostatistical package that is mostly used for spatial data analysis and prediction. It provides an advanced set of tools for variogram modelling, kriging, IDW, and conditional simulation. These methods are particularly useful in geological applications, where spatially distributed data is often collected, and accurate predictions at unobserved locations are critical for decision-making or risk assessment.

The possible geological applications are numerous and include:

Mineral Resources Studies: The interpolation to estimate mineral concentrations and their uncertainties at unobserved locations are a striking example of the possible application of geostatistical studies. These data interpretations can guide further exploration efforts and inform decision-making processes for mine planning and extraction. For example, Du et al. (2017) used kriging to estimate the thickness of cobalt-rich crust deposits around seamounts. Akbar (2012) estimated the grade from the Choghart north anomaly iron ore deposit in Yazd province, Iran, using an ordinary kriging (OK) method.

Soil Contamination: The spatial distribution of contaminants can be assessed and predicted in soil, such as heavy metals, based on sampled data. The interpolation of chemical elements' concentrations at unsampled locations aids in the identification of contaminated areas and guiding remediation efforts. In a remarkable example, Wang et al. (2023) evaluated the human health risk of soil metal pollution and interpolated their data using an empirical Bayesian kriging model.

Groundwater Modelling: Estimation of the spatial patterns in groundwater levels, hydrogeological properties, and contaminant plumes are other examples of the possible applications of interpolating data. Spatial interpolation or conditional simulation can be used to estimate these variables and their uncertainties, providing valuable input for groundwater flow and transport models. As an example, Kumar et al. (2023) applied kriging to estimate the spatial variation of permeability in soils.

Geological Mapping: Rock type, fault density, and geological unit thickness often exhibit spatial dependence. Geostatistical methods can be used to infer their continuity, aiding in the interpretation of geological structures and guiding further data collection. Zeng et al. (2022) defined the spatial distribution of lithologies for reservoir modelling, using a prediction method to provide a solution that is data supported, for production and decision-making in petroleum engineering.

Geothermal Resource Assessment: Estimating the subsurface temperature distributions and heat flow based on sparse measurements is an important task in geothermal assessments. The 'gstat' package can be applied to interpolate these variables and estimate uncertainties, informing decision-making for geothermal energy exploration and development. Bonté et al. (2020) estimated subsurface temperature in sedimentary basins using interpolation methods.

Among the methods available from the 'gstat' package, we highlight the following:

IDW: As stated above, this is a method for spatial interpolation that weights the data points according to their distance from the interpolation point.

Variogram Modelling: Exponential, Gaussian, and spherical models can be fitted to the empirical variogram to estimate spatial dependence structure.

Kriging: The package supports OK, simple kriging (SK), and universal kriging (UK) for spatial interpolation and prediction.

Cross-validation: Leave-one-out cross-validation can be performed to assess the performance of the kriging model.

Conditional Simulation: Gaussian and indicator simulations can be used to generate realisations of the spatial process, incorporating the uncertainty associated with the estimation.

For these spatial analysis, 'gstat' has, among others, the following functions that can be used in interpolation projects:

gstat(): Creates a 'gstat' object for variogram modelling and kriging.
gstat.formula(): Creates a 'gstat' object using formula interface.
idw(): Performs Inverse Distance Weighted interpolation.
vgm(): Defines a variogram model.
variogram(): Computes the empirical variogram.
variogramLine(): Computes the empirical variogram for a line segment.
variogramSurface(): Computes the empirical variogram for a grid.
fit.variogram(): Fits a variogram model to the empirical variogram.
krige(): Performs kriging (ordinary or universal).
krige.cv(): Performs cross-validation for kriging.
predict.gstat(): Predicts values at new locations using a 'gstat' object.

References

Akbar, D. (2012). Reserve estimation of central part of Choghart north anomaly iron ore deposit through ordinary kriging method. International Journal of Mining Science and Technology, 22(4), 573–577.

Atkinson, P. (2005). Spatial prediction and surface modeling. Geographical Analysis, 37(2), 113–124.

Bivand, R., Pebesma, E., & Gomez-Rubio, V. (2013). Applied Spatial Data Analysis with R, Second Edition. Springer, NY.

Bonté, D., Guillou-Frottier, L., Garibaldi, C., Bourgine, B., Lopez, S., Bouchot, V., & Lucazeau, F. (2010). Subsurface temperature maps in French sedimentary basins: new data compilation and interpolation. Bulletin de la Société géologique de France, 181(4), 377–390.

Du, D., Wang, C., Du, X., Yan, S., Ren, X., Shi, X., & Hein, J. (2017). Distance-gradient-based variogram and kriging to evaluate cobalt-rich crust deposits on seamounts. Ore Geology Reviews, 84, 218–227.

Goovaerts, P. (1997). Geostatistics for Natural Resources Evaluation. Applied Geostatistics Series. Oxford University Press, New York.

Gräler, B., Pebesma, E., & Heuvelink, G. (2016). Spatio-temporal interpolation using gstat. The R Journal, 8(1), 204–218.

Hengl, T. (2007). A Practical Guide to Geostatistical Mapping of Environmental Variables. Publications Office, Luxembourg. OCLC: 758643236.

Kumar, P., Rao, B., Burman, A., Kumar, S., & Samui, P. (2023). Spatial variation of permeability and consolidation behaviors of soil using ordinary kriging method. Groundwater for Sustainable Development, 20, 100856.

Lado, L., Hengl, T., & Reuter, H. (2008). Heavy metals in European soils: A geostatistical analysis of the FOREGS Geochemical database. Geoderma, 148(2), 189–199.

Lu, G., & Wong, D. W. (2008). An adaptive inverse-distance weighting spatial interpolation technique. Computers & Geosciences, 34(9), 1044–1055.

Shepard, D. (1968). A two-dimensional interpolation function for irregularly-spaced data. In: Blue, R. B. S., & Rosenberg, A. M. (Eds.), Proceedings of the 1968 ACM National Conference. ACM Press, New York, pp. 517–524.

Stojdl, J., Matys Grygar, T., Elznicova, J., Popelka, J., Vachova, T., & Hosek, M. (2017). Kriging - a challenge in geochemical mapping. In: EGU General Assembly Conference Abstracts, p. 3615. https://ui.adsabs.harvard.edu/abs/2017EGUGA..19.3615S/abstract

Wang, L., Liu, R., Liu, J., Qi, Y., Zeng, W., & Cui, B. (2023). A novel regional-scale human health risk assessment model for soil heavy metal (loid) pollution based on empirical Bayesian kriging. Ecotoxicology and Environmental Safety, 258, 114953.

Zeng, L., Ren, W., Shan, L., Huo, F., & Meng, F. (2022). Lithology spatial distribution prediction based on recurrent neural network with Kriging technology. Journal of Petroleum Science and Engineering, 214, 110538.

22

Inverse Distance Weighting (IDW)

Inverse distance weighting (IDW)[1] is an interpolation method that estimates values at unsampled locations based on the weighted average of nearby points. The 'gstat' and the 'spatstat' packages both provide the function idw() for performing IDW interpolation.

The idw() function assigns weights to nearby points based on their distances to the unsampled location. The closer a point is to the unsampled location, the higher weight it is given in the interpolation. The weights are calculated using a power function, where the exponent (p) determines the rate of decay of the weights with distance (Figure 22.1). A larger exponent results in a more rapid decay of the weights, giving greater emphasis to the values of the closest points.

The idw() function also allows for the specification of a minimum and maximum number of points to use in the interpolation as well as a maximum search distance for nearby points. This enables the user to control the extent to which distant points contribute to the interpolation and the level of smoothing in the resulting surface.

To use idw() from the 'gstat' package, the user must first create a data frame containing the known data points and their corresponding values and define the spatial coordinates of the data points using the 'sp' package. The idw() function takes the 'data.frame' object as input and outputs a 'SpatialPixelsDataFrame' object of the interpolated surface.

To use idw() from the 'spatstat' package, the user must first create a point pattern object using the ppp() function. The point pattern object should contain the known data points and the location where the values are to be estimated. The function then takes the point pattern object and outputs a pixel image of the interpolated surface.

While IDW interpolation is a relatively simple geospatial interpolation method, it does have some caveats and weaknesses, particularly when used in geology. Among the considerations that must be taken, I would list the following:

Sensitivity to data distribution: IDW is highly sensitive to the distribution of data points. If data points are unevenly distributed or clustered, such as when a sampling grid is clustered or biased, IDW can

[1] For more, see https://gisgeography.com/inverse-distance-weighting-idw-interpolation/ and https://www.e-education.psu.edu/natureofgeoinfo/c7_p9.html

DOI: 10.1201/9781032651880-25

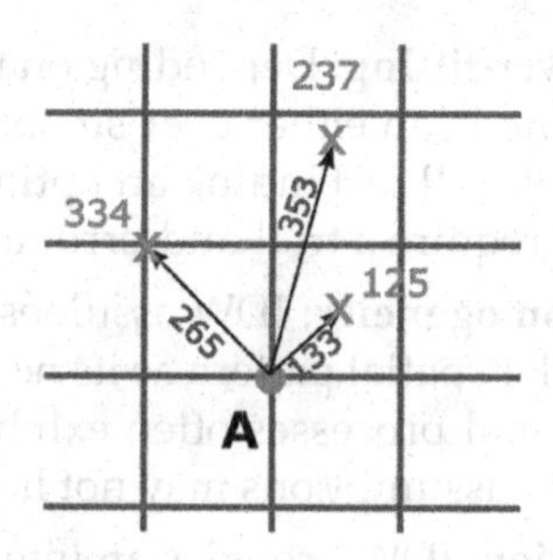

$$Z_A = \frac{\sum_{i=1}^{n}\left(\frac{z_i}{d_i}\right)}{\sum_{i=1}^{n}\left(\frac{1}{d_i}\right)} = \frac{\frac{125}{133} + \frac{334}{265} + \frac{237}{353}}{\frac{1}{133} + \frac{1}{265} + \frac{1}{353}}$$

$$Z_x = \frac{\sum_{i=1}^{n}\left(\frac{z_i}{d_i^p}\right)}{\sum_{i=1}^{n}\left(\frac{1}{d_i^p}\right)}$$

FIGURE 22.1
Example for calculating the value (Z_A) of a point A using an inverse distance weighted algorithm.

produce unrealistic results. It tends to create artefacts, such as sharp edges and abrupt changes, where data points are sparse.

Boundary effects: IDW does not consider any information outside the convex hull of the data[2] points. This can lead to boundary effects where the interpolation is less accurate near the edges of the study area.

No accounting for anisotropy: IDW assumes isotropic spatial correlation, meaning it treats all directions equally. Sometimes, the geological structures or the geochemical distribution of elements, e.g., may exhibit anisotropy, where the correlation structure varies with direction. IDW cannot capture such anisotropic behaviour.

---

[2] The convex hull of a set of data points is the smallest convex polygon that encloses all of the points in the set. A convex polygon is a polygon in which any line segment connecting two points in the polygon lies entirely within the polygon.

Over-smoothing and over-fitting: Depending on the value of the IDW exponent (p), the method can either over-smooth the surface[3] (large p) or over-fit the data (small p). Finding an optimal value of p can be challenging and may require a trial-and-error approach.

Assumption of local homogeneity: IDW assumes that each interpolation location has similar spatial pattern as its neighbours. In geology, geological structures and processes often exhibit spatial variability, and local homogeneity assumptions may not hold true.

No uncertainty estimation: IDW provides an interpolated surface, but it does not provide any uncertainty estimation. Users cannot assess the reliability or confidence of the interpolated values.

Outlier sensitivity: IDW is sensitive to outliers. A single extreme value can exert significant influence on nearby interpolated values.

Dependence on neighbourhood size: The choice of neighbourhood size (number of nearest neighbours to consider) can significantly impact the results of IDW interpolation. Selecting an appropriate neighbourhood size can be subjective and may affect the final output.

Given these limitations, it's essential to use IDW with caution and consider its appropriateness based on the specific characteristics of the geospatial data and the goals of the analysis. In cases where data distribution is uneven, spatial correlation is anisotropic, or uncertainty estimation is essential; alternative interpolation methods such as kriging or spline-based approaches may be more suitable in geology applications.

22.1 Timor-Alor Earthquakes

The Timor-Alor region is set in the border between the Australian and Eurasian plates (Coudurier-Curveur et al., 2021). This tectonic setting implies a series of earthquakes with changing depth depending on the subducting plate position. In this example, we will look at the earthquake's spatial distribution and its corresponding depth in the Timor-Alor region.

#22-01

```
# Load the libraries
library(readr)
library(sf)
library(raster)
```

[3] This means that if the exponent 'p' is set to a very high value, it might give too much emphasis to the closest points and may over-smooth the surface by making it overly uniform. This can result in the loss of fine-grained details in the interpolated surface.

```
library(leaflet)
library(ggplot2)
library(gstat)

# Read in earthquake data
base_query = "https://earthquake.usgs.gov/fdsnws/event/1/
query?format=csv&"
time_interval = "starttime=1900-01-01&endtime=2020-12-31&"
mag_interval = "minmagnitude=3&"
lat_interval = "maxlatitude=-7.0&minlatitude=-11.0&"
lon_interval = "maxlongitude=128.0&minlongitude=123.0"
query = paste0(base_query, time_interval, mag_interval, lat_
interval, lon_interval)
timor_eq = read_csv(query)

# Plot the earthquakes locations on an interactive map
leaflet(timor_eq) %>%
  addTiles() %>%
  addCircleMarkers(lng = ~longitude, lat = ~latitude,
     popup = paste0("date: ",timor_eq$time,"<br> mag: ",
timor_eq$mag),
     radius = ~exp(mag) * 0.02, color = "red", fillOpacity = 0.3)
```

The resulting plot comprises 3028 earthquakes in the selected time interval and magnitude and is shown in Figure 22.2.

In this example, the interpolation is made for a grid of points that is created, dividing the extent of the area in 100 × 100 cells. For this, the min() and max() functions are used to extract the limits, the seq() function is used to create a sequence of values, and the expand.grid() function to create the grid.

#22-02/01 2 stars

```
# Convert the earthquakes object to a data.frame
timor_eq_df = as.data.frame(timor_eq)

# Convert the earthquakes to a spatial object
coordinates(timor_eq_df) = c("longitude", "latitude")
timor_eq_df$y = timor_eq$latitude
timor_eq_df$x = timor_eq$longitude

# Create the grid
x_seq = seq(min(timor_eq_df$longitude), max(timor_eq_
df$longitude), length.out = 100)
y_seq = seq(min(timor_eq_df$latitude), max(timor_eq_
df$latitude), length.out = 100)
timor_eq_grid = expand.grid(x = x_seq, y = y_seq)

# Make the grid a SpatialPoints object
timor_eq_grid = SpatialPoints(timor_eq_grid)
```

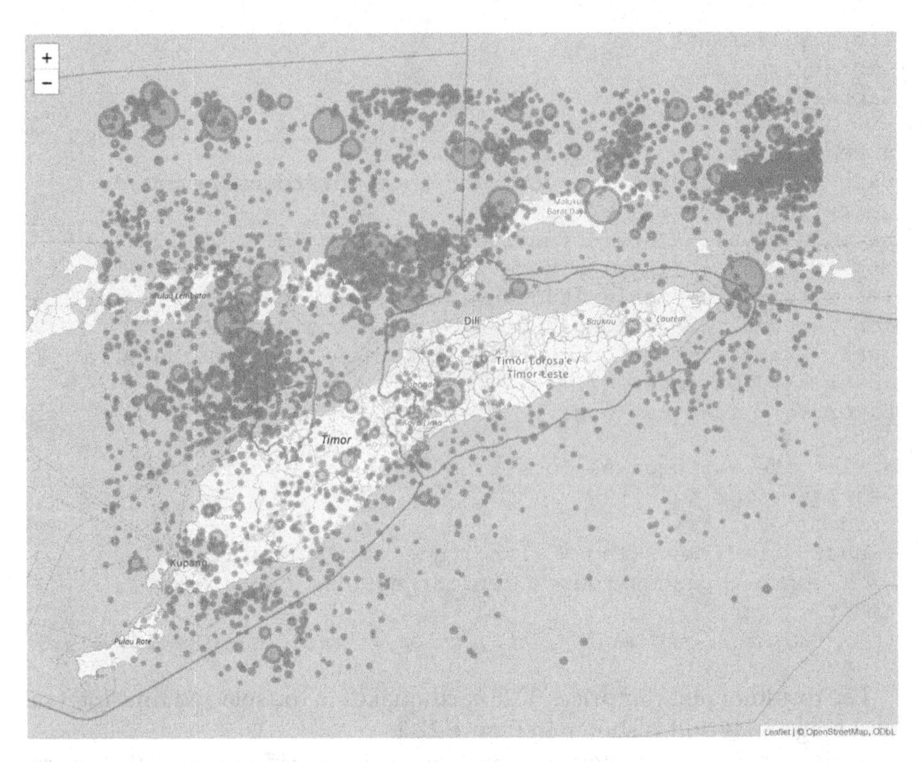

FIGURE 22.2
The earthquakes in the Timor-Alor region.

The grid is converted to a 'SpatialPointsDataFrame' variable that is used as an input to the idw() function. In this case, we will be using the function from the 'gstat' package, and that is why the 'gstat::' is used.

#22-02/02

```
# Perform inverse distance weighting interpolation using idw()
timor_eq_idw = gstat::idw(formula = depth ~ x + y,
                          locations = timor_eq_df,
                          newdata = timor_eq_grid,
                          maxdist = 1, nmin = 2, idp = 2)
```

The idw() function parameters, from the 'gstat' package,[4] can be used to fine-tune the interpolation results.

'maxdist': It controls the maximum distance between the target point and the sampled points that should be included in the interpolation. Increasing this value can lead to longer computation times and

[4] Note the use of gstat::idw() for clarity.

potentially poorer interpolation results due to the inclusion of more distant points.

'nmin': It controls the minimum number of sampled points that must be within the 'maxdist' distance of the target point for the interpolation to be computed. If there are fewer than 'nmin' sampled points within the 'maxdist' distance, the interpolated value will be set to not applicable (NA). Increasing this value can lead to sparser interpolation results.

'idp': It is the IDW exponent (p value) that controls the degree of smoothing in the interpolation. A higher 'idp' value results in a smoother interpolation while a lower value results in a more jagged interpolation that closely follows the sampled points. It is important to note that increasing 'idp' can lead to greater interpolation error if the data is too sparse.

The gridded result is better visualised as a raster. Therefore, it is transformed to a raster with the rasterFromXYZ() function and plotted as a raster.

#22-02/03 2 stars

```
# Create the raster
r = rasterFromXYZ(timor_eq_idw)

# Define the CRS of the raster
crs(r) = CRS("+proj=longlat +datum=WGS84")

# Make the plot
plot(r, main="Timor-Alor earthquakes (idw)")
points(timor_eq_df@coords, cex = 0.4, col="red")
```

The resulting plot is shown in Figure 22.3.

It is readily visible that there is a trend of increasing depth of the earthquakes from the South-Southeast to the North-Northwest, where the deeper earthquakes occur. Naturally, the fitting is not perfect, and the interpolated surface has negative depth earthquakes, i.e. above ground earthquakes in the southern part. This is one example of the boundary effects issues that are raised by the use of IDW. One solution to this might be to clip the raster 'r' to the limits of the earthquakes.

Changing the formula in the idw() function allows the visualisation of other variables. For example, one can try to understand the distribution of the earthquake's magnitudes. In this case, we used a 'nmin = 10' in order to have a higher number of earthquakes included in the evaluation of the magnitude.

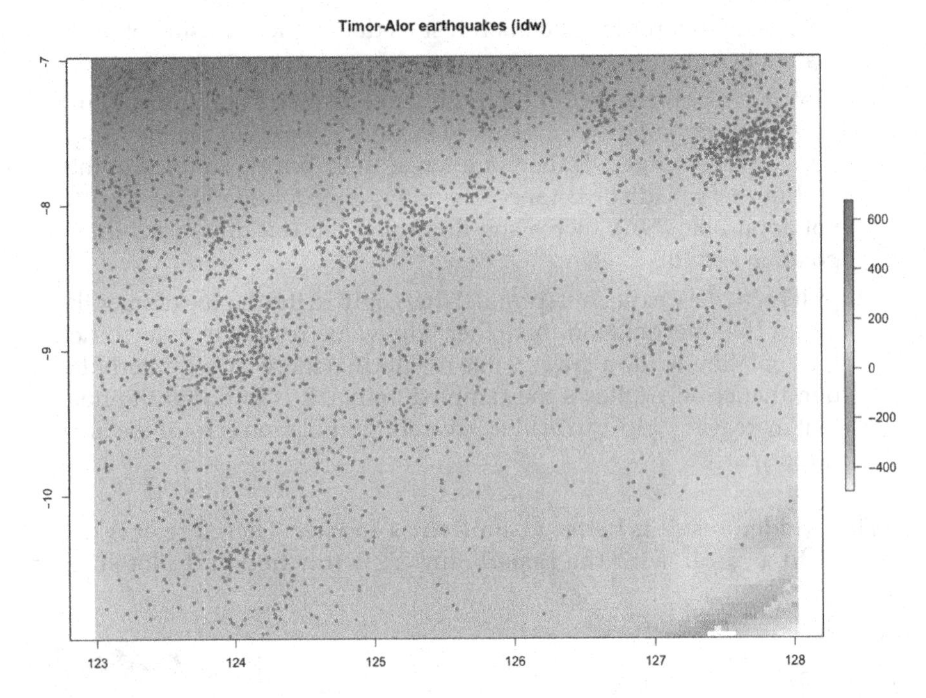

FIGURE 22.3
The resulting plot of the inverse distance weighting (IDW) interpolation.

#22-03/01

```
# Perform inverse distance weighting interpolation using idw()
timor_eq_mag_idw = gstat::idw(formula = mag ~ x + y,
                        locations = timor_eq_df,
                        newdata = timor_eq_grid,
                        maxdist = 1, nmin = 10, idp = 2)
```

The resulting surface is a function of the location (x, y) and magnitude of the earthquakes. For a better visualisation, the pretty() function from the 'raster' package is used to create a visually more appealing colour scheme for the values of the magnitude.

#22-03/02

2 stars

```
# Create the raster
r = rasterFromXYZ(timor_eq_mag_idw)

# Define the CRS of the raster
crs(r) = 4326

# Set pretty breaks
brks = pretty(r[], n = 10)
```

```
# Make the plot
plot(r, breaks= brks, col = terrain.colors(length(brks) - 1),
      main="Timor-Alor Earthquakes magnitude (idw)")
points(timor_eq_df@coords,cex =0.2, col="red")
```

The resulting plot is shown in Figure 22.4.

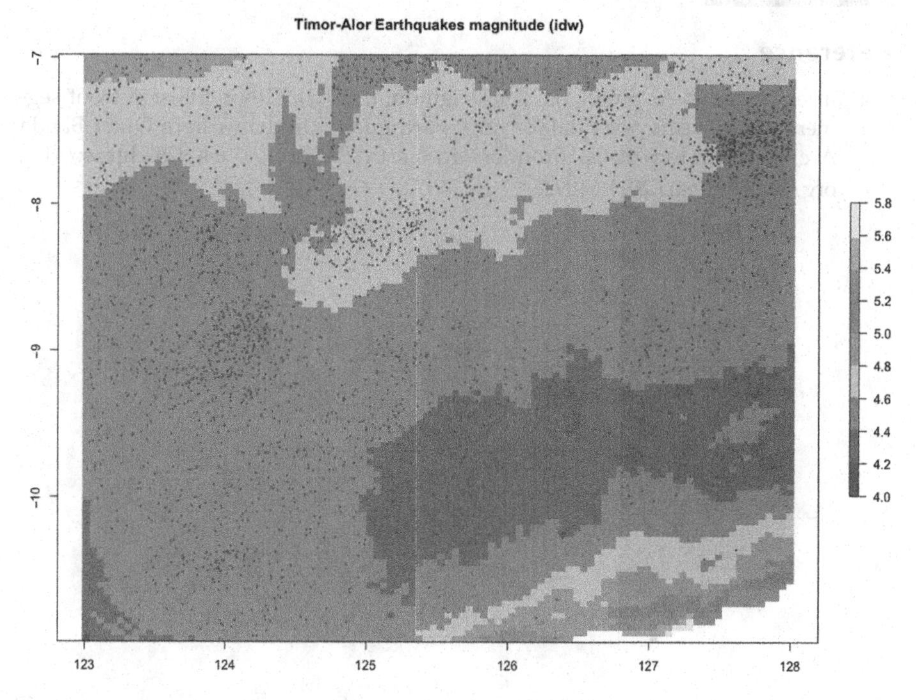

FIGURE 22.4
The magnitude distribution of earthquakes in the Timor-Alor region.

Notice that the pattern is not continuous as in the case of the depth of the earthquakes. In this case, the South coast of East Timor has earthquakes with lower magnitudes (dark grey) than the rest of the area (light grey).

22.2 Concluding Remarks

Interpolations are beautiful! The transition to a higher number of dimensions, marked by the progression from point data to a surface, comprises a venture into uncharted territory. In such instances, it is the expertise and contextual understanding of the analyst that should guide the decisions regarding the parameters to be employed.

IDW is easy to understand and does not imply a lot of parameterising or twitching to obtain usable results. As it is fast and simple, it is always a 'go for it' method before diving into more complex calculations.

Reference

Coudurier-Curveur, A., Singh, S.C., & Deighton, I. (2021). Timor collision front segmentation reveals potential for great earthquakes in the western Outer Banda Arc, Eastern Indonesia. Frontiers in Earth Science, 9, 640928. https://doi.org/10.3389/feart.2021.640928

23

Kriging

Kriging is a geostatistical interpolation method that provides optimal, unbiased predictions of spatially distributed variables based on sampled data. It was first developed by the South African mining engineer Danie Krige[1] and later formalised by the French mathematician Georges Matheron.[2] Kriging has since become a standard tool in geostatistics and is widely used in various geological applications.

The basic principle behind kriging is to use a weighted combination of nearby observations to predict the value of a variable at an unsampled location. The weights are determined based on the spatial autocorrelation structure, as described by the variogram, which quantifies the spatial dependence between pairs of observations as a function of their separation distance (Hengl, 2007; Chilès & Delfiner, 2012).

23.1 Types of Kriging

There are several types of kriging that can be used to model spatial data, including:

Ordinary kriging (OK): This is the most basic type of kriging, which assumes that the mean of the underlying process is constant across the spatial domain. It is used to make predictions at unsampled locations based on the spatial patterns in the observed data.

Simple kriging (SK): It is similar to OK but allows for a linear trend in the mean of the underlying process.

Universal kriging (UK): This type of kriging allows for a more general mean function, including polynomial trends of any degree, and can be used to model complex spatial patterns.

Co-kriging (CK): This type of kriging is used when there are multiple correlated variables, and the goal is to make predictions for one variable based on measurements of the other variables.

[1] https://en.wikipedia.org/wiki/Danie_G._Krige
[2] https://en.wikipedia.org/wiki/Georges_Matheron

DOI: 10.1201/9781032651880-26

These are some of the most commonly used types of kriging. The choice of which type of kriging to use will depend on the characteristics of the data and the geological question being addressed.

As a tool to help on the decision of which method to apply, we suggest the strength, weakness, opportunities, and threats (SWOT) analysis of these four kriging methods presented in Table 23.1.

23.2 Variograms and Semi-Variograms

A variogram is a tool used to assess the spatial continuity of a variable. This spatial continuity refers to the interdependence of adjacent samples within a certain distance and is represented mathematically through a variogram model. The variance between the values of two points at a distance L from each other can demonstrate the correlation between the values. In the presence of spatial structure, it is expected that the dependence between the values of close points is greater than that of distant points. This distance-dependent variance, known as a variogram, can be depicted through $2\gamma(L)$, where $\gamma(L)$ is a semi-variogram (Goovaerts, 1997).

The first step to analyse the spatial data is to create semi-variograms to interpret the spatial variation of data. The semi-variogram is a graphical representation of the spatial autocorrelation of a variable. It shows the relationship between the difference in values of the variable at two different locations as a function of the distance between those locations.

A semi-variogram plot (Figure 23.1) typically consists of a set of points and a smoothed curve that fits the points. Each point represents the average difference in values of the variable for all pairs of observations that are separated by a certain distance. The x-axis of the plot represents the distances between pairs of observations, and the y-axis represents the average difference in values.

Points that are close to each other are likely to be more similar than points that are far apart. The range represents the distance over which points show spatial similarity, beyond which points are no longer spatially related. The nugget represents the random error or uncertainty.

The shape of the semi-variogram curve provides insights into the spatial structure of the variable being studied. For example:

- A semi-variogram that increases rapidly at small distances and then levels off at larger distances suggests a strong spatial autocorrelation at small distances and a relatively weak autocorrelation at larger distances. This could indicate a clustered pattern in the data, where similar values tend to occur close together.
- A semi-variogram that increases slowly or not at all with distance suggests a weak or absent spatial autocorrelation. This could indicate

TABLE 23.1

SWOT Analysis of the Different Kriging Methods

	Strength	Weakness	Opportunities	Threats
Ordinary kriging	Accounts for spatial variability by estimating both the mean and spatially varying variance. Assumes stationarity, making it suitable for situations with no systematic trends. Provides optimal linear unbiased estimates when underlying assumptions are met. Handles missing data gracefully through the estimation process.	Assumes a constant mean, which may not be appropriate for non-stationary data. Sensitive to outliers and extreme values, potentially affecting the interpolation results. Computationally intensive for large datasets due to the estimation of the variogram. Requires a well-defined variogram model, which might be challenging to determine for complex datasets.	Can be combined with auxiliary variables for external drift kriging to improve interpolation results. Continued development of spatial statistical techniques may address some of the limitations.	Inappropriate application on non-stationary datasets may lead to biased results. Computationally demanding nature may limit its use in large-scale applications.
Simple Kriging	Provides optimal linear unbiased estimates when the constant mean is known or accurately estimated. Suitable for situations where the mean is constant across the study area. Handles missing data gracefully through the estimation process.	Assumes a known constant mean, which might not always be available or accurate. Less flexible than other kriging methods in handling spatially varying trends. Sensitive to outliers and extreme values.	Can be combined with auxiliary variables for external drift kriging to improve interpolation results. Potential for improved performance when the constant mean is accurately known.	Inappropriate application on non-stationary datasets or datasets with uncertain mean estimates. Limited suitability for scenarios with spatially varying trends.
Universal kriging	Incorporates a trend model to account for spatially varying trends and non-stationarity. Flexible in capturing complex spatial patterns and trends. Provides optimal linear unbiased estimates when underlying assumptions are met.	Requires the specification of a trend model, which may be challenging in practice. May become computationally intensive for large datasets with complex trends. Sensitive to the choice of trend function, which may impact interpolation results.	Provides improved performance when spatial trends are present and need to be accounted for. Potential for enhanced accuracy and robustness in non-stationary environments.	Mis-specification of the trend model may lead to biased results. Computationally demanding nature may limit its application in large-scale projects.
Co-kriging	Utilises the spatial correlation between multiple variables to improve interpolation. Provides opportunities to combine information from various datasets. Can handle situations where spatially correlated variables are available.	Requires cross-variograms and may become computationally intensive for multiple variables. Sensitive to the spatial relationships between the variables, which may not always be well understood. May introduce additional complexity and uncertainty due to the integration of multiple datasets.	Can enhance interpolation results by leveraging the spatial information from multiple sources. Potential for improved accuracy and reliability in areas with data overlap.	Complex data integration may lead to increased uncertainty and potential biases. Appropriate cross-validation and validation strategies are essential to assess the validity of co-kriging results.

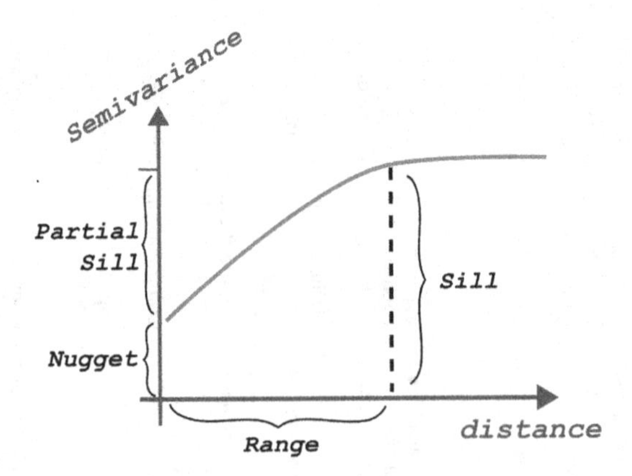

FIGURE 23.1
The elements of a semi-variogram.

a random or uniform pattern in the data, where values are not associated with each other based on their location.

- A semi-variogram that increases rapidly with distance, but does not level off, suggests a strong spatial autocorrelation that extends over a large spatial scale. This could indicate a trend in the data, where values tend to increase or decrease systematically with increasing distance from a certain location.

The semi-variogram graphical shape can also be used to select an appropriate variogram model to be applied in the kriging analysis. For this, the semi-variogram curve is compared to various theoretical variogram models (such as spherical, exponential, and Gaussian) and the best-fitting model is selected based on the shape of the semi-variogram (Figure 23.2). The parameters of the selected model can then be estimated using the semi-variogram points and used as input to the kriging algorithm.

Clustered, random (or uniform), and trended patterns of data are the most commonly found in geological data. For example, when analysing the geochemistry from stream sediments, there can be a nugget effect if one is looking for gold. When evaluating the depth of the earthquakes, a trended surface can be present. In the case of looking at the location of mineral deposits, a clustered pattern can be present.

For illustration purposes, coded examples for generating semi-variograms that correspond to each of these three patterns are present in the following examples.

23.2.1 Clustered Pattern

For creating this type of pattern, we adapted the previous while() loop (snippet #20-01) to create a clustered distribution of points. For starting, 20 points are uniformly distributed in a region of extension min(25, 25) and max(75, 75).

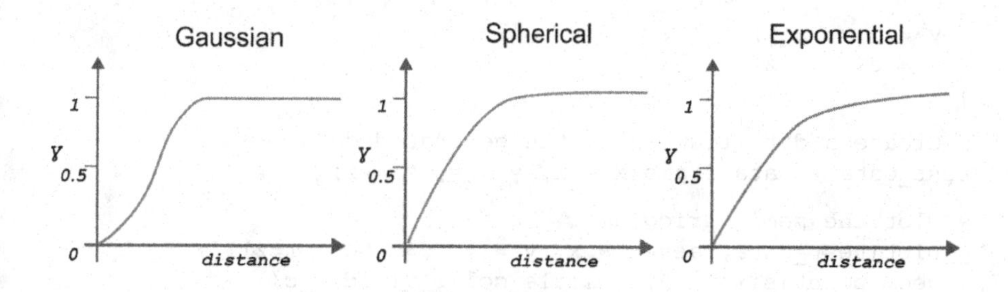

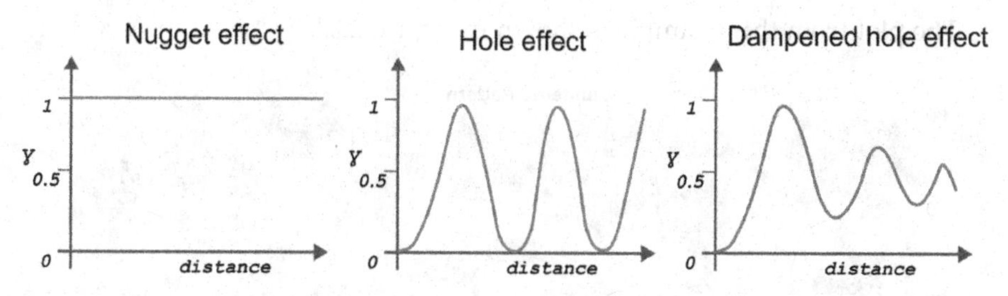

FIGURE 23.2
Theoretical variograms and its possible interpretation.

The 'z' values will vary between 80 and 220 in a random uniform distribution with the runif() function. Then, the loop is used to create the 500 points with variations using the jitter() function. Finally, the data is transformed to 'data.frame' to be plotted.

#23-01

```
# Load the Libraries
library(ggplot2)
library(gstat)
library(sf)

# Generate some artificial data with a clustered pattern
set.seed(123)
x = c(runif(20, min = 25, max = 75))
y = c(runif(20, min = 25, max = 75))
z = c(runif(20, min = 80, max = 220))

# Create 100 clustered points with a uniform distribution
npoint = 20
while(npoint < 500) {
  npoint = npoint + 1
  random_point = sample(1:20, 1)
  p_x = jitter(x[random_point], factor = 2)
  p_y = jitter(y[random_point], factor = 2)
  p_z = jitter(z[random_point], factor = 2)
  x = c(x, p_x)
```

```
  y = c(y, p_y)
  z = c(z, p_z)
}

# Create a data.frame from the generated data
test_data = data.frame(x = x, y = y, z = z)

# Plot the semi-variogram
ggplot(test_data, aes(x = x, y = y, z=z, color=z)) +
  geom_point(size = 2) + scale_color_viridis_c() +
  ggtitle("Clustered Pattern") + xlab("X") + ylab("Y")
```

The plot from this example is shown in Figure 23.3.

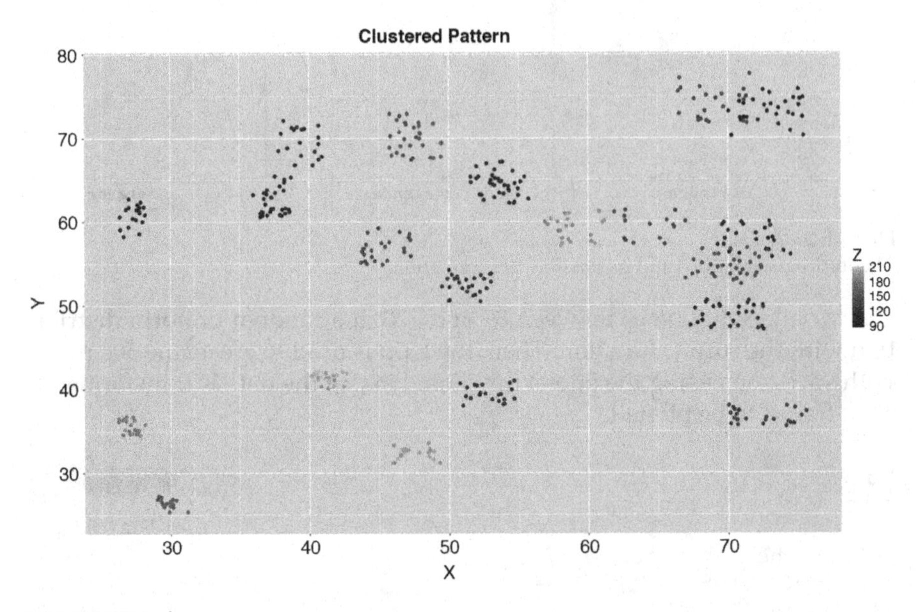

FIGURE 23.3
An example of a clustered pattern of data.

The next step creates a 'sp' object and uses the variogram() function to create the variogram curve. The result is plotted using the geom_smooth() function from the 'ggplot2' package.

#23-02 3 stars

```
# Convert the  data to a spatial object
test_data = st_as_sf(test_data, coords =  c("x", "y"))

# Calculate the semi-variogram
sv = variogram(z ~ 1, data = test_data, width = 0.1, cutoff = 60)

# Plot the semi-variogram
ggplot(sv, aes(x = dist, y = gamma)) +
```

```
geom_point() + geom_smooth() +
ggtitle("Semi-variogram for Clustered Pattern") +
xlab("Distance") + ylab("Semivariance")
```

The resulting semi-variogram is shown in Figure 23.4. As expected, it has a step growth in the lower distances and tends to stabilise at higher distances; this is an indication of a clustered pattern. In this case, the smoothed curve is under-fitted and therefore should not be considered for interpretation purposes.

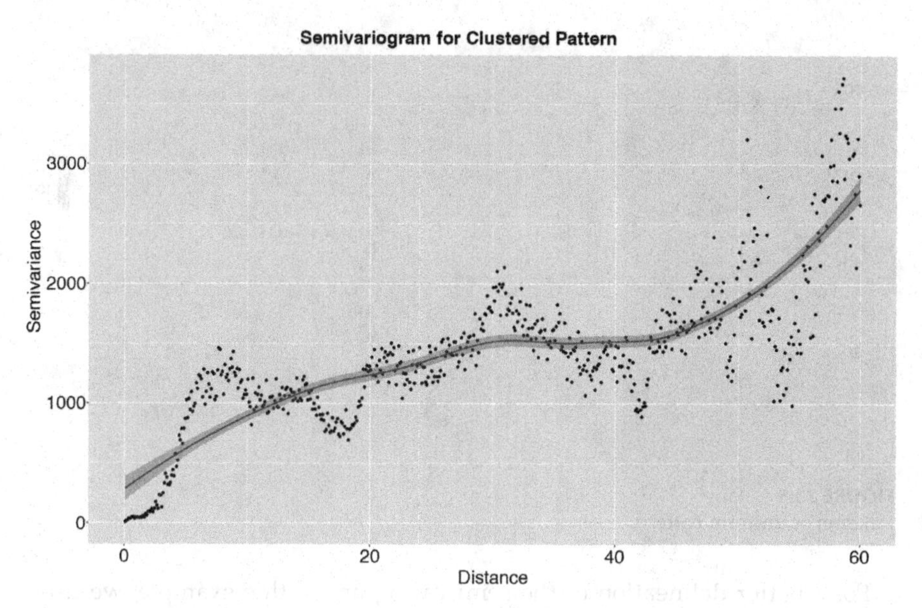

FIGURE 23.4
The semi-variogram of a clustered pattern.

23.2.2 Random or Uniform Pattern

A random or uniform pattern of 500 points is created in the following example. The 'z' values vary with a mean of 150 and a standard deviation of 70.

#23-03

```
# Generate some artificial data with a random or uniform pattern
set.seed(123)
x = runif(500, min = 25, max = 75)
y = runif(500, min = 25, max = 75)
z = rnorm(500, mean = 150, sd = 70)

# Create a data.frame from the generated data
test_data = data.frame(x = x, y = y, z = z)
```

```
# Plot the data
ggplot(test_data, aes(x = x, y = y, z=z, color=z)) +
  geom_point(size = 2) + scale_color_viridis_c() +
  ggtitle("Random Pattern") + xlab("X") + ylab("Y")
```

The result is illustrated in Figure 23.5.

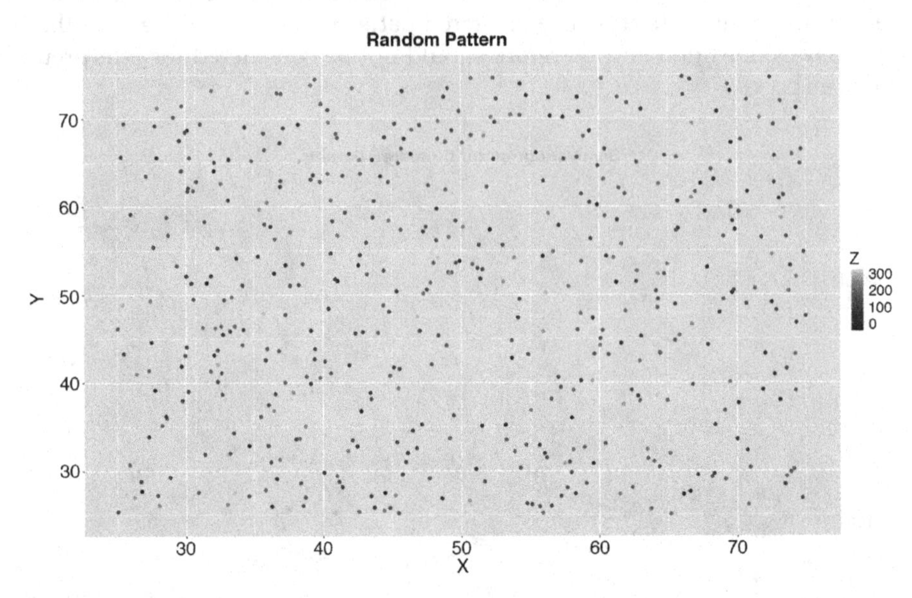

FIGURE 23.5
Random or uniform pattern.

For a better delineation of the semi-variogram in this example, we use a width of 0.5; that is, the distances are calculated at every 0.5 units.

#23-04 3 stars

```
# Convert the  data to a spatial object
test_data = st_as_sf(test_data, coords =  c("x", "y"))

# Calculate the semi-variogram
sv = variogram(z ~ 1, data = test_data, width = 0.1, cutoff = 60)

# Plot the semi-variogram
ggplot(sv, aes(x = dist, y = gamma)) +
  geom_point() + geom_smooth() +
  ggtitle("Semi-variogram for Random or Uniform Pattern") +
  xlab("Distance") + ylab("Semivariance")
```

The resulting semi-variogram is a line that does not increase and fluctuates with the increasing distance. This is because of the random data, there is very little auto correlation, so the range is very small and the sill is reached

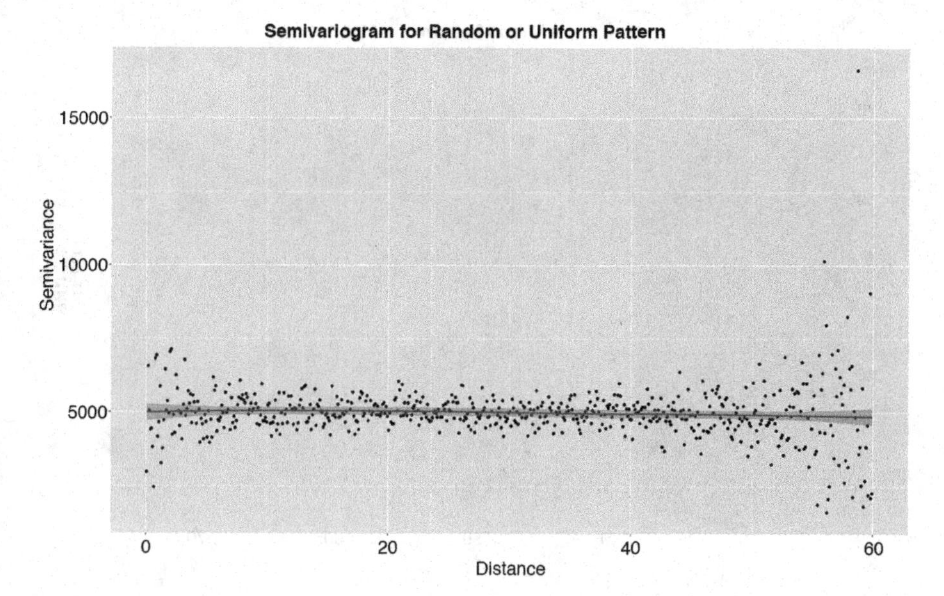

FIGURE 23.6
Semi-variogram for a uniform or random distribution.

very quickly. This is what is expected for a uniform or random distribution of data. The resulting plot is shown in Figure 23.6.

23.2.3 Trend Pattern

This example creates a uniform distribution of points, and the 'z' value increases with x and y, i.e. from bottom left to top right. For this example, a mean of 10 and a standard variation of 2 for the 'z' values is used.

#23-05 3 stars

```
# Trend Pattern
# Generate some artificial data with a trend pattern
set.seed(123)
x = runif(500, min = 25, max = 75)
y = runif(500, min = 25, max = 75)
z = x + y + rnorm(500, mean = 10, sd = 2)

# Create a data.frame from the generated data
test_data = data.frame(x = x, y = y, z = z)

# Plot the data
ggplot(test_data, aes(x = x, y = y, z=z, color=z)) +
  geom_point(size = 2) + scale_color_viridis_c() +
  ggtitle("Trend Pattern") + xlab("X") + ylab("Y")
```

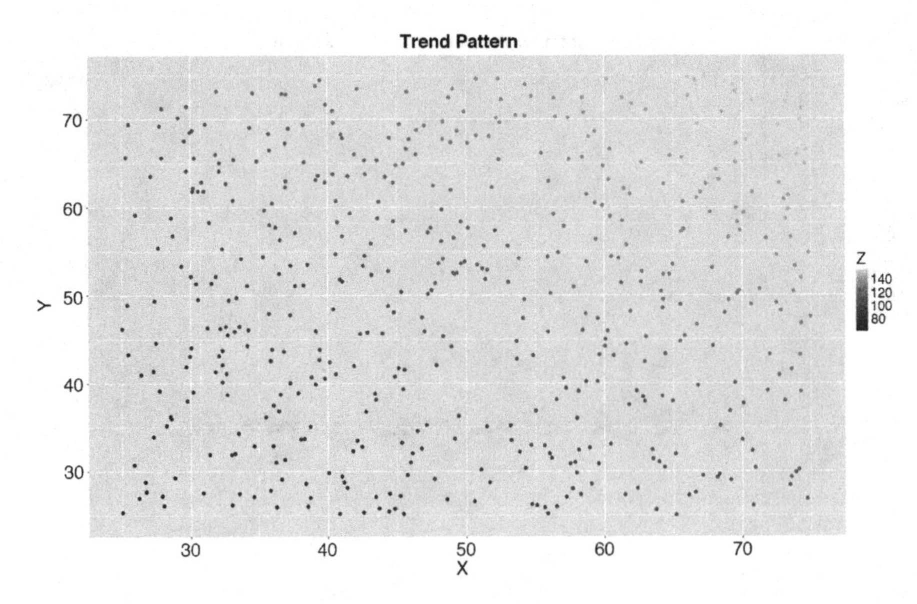

FIGURE 23.7
The trend pattern of data.

The obtained data pattern is shown in Figure 23.7.

For this example, a width of 0.1 and a cut-off at a distance of 60 are used to a better visualisation of the semi-variogram behaviour.

#23-06

3 stars

```
# Convert the data to a spatial object
test_data = st_as_sf(test_data, coords =  c("x", "y"))

# Calculate the semi-variogram
sv = variogram(z ~ 1, data = test_data, width = 0.1, cutoff = 60)

# Plot the semi-variogram
ggplot(sv, aes(x = dist, y = gamma)) +
  geom_point() + geom_smooth() +
  ggtitle("Semi-variogram for Trend Pattern") +
  xlab("Distance") + ylab("Semivariance")
```

The resulting semi-variogram is shown in Figure 23.8.

Notice that the values increase in small distances but do not level up, an indication of a trended pattern of data.

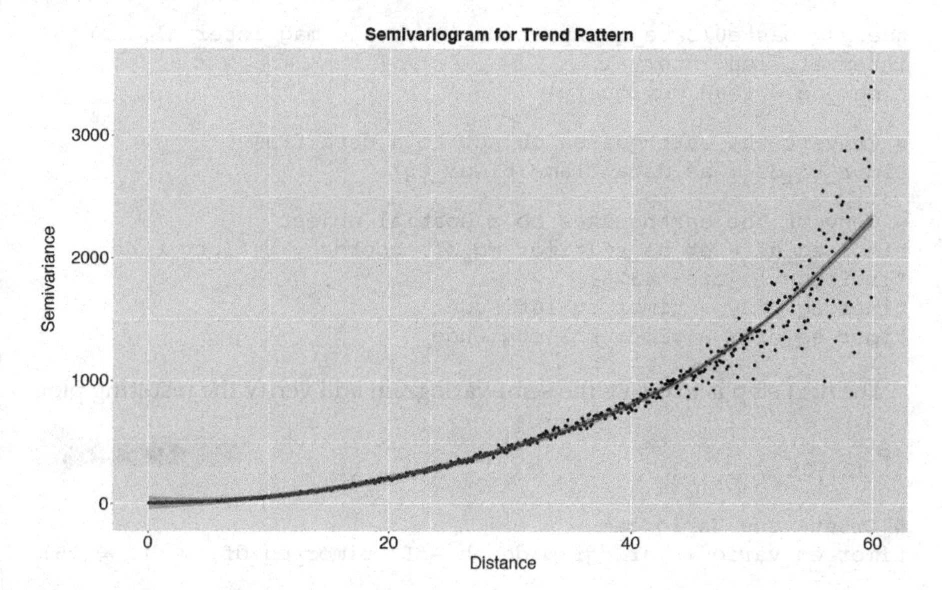

FIGURE 23.8
Semi-variogram for a trend pattern of data.

23.3 Kriging Earthquakes

Let us return to the code snippet of the study of the earthquakes in the Timor-Alor region. For reading the data and creating the variables, the following code is used (code snippet #22-01 and #22-02/01).

#23-07

```
# Load the libraries
library(readr)
library(sf)
library(raster)
library(leaflet)
library(ggplot2)
library(gstat)

# Read in earthquake data
base_query = "https://earthquake.usgs.gov/fdsnws/event/1/
query?format=csv&"
time_interval = "starttime=1900-01-01&endtime=2020-12-31&"
mag_interval = "minmagnitude=3&"
lat_interval = "maxlatitude=-7.0&minlatitude=-11.0&"
lon_interval = "maxlongitude=128.0&minlongitude=123.0"
```

```
query = paste0(base_query, time_interval, mag_interval, lat_
interval, lon_interval)
timor_eq = read_csv(query)

# Convert the earthquakes object to a data.frame
timor_eq_df = as.data.frame(timor_eq)

# Convert the earthquakes to a spatial object
timor_eq_df = st_as_sf(timor_eq_df, coords = c("longitude",
"latitude"), crs= 4326)
timor_eq_df$y = timor_eq$latitude
timor_eq_df$x = timor_eq$longitude
```

The next step is to create the semi-variogram and verify the resulting plot.

#23-08

4 stars

```
# Create the Variogram
timor_eq_vario = variogram(depth ~ 1, timor_eq_df, width = .5)

# Plot the semi-variogram and fitted curve
ggplot(data.frame(timor_eq_vario), aes(x = dist, y = gamma)) +
  geom_smooth() + geom_point() +
  xlab("Distance") + ylab("Semivariance") +
  ggtitle("Semi-variogram of EQ in the Timor-Alor region")
```

Figure 23.9 shows the results for the semi-variogram.

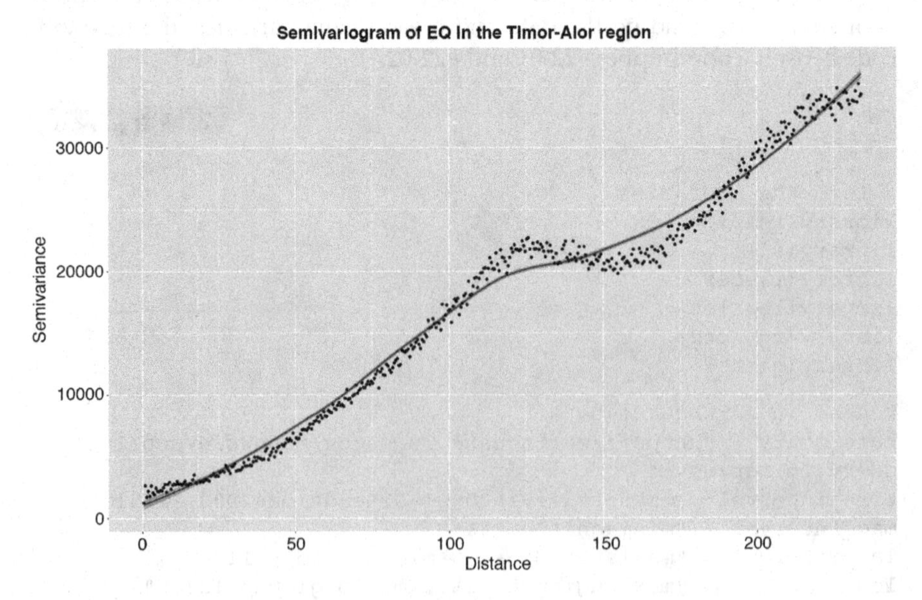

FIGURE 23.9
Semi-variogram obtained for the earthquakes in the Timor-Alor region.

The values increase rapidly but don't reach a sill, i.e. they don't level, indicating that there is a clear trend in the data (cf. Figure 23.8).

Based on the shape of the semi-variogram plot, different kriging methods can be chosen (cf. Section 23.1 and Table 23.1). For example, if the semi-variogram plot shows a linear relationship with a clear sill, SK may be the most appropriate method. If the semi-variogram plot is more complex and does not exhibit a clear sill, OK may be more appropriate. UK and CK may be used when there are additional covariates or when the data exhibits nonstationary, such as trend or drift.

Hence, for this example, two kriging models will be tested: a) the OK and the UK.

23.3.1 Ordinary Kriging

For OK, no model is necessary and the krige() function is used directly, specifying the relation between the variables.

#23-09

```
# Temporary variable for prediction
timor_df_new = timor_eq_df

# Fit an ordinary kriging model to the Timor-Alor Earthquakes
timor_eq_ok_pred =  krige(depth ~ y + x, timor_eq_df, newdata
= timor_df_new)

# Rename the variables from the Ordinary Kriging
names(timor_eq_ok_pred) = c("depth", "var1", "geometry" )

# Plot the semi-variogram and fitted curve
ggplot(timor_eq_ok_pred, aes(color = depth)) +
  geom_sf() + scale_color_viridis_b() +
  xlab("Longitude") + ylab("Latitude") +
  ggtitle("Predicted Depth Ordinary Kriging")
```

The resulting plot is shown in Figure 23.10.

Notice that the trend of deepening earthquakes to the North-Northwest direction is quite clear. However, there are earthquakes with negative depth, i.e. above the ground!!! In this case, the kriging is not perfect.

23.3.2 Universal Kriging

UK uses a model for predicting the data values. For this model, the vgm() function is used to create a variogram model object, which is then passed to

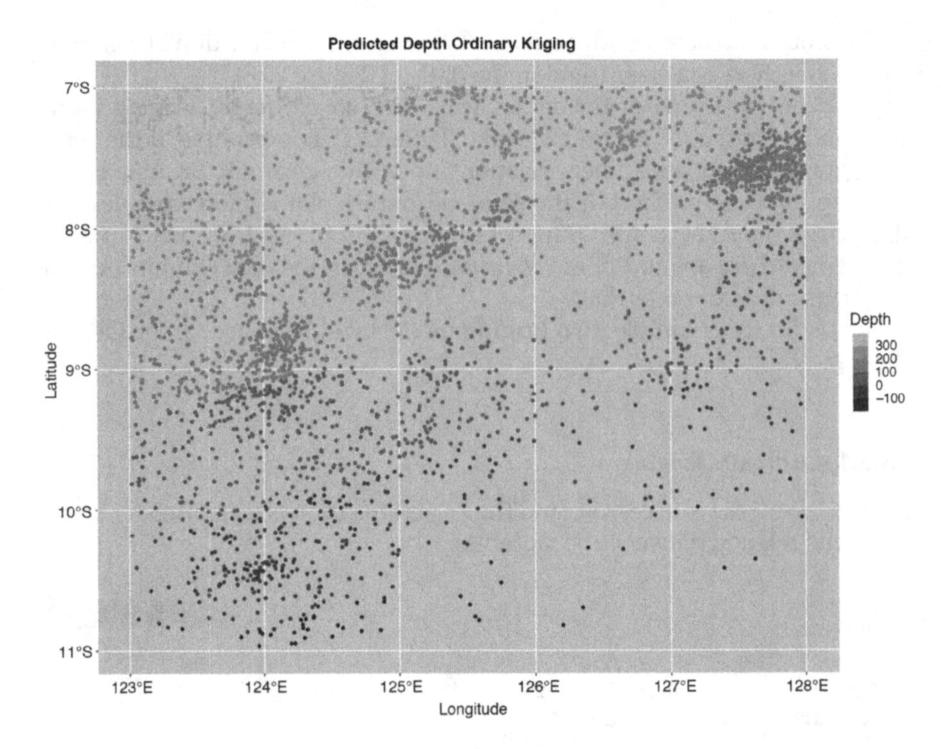

FIGURE 23.10
The prediction of depth based on ordinary kriging.

the krige() function to describe the spatial structure of the data. The parameters of vgm() used are as follows (cf. Figure 23.1):

'psill': It is the partial sill, which represents the difference between the total variance of the data and the nugget effect. The partial sill describes the spatial dependence of the data and determines the strength of the spatial correlation.

'model': It is the type of variogram model to be fit to the data. The options include "Exp", "Gau", "Spher", "Lin", and "Mat". These models describe different shapes of the variogram, and the appropriate model depends on the spatial structure of the data.

'range': It is the range of the variogram model, which represents the distance at which the spatial autocorrelation of the data is negligible. The range is used to determine the bandwidth of the kriging weights and has a significant impact on the accuracy of the prediction.

'nugget': This is the nugget effect, which represents the amount of variance in the data that is not explained by spatial dependence. The nugget effect accounts for measurement error, small-scale variability, and other sources of spatial heterogeneity.

The code for the UK is quite similar, but the vgm() function is introduced.

#23-10

4 stars

```
# Create the Variogram Model
timor_eq_vario_model = vgm(psill = 0.1, model = "Exp", range =
1, nugget = 0)

# Temporary variable for prediction
timor_df_new = timor_eq_df

# Fit an simple kriging model to the Timor-Alor Earthquake data
timor_eq_uk = krige(depth ~ 1 + y + x,
                       timor_eq_df, newdata = timor_df_new,
                       model = timor_eq_vario_model)

# Rename the variables from the Ordinary Kriging
names(timor_eq_uk) = c("depth", "var1", "geometry" )

# Plot the semi-variogram and fitted curve
ggplot(timor_eq_uk, aes(color = depth)) +
  geom_sf() + scale_color_viridis_b() +
  xlab("Longitude") + ylab("Latitude") +
  ggtitle("Predicted Depth Universal Kriging")
```

The resulting prediction plot is shown in Figure 23.11.

In this case, the results have a more meaningful result, without values above ground and the prediction better suited to what is expected.

23.4 Creating the Rasters

A kriged version of the depth map can be created using a grid for interpolation of the depth values. This grid can be transformed to a raster and plotted or exported to a file.

As an alternative to have the grid transformed to a 'SpatialPointsDataFrame' as in Chapter 22, this time we will transform the grid to a 'sf' object, called 'timor_grid_sf'. The 'new_data' parameter has to have an object that has the same variables as the prediction ones in the formula, i.e. 'x' and 'y'.

The resulting prediction named 'timor_grid_ok' is converted to a raster using the rasterFromXYZ() function.

In this example, we create a leaflet map to display the resulting raster.

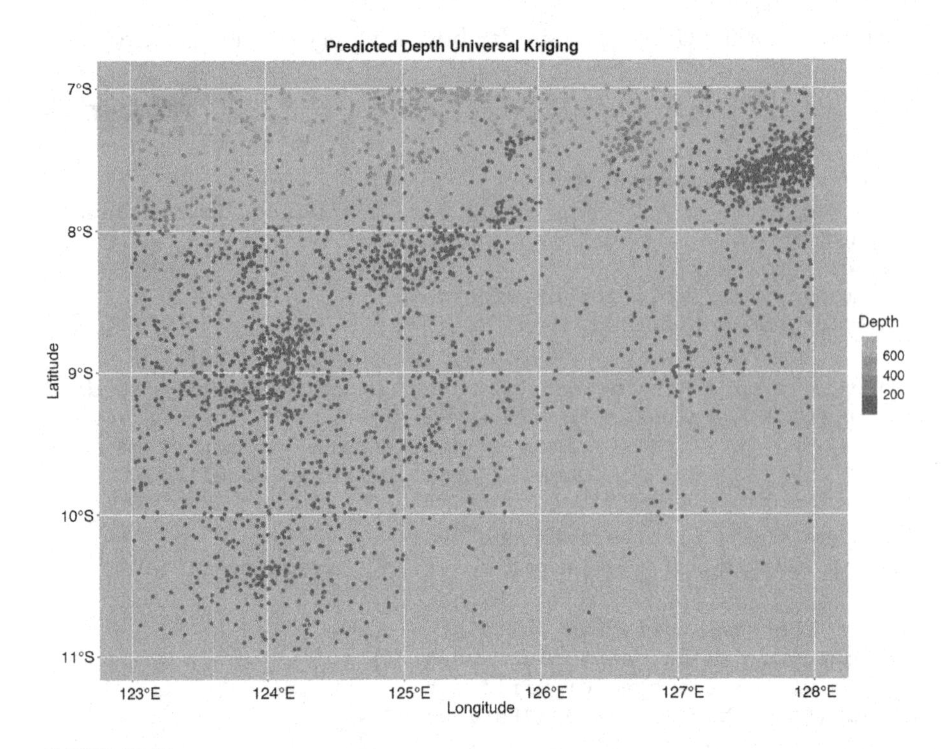

FIGURE 23.11
The Universal Kriging model for the Timor-Alor earthquakes.

#23-11

3 stars

```
# Create the grid
x_seq = seq(min(timor_eq_df$x), max(timor_eq_df$x), length.out
= 100)
y_seq = seq(min(timor_eq_df$y), max(timor_eq_df$y), length.out
= 100)
timor_grid = expand.grid(x = x_seq, y = y_seq)

# Create the sf grid
timor_grid_sf = st_as_sf(timor_grid, coords = c("x", "y"), crs
= st_crs(timor_eq_df))
timor_grid_sf$x = timor_grid$x
timor_grid_sf$y = timor_grid$y

# Fit an ordinary kriging model to the Timor-Alor Earthquake data
timor_grid_ok = krige(depth ~ y + x, timor_eq_df, newdata =
timor_grid_sf)

# Extract x, y, z coordinates
coordinates = st_coordinates(timor_grid_ok)
```

```
grid_x = coordinates[, "X"]
grid_y = coordinates[, "Y"]
grid_z = timor_grid_ok$var1.pred

# Create a prediction data frame
timor_prediction_grid = data.frame(x = grid_x, y = grid_y, z =
grid_z)

# Transform the grid to a raster
r_ok = rasterFromXYZ(timor_prediction_grid, crs = 4326)

# Plot the raster in a leaflet map
leaflet(timor_eq_df) %>%
  addTiles() %>%
  addRasterImage(r_ok, group ="Interpolation", opacity = 0.7)
%>%
  addCircleMarkers(group = "Earthquakes",
                   data = timor_eq_df,
                   lng = ~st_coordinates(timor_eq_df)[, "X"],
                   lat = ~st_coordinates(timor_eq_df)[, "Y"],
                   color = "black", radius = 0.5, fillOpacity =
0.7) %>%
  addLegend(group ="Legend", pal = colorNumeric(palette =
"RdYlGn", domain = values(r_ok)), values = values(r_ok), title
= "Depth (km)", position = "bottomright") %>%
  addScaleBar(position = "bottomleft", options = scaleBarOptio
ns(imperial=FALSE)) %>%
  addLayersControl(overlayGroups = c("Interpolation",
"Earthquakes", "Legend"))
```

The result displayed on top of a leaflet map is presented in Figure 23.12.

This data does have a trend from southeast (SE) to northwest and therefore does not comply with the assumptions for the OK to have a stationary mean. That is why in the SE part of the image the earthquakes are above the surface.

The same procedure can be used to create the raster with the UK method.

This type a variogram model is created considering an exponential model 'model = "Exp"' a sill of 0.1 'psill = 0.1', a range of 5. These parameters can be fine-tuned to improve the results.

The following code illustrates the creation of a UK raster.

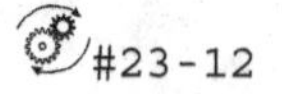
#23-12

```
# Create the grid
x_seq = seq(min(timor_eq_df$x), max(timor_eq_df$x), length.out
= 100)
y_seq = seq(min(timor_eq_df$y), max(timor_eq_df$y), length.out
= 100)
timor_grid = expand.grid(x = x_seq, y = y_seq)
```

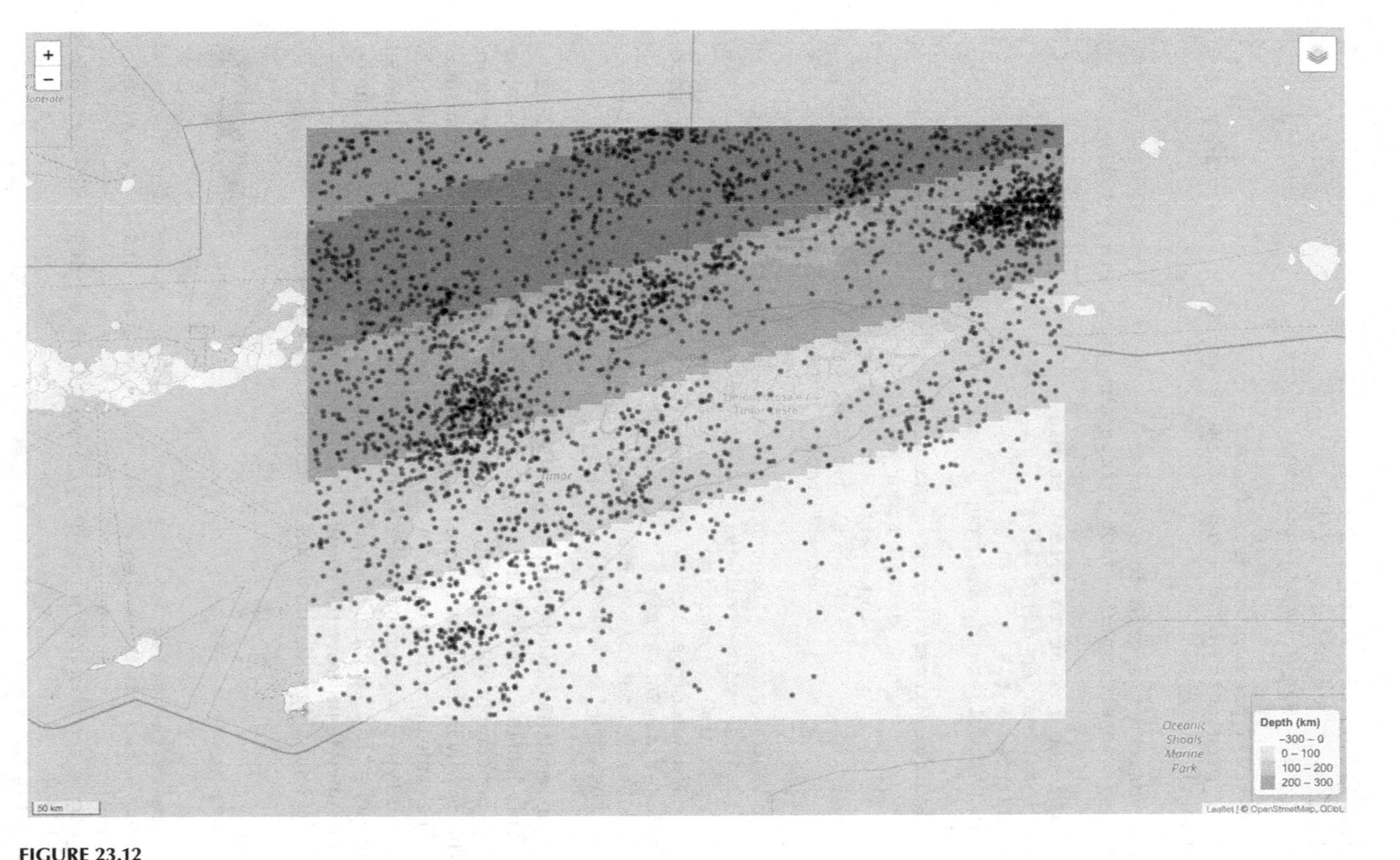

FIGURE 23.12
Depth map for earthquakes based on the ordinary kriging for the Timor-Alor region.

```
# Create the sf grid
timor_grid_sf = st_as_sf(timor_grid, coords = c("x", "y"), crs
= st_crs(timor_eq_df))
timor_grid_sf$x = timor_grid$x
timor_grid_sf$y = timor_grid$y

# Create the Variogram Model
timor_eq_vario_model = vgm(psill = 0.1, model = "Exp", range = 5)

# Fit an ordinary kriging model to the data
timor_grid_uk = krige(depth ~ y + x, timor_eq_df, newdata =
timor_grid_sf, model = timor_eq_vario_model)

# Extract x, y, z coordinates
coordinates = st_coordinates(timor_grid_uk)
grid_x = coordinates[, "X"]
grid_y = coordinates[, "Y"]
grid_z = timor_grid_uk$var1.pred

# Create a prediction data frame
timor_prediction_grid = data.frame(x = grid_x, y = grid_y, z =
grid_z)

# Transform the grid to a raster
r_uk = rasterFromXYZ(timor_prediction_grid, crs = 4326)

# Plot the raster in a leaflet map
leaflet(timor_eq_df) %>%
  addTiles() %>%
  addRasterImage(r_uk, group ="Interpolation", opacity = 0.7)
%>%
  addCircleMarkers(group = "Earthquakes",
                   data = timor_eq_df,
                   lng = ~st_coordinates(timor_eq_df)[, "X"],
                   lat = ~st_coordinates(timor_eq_df)[, "Y"],
                   color = "black", radius = 0.5, fillOpacity =
0.7) %>%
  addLegend(group ="Legend", pal = colorNumeric(palette =
"RdYlGn", domain = values(r_uk)), values = values(r_uk), title
= "Depth (km)", position = "bottomright") %>%
  addScaleBar(position = "bottomleft", options = scaleBarOptio
ns(imperial=FALSE)) %>%
  addLayersControl(overlayGroups = c("Interpolation",
"Earthquakes", "Legend"))
```

Figure 23.13 shows the results, with the earthquakes plotted to better observe the subtleties of the model, that better capture the variations of the data.

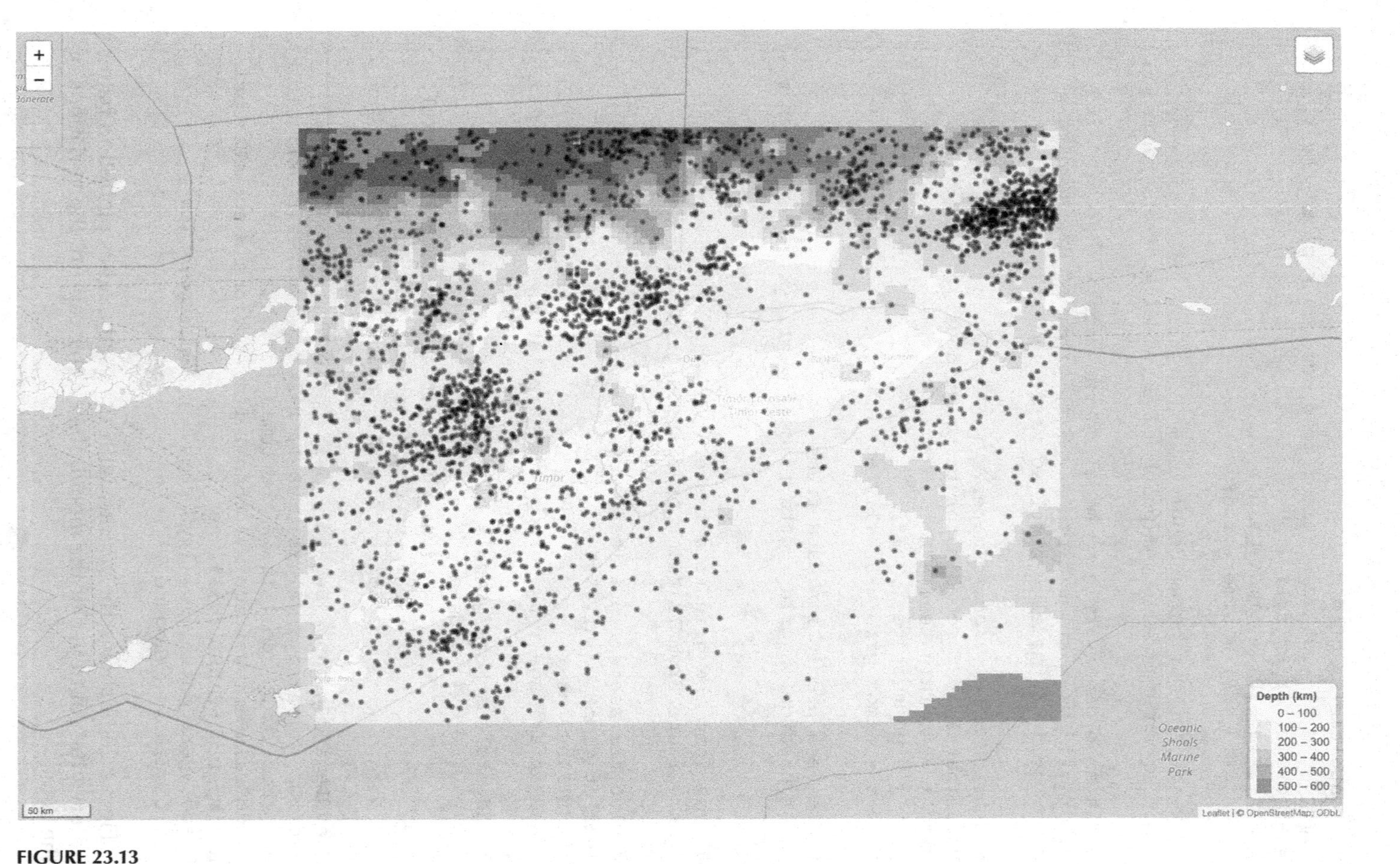

FIGURE 23.13
Depth map for earthquakes based on the simple kriging for the Timor-Alor region.

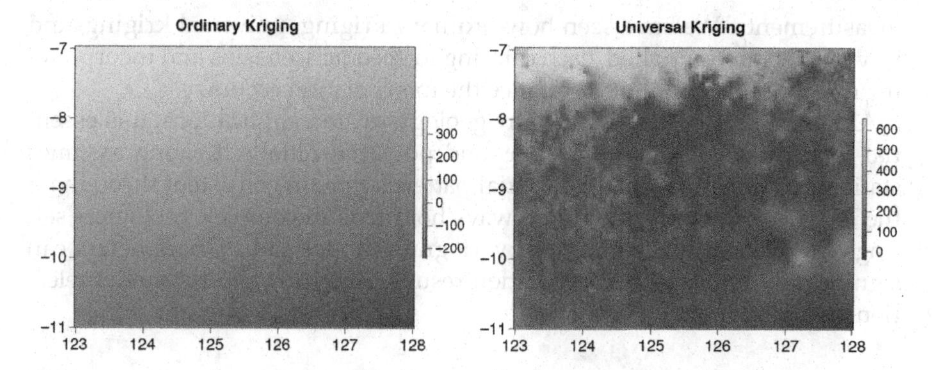

FIGURE 23.14
Comparison of the depth map results for both kriging methods.

As a final exercise, one can compare both ordinary and universal kriged maps by plotting them side by side.

Notice that the OK presents a very simple solution, similar to the one from the inverse distance weighted example (Chapter 22). The UK solution is more complex and captures better the structure of the data.

#23-13

```
# Compare both plots
par(mfrow = c(1, 2))
plot(r_ok, main = "Ordinary kriging")
plot(r_uk, main = "Universal kriging")
```

Figure 23.14 shows the results.

23.5 Concluding Remarks

Throughout this chapter, we explored examples of applications of kriging in geology, namely in its application for the outlining the earthquake pattern in a region, but its possible applications are much more far-reaching and range from mineral exploration to environmental monitoring.

One of the key strengths of kriging lies in its ability to account for spatial autocorrelation, taking into consideration the underlying spatial structure of geological data. By capturing the spatial dependence between data points, kriging provides reliable estimates at unsampled locations and produces continuous surfaces that reflect the natural geological variations.

Moreover, the versatility of kriging allows it to handle different geological data types, such as point data, spatial grids, or irregularly spaced measurements. We have seen how ordinary kriging, universal kriging, and its variants can be applied, each catering to specific scenarios and incorporating external information to enhance the interpolation accuracy.

Although kriging excels in many geology-related applications, it is essential to acknowledge its limitations and potential pitfalls. Kriging assumes stationarity, meaning that the spatial patterns remain consistent throughout the study area, which may not always hold true in complex geological settings. Additionally, the choice of variogram model and its parameters can significantly impact the interpolation results, requiring careful model selection and validation.

References

Chilès, J. & Delfiner, P. (2012). Geostatistics: Modelling Spatial Uncertainty. John Wiley & Sons, p. 699.

Goovaerts, P. (1997). Geostatistics for Natural Resources Evaluation. Applied Geostatistics Series. Oxford University Press, New York, Oxford, pp. 496.

Hengl, T. (2007). A Practical Guide to Geostatistical Mapping of Environmental Variables. Publications Office, Luxembourg. OCLC: 758643236.

24

The Way Forward

If you arrived here by reading all the chapters – congrats – I am sure it has been a long journey. If you arrived here by jumping between parts of the book, just following your interests – congrats – the job is done.

This is where I will pour some thoughts on what to do next. I follow the Niels Bohr vision that once said as follows:

> *Prediction is very difficult, especially if it's about the future.*

As a geology professor with more than 20 years trying to convince colleagues and students that using computers is a good idea, I designed this book as a handbook for geologists that dare to go deeper than word processing or spreadsheet use. The three parts division is meant to introduce R in Part I, with some computer programming insights, an introduction to spatial analysis in Part II, with familiar parts to who already use geographic information system (GIS), and a more advanced daily geologist problem solving in Part III, dealing with point pattern analysis and interpolations, which require some more advanced understanding of data science and geostatistics.

The progressive difficulty of the subjects presented combined with natural pedagogical concerns, hopefully, will primarily make this book a valuable instrument in any graduate course of spatial analysis or advanced GIS for earth sciences. Furthermore, Part I that introduces the R programming language can also be used in undergraduate courses in geology to provide guidance in the R language and in computer programming lessons. The book includes three comprehensive chapters (8, 13, and 18) that equip readers with in-depth examples, facilitating the integration of acquired knowledge. These illustrative examples not only reinforce the concepts but also introduce novel functions and applications, nurturing a more complete understanding of spatial science applied to geology.

Geology, or broadly the earth sciences, is definitely transmuting from a natural history perspective to more data-oriented visions. This strengthens the necessity of geologists to become more and more acquainted with the tools of data science. The perspective of a data science approach to geological, geochemical, or mineralogical problems was also in my mind when presenting the examples of the book. Firstly, the attention of the reader is drawn to the paramount necessity to have a convenient dataset to work with by cleaning, substituting, and subsetting data. The subsequent step implies a detailed

DOI: 10.1201/9781032651880-27

examination of the data through exploratory data analysis in a 'know your data' phase. The text, intentionally, does not provide any profound interpretation of the results obtained in the case studies, leaving that for the specialists in each field subject or region.

Science as a whole is also suffering a transformation from a realm of scarce, hard to access, and expensive data to free, big, and open-access data. This book benefits and explores this movement. All of the examples presented are built upon data that is freely available. This is a Janus-faced question because some of these servers may undergo updates or modifications, eventually necessitating adjustments to the examples to accommodate the new circumstances. However, the open science movement follows a unidirectional path, and the necessary adaptations will naturally be available on the companion website of this book.

Another ambivalence rises from the usage of certain packages and not others. R has a great community of contributors who are always available to help and share knowledge, and many of its greatest packages rely on this tribe. However, as in any dynamic community, things are always changing entailing the necessity to adapt. Throughout the writing process, some of my preferred packages underwent updates, while others were deprecated or discontinued. While initially frustrating, these changes necessitated adaptations and modifications to existing code, ultimately fostering a deeper understanding of the code and leading to a more refined outcome.

It is my aspiration that the topics explored in this book will serve as a springboard for new frontiers in spatial science, empowering geologists with innovative tools for ground-breaking investigations. However, novel methodologies are emerging, primarily involving big data, machine learning, and artificial or spatial intelligence. Indeed, these are the directions I am pursuing in the coming years. I anticipate our paths will converge there.

Index

Note: Locators in *italics* represent figures; **bold** indicate tables in the text. Page numbers followed by 'n' denote notes.

A

Akbar, D., 384
Analysis of variance (ANOVA), 242, 251
Apply family operators, **16**
apply() functions, 66–69
Arendt, C., 36
Arithmetic operators, 14, **15**
Array, defined, 12
as.*() functions, 50
Aspect, 291, 293–294
assignment operators, 14, **15**
ASTER (Advanced Spaceborne Thermal Emission and Reflection Radiometer), 301n7

B

Band combinations, 327–329
  geological features, 329
  IR false colour, 327
  short-wave infrared (SWIR) spectral bands, 328
  SWIR (B12, B8, and B4), 328
  vegetation, mapping and monitoring, 328
Bare Soil Index (BSI), 331, *334*
Bathymetry, retrieving, 266–270
Bivand, R., 136, 151
Bivariate spline, 384
Bonté, D., 385

C

Cadastral data, 160
Canonical correlation analysis (CCA), 120
Cartesian coordinate system, 147
Categorical grid, 258
Central tendency, 74–75
Chlorophyll Vegetation Index (CVI), 331, *334*
Cluster analysis, 120
colnames() function, 111
colorFactor() function, 232
Comma-separated values (CSV) files, 27–28
  reading, 27–28
Comparison operators, 14, **15**
Complementary-to-basic functions, 70
Console panel, 9
Continuous grid, 258
Coordinate reference systems (CRS), 31, 136, 145–147, 164, 178, 185, 190, 194, 196, 197, 202, 216, 224, 226, 280, 326
  defining, 193–194
Coordinate system, 145, 158
Copernicus Data Space Ecosystem, 301
corrgram() function, 124–126
'corrgram' package, 122, 124
CRAN (Comprehensive R Archive Network), 19, 23, 24, 167
  overview, 21–22

D

Data, 100
  attributes() function, 40
  Bureau of Mineral Resources Subset, 109–113
  class() function, 38–39
  cleaning and manipulation, 38–59
  in CSV files, 27–28
  insights, 102–104
  knowing, 38–40
  length() function, 39–40
  other types, 37
  plotting, 107–109
  reading, 27–37, 100–102
  reshaping, 56–59
  retrieving from excel files, 35–36
  retrieving from web server, 33–35

retrieving from website, 28–33
str() function, 39
subsetting, 40–41, 104–106 (*see also* subsetting data)
transforming, 106–107
typeof() function, 39
writing, 27–37
Data analysis, 3, 4
popular packages for, 22–23
Data analytics
apply() functions, 66–69
arrays and vectors, 61
correlating variables, 79–80
creating functions, 69–70
data frame-related functions, 64–66
functions, 61–70
gsub() function, 63
inferential statistics, 80–89
paste0() function, 62–63
paste() function, 62–63
sort() function, 61–62
substr() function, 63
unique() function, 62
Data cleaning, 59
Data frame-related functions, 64–66
colnames() and rownames() functions, 65–66
head() function, 64–65
ncol() function, 64
nrow() function, 64
tail() function, 65
Data frames, 12, *13*
Data structures, 10–14, *11*, 14
Data visualisation, 91–99
bar plots, 91, *92*
box plots, 93, *94*
histograms, 92, *93*
layout() function, 97, *98*
par() function, 96
plots, organising, 95–97
popular packages for, 22–23
save plots, 97–99
scatter plots, 94–95
Datum-based coordinate system, 146
'.dbf' extension, 139
describeBy() function, 240
describe() function, 122
Descriptive statistics, 113–126
descriptive parameters, 122–124
dev.off() function, 99
Digital elevation models (DEMs), 291, 341, 344, 346, 354
Dispersion, 75–76
Distribution patterns, 361–369
clustered points, 365–367
K-function, 361–365
randomly distributed points, 367–369
regular points, 367
'dplyr' package, 22, 254
Dragulescu, A., 36

E

Earthquake patterns, 362
Element-wise operators, **16**
Elevation, retrieving, 263–265
'elevatr' package, 263, 278
'EMODnet' package, 278
EMODnet (European Marine Observation and Data Network) Bathymetry project, 266–267
Environment panel, 9
EPSG (European Petroleum Survey Group) system, 146
ESA (European Space Agency) Copernicus website, 300
Excel files, data, 35–36
reading, 35
writing, 35–36
Exploratory data analysis (EDA), 71, 89, 359, 420
basic functions, 72–76
central tendency, 74–75
dispersion, 75–76
'meuse' dataset, 71–72

F

Factor analysis, 120
Fault patterns, 362
filter() function, 16
Flow control functions, 17
for() loops, 18–19
if()-else statements, 17–18
while() loops, 19
for() loops, 18–19
Friendly, M., 61, 124
Fry, N., 376

G

GCDkit package, 23
Gentleman, Robert, 3
Geocentric coordinate system (GEOCS), 146
Geochemical survey, 160
Geographic coordinate system (GCS), 146
Geographic information system (GIS), 419
Geometry, defining, 194–197
GeoPackages, 140
Geophysical survey, 160
'geophys' package, 23
Geothermal system patterns, 362–363
GeoTIFF (Geographic Tagged Image File Format), 259
get_elev_raster() function, 265
'ggnewscale' package, 341
'ggplot2' package, 23, 273
  Grammar of Graphics, 173–175, 183
ggplot() function, 341
'ggtern' package, 23
'GLG' variable, 186
Goldschmidt classification, 120
GPS (Global Positioning System) data, 160
'graphics' package, 91
'grDevices' package, 97, 98
grep() function, 45, 45n4
Grid ASCII, 259
Gridded data, 143–145, 158, 257, 278
Groundwater resources management, 133
Grouping operators, **16**
Group statistics, 240
  correlation, geological unit, 241–242
  descriptive statistics, geological unit, 240–241
  plotting independence, 246–249
  testing independency, geological unit, 242–246
'gstat' package, 384–386
  geological mapping, 385
  geothermal resource assessment, 385
  groundwater modelling, 385
  mineral resources studies, 384
  soil contamination, 384–385

H

Hanna, S., 376
HDF (hierarchical data format), 259
Hillshade, 291, 294–296, 343
History panel, 9

I

idw() function, 388, 392, 393
if()-else statements, 17–18
Ihaka, Ross, 3
Inferential statistics, 80–89
  ANOVA (analysis of variance), 83–85
  regression analysis, 85–89
  t-test, 80–82
  Welch two-sample t-test, 82–83
install.packages() function, 21, 22
Interquartile range (IQR), 77, 116n3
Inverse distance weighting (IDW), 383, 385, 388–396
  anisotropy, no accounting, 389
  boundary effects, 389
  data distribution, sensitivity, 388–389
  local homogeneity, assumption, 390
  neighbourhood size, dependence, 390
  no uncertainty estimation, 390
  outlier sensitivity, 390
  over-smoothing and over-fitting, 390

J

JPG (Joint Photographic Experts Group), 259

K

K-function, 361–364, 376
  interpretation, 364–365
Klischies, M., 273
Krige, Danie, 397
Kriging, 383, 385, 397–418
  co-kriging, 397
  earthquakes, 407–412
  ordinary kriging, 397, 409
  semi-variograms, 398–407
  simple kriging, 397
  universal kriging, 397, 409–411
  variograms, 398–407
Kurtosis, 124

L

Landsat-8 images
download data, 314–315
know your data, 315–318
NDVI, 320–322
plotting composite RGB image, 318–320
read data, 318
Landsat program, 300n5
lapply() function, 114
'leaflet' package, 189
Libraries, 19–21
LiDAR (Light Detection and Ranging) data, 160
Limit of Detection (LOD), 46
Lindsay, John, 338
lithology dataset, 103
Local coordinate system (LCS), 146
Logical operators, 14, **15**

M

Maps, preparing, 228–230
ggplot() map, 230–231
'leaflet' map, 231–235
Martín-Fernández, J. A., 47
mask() function, 290
Matheron, Georges, 397
Matloff, N., 17
Matrices, 12
Matrix operators, **16**
Median absolute deviation, 123
'meuse' dataset, 71–72, *88*, *178*, *266*, *287*
Mineral deposit patterns, 362
Mirai Solutions GmbH, 36
Missing values, 46–47
handling, 47–48
is.na() function, 47
na.omit() function, 48
na.rm() argument, 48
replacing, 48–49
MODIS satellite, 300n6
Moisture Stress Index (MSI), 331, *334*
Multi-layer raster, 258
MultiSpectral Instrument (MSI), 301n9

N

NDVI (Normalised Difference Vegetation Index), 320–322, 330–331, *333*
NDWI (Normalised Difference Water Index), 331, 332, *333*
Network Common Data Form (NetCDF), 259

O

OpenStreetMap (OSM), 167–172, 189
'osmdata' Package, 167–172
Operators, 14
basic operators, 14
other operators, 14–15
special operator, pipe, 15–17
'osmdata' package, 167–172
Outliers, 76
data with, *77*
data without, *78*
handling, 76–79

P

Packages, 19–21
for geologists, 23–24
Palarea-Albaladejo, J., 47
Patterson plot, 376
Pawlewicz, M. J., 186
Pebesma, E., 136
Pipe operator, 15–17
plot() function, 99
Plot panel, 9
plotRGB() function, 313
Plotting tool, 9
PNG (Portable Network Graphics), 259
png() function, 99
Point data interpolation, 383–386
Point pattern analysis (PPA), 358, 361–381
distribution patterns, 361–369
Earthquakes Data, 373–376
Fry method, 376–379
nearest neighbour, 379–380
Northern Territory Dataset, 369–371
point density analysis, 371–373
Point process models, 358

Poisson distribution, 380
Pre-installed packages, 20
previewColors() function, 232, *233*
Principal component analysis (PCA), 120
'.prj' extension, 139
Projected coordinate system (PCS), 146
'provenance' package, 23–24
'psych' package, 121, 122

R

Raster data, 143–145, 257, 278
  basic descriptive statistics of, 277
  files, types, 258–259
  filtering, 284–285
  masking, 286–290
  objects, converting between, 262–263
  objects, information about, 274
  reading and writing, 260–262
  resampling, 281–283
  types, 258
rasterize() function, 286
Raster operations, 279–290
  aggregating and disaggregating, 283–284
  cropping, 281
  reprojection, 279–280
'raster' package, 273, **275**
Rasters, creating, 412–417
rasterToPoints() function, 341
R community, 20–21
R console, 21
'RCurl' package, 37
read.csv2() function, 27n3
read.csv() function, 27n2
Reading spatial data files
  KML files, 162
  shapefiles, 161–162
Regression analysis, 85–89, 120
Regular expressions, 45n4
Remote sensing raster, 258
Reshaping data, 56–59
  data frame, transposing, 56–57
  long to wide and vice versa, 57–59
  pivot_longer() function, 58
  pivot_wider() function, 59
Rio Tinto area, *319*, 324, *342*, *345*, *348*, *349*
'RJSONIO' package, 37
R programming language, 3, 4; *see also individual entries*
  advanced statistical analysis, 5
  advantage of, 3–4
  data handling and manipulation, 4
  downloading and installing, 8–9
  extensive support and resources, 6
  geospatial analysis, 4–5
  packages and libraries, 19–21
  reproducibility and automation, 5–6
  resources for learning, 24–25
  syntax and data structures, 10–14
  transforming variables in, 50–55
  versions and updates, 24
  visualisation, 5
RStudio, 21
  downloading and installing, 8–9
  environment and workspace, 9–10
  versions and updates, 24

S

sapply() function, 115
Satellite images, 299–322
Satellite packages, 299
  'openeo' package, 299–300
  'rasterVis' package, 300
  'rgee' package, 300
Scalars, 11
Schauberger, P., 36
Scripts, 9
Semi-variograms, 398–407
  clustered pattern, 400–403
  random or uniform pattern, 403–404, *405*
  trend pattern, 405–407
Sentinel-2
  grid structure, 306
  overlap, 307
  tile naming convention, 307
  tile size, 306
  UTM zones, 307
Sentinel images, 301–314
  available images, verifying, 303–307
  band combinations, 327–329
  basins and rivers, 349, *349*
  calculating indices, 329–332, *334*
  Copernicus program, 302–303

downloading data, 308–310
environment, setting, 324–327
geomorphological analysis, 338–353
geomorphs tool, 349–353
installation, 'whitebox' package, 340
level-0 (L0), 301
level-1C (L1C), 302
level-2A (L2A), 302
Rio Tinto area, case study, *319*, 324, *342*, *345*, *348*, *349*
Strahler classification, 347–348, *348*
streams and waterflow, 346–348
thematic map, 333–338
visualising, 310–314, *314*
'whitebox' package, 338–340
'sf' package, 136
Shapefiles, 139
'.sbn' and '.sbx' extensions, 139
'.shp' extension, 139
'.shp.xml' extension, 139
'.shx' extension, 139
Single-layer raster, 258
Skewness, 123, *124*
Slope, 291, 292–293
Spatial analysis
fundamentals of, 131–134
geology and earth sciences, applications, 132–134
geostatistics, 132
GIS, 132
multivariate spatial analysis, 132
point pattern analysis, 132
remote sensing, 132
spatial interpolation, 132
spatial statistics, 132
terrain analysis, 132
Spatial analysis packages, 147–157
'raster' package, 153–155
'sf' package, 151–153
'sp' package, 148–151
'terra' package, 156–157
web pages, R packages, 157
Spatial calculations (numeric), 216–217
st_area() function, 217–218
st_centroid() function, 217, *218*
st_distance() function, 219
st_length() function, 218–219
Spatial covariates, 359
Spatial data
buffering, 190, 198–199
clipping, 203–207
cropping, 200–203
geometries and projection, 193–198
'ggplot2' package, Grammar of Graphics, 173–175
handling, 190–224
'leaflet' package, 175–177
plotting spatial lines, 179–183
plotting spatial points, 177–179
plotting spatial polygons, 183–189
querying, 190
reprojection, 190, 197–198
'sf' object, 191–193
'sp' package object, 191–193
subsetting, 190, 199–207
visualising, 172–189
Spatial file types
data conversion, 163–166
Data Frame to Spatial* Object, converting, 164–165
'sf' and GPX, converting between, 166
'sf' and JSON, converting between, 166
'sf' and KML, converting between, 167
'sp' and 'sf,' converting between, 165
'SpatialLinesDataFrame' object, 180
Spatial objects, 135–158
coordinate reference systems (CRS), 145–147
gridded and raster objects, 143–145
grids or rasters, 136, *137*
lines, 136
points, 136
polygons, 136
vectorial objects, 137–143
Spatial operations (geometric), 219–220
st_difference() function, 221–222
st_intersection() function, 222–224, *223*
st_sym_difference() function, 220
st_union() function, 220–221
SpatialPointsDataFrame object, 164, 195
Spatial queries (logical), 207
st_contains() function, 207–209
st_crosses() function, 210–211
st_intersects() function, 214–216
st_overlaps() function, 212–213
st_touches() function, 209–210
st_within() function, 213–214
Spatial statistics, 357–359

'SpatRaster' objects, 273
'spatstat' package, 358–359
st_as_sf() function, 249
Statistical techniques, 71
st_buffer() function, 249
Stream sediments, subsetting, 235–240
  answering questions, 238–240
  geological unit dataset, 236–237
  naming geological units, 238
Strength, weakness, opportunities, and threats (SWOT) analysis, 398
Structural geology, 133
Subsetting data, 40–41, 104–106
  environmental studies, 41
  filtering data, 41
  numeric data, 41–44
  region of interest analysis, 41
  resource exploration, 41
  spatial analysis, geological feature, 41
  temporal analysis, 41
  text data, 44–46
Subsetting operators, **16**
Syntax, 10–14

T

Terrain operations, 291–298
  aspect, 293–294
  hillshade, 294–296
  slope, 292–293
  Terrain Ruggedness Index (TRI), 296–297
Terrain Ruggedness Index (TRI), 296–297
'terra' package, 273, **275**
'tidyr' package, 23
time series grid, 258
time series raster, 258
Timor-Alor Earthquakes, 390–395
transforming variables, 50–55
  appending columns, 54–55
  appending rows, 52–54
  categorical to continuous, converting, 51–52
  character to numeric, converting, 50
  continuous to categorical, converting, 51
transpose function t(), 56–57

U

United States Geological Survey (USGS), 300
URL (Uniform Resource Locator), 29n5
UTM (Universal Transverse Mercator), 147

V

variables, 11
variograms, 398–407, 418
  modelling, 385
vector data, 159–189
vectorial objects, 137–143
  GeoJSON, 141
  GeoPackages, 140
  GPX (GPS Exchange Format), 141–142
  KML (Keyhole Markup Language), 142–143
  shapefiles, 139
vectorial spatial data, 159
  sources of, 159–160
vectors, 11
Verzani, J., 9

W

Walker, A., 36
web server, defined, 33n10
WFS Service, 270–274
WGS84 (World Geodetic System 1984), 147, 185, 326
whitebox package, 24
Wickham, H., 15
Wilcoxon rank-sum test, 251–253
Wilkinson, Leland, 173
window functions, 359
WKT-2 (Well-Known Text, version 2), 194n2
write.csv2() function, 28n4
writing spatial data files, 162
  KML file, 162–163
  shapefiles, 162

Z

Zeng, L., 385

Printed in the United States
by Baker & Taylor Publisher Services

Printed in the United States
by Baker & Taylor Publisher Services